T0173819

Pharmaceutical Photostability and Stabilization Technology

DRUGS AND THE PHARMACEUTICAL SCIENCES
A Series of Textbooks and Monographs

Executive Editor

James Swarbrick
PharmaceuTech, Inc.
Pinehurst, North Carolina

Advisory Board

Larry L. Augsburger
University of Maryland
Baltimore, Maryland

Harry G. Brittain
Center for Pharmaceutical Physics
Milford, New Jersey

Jennifer B. Dressman
Johann Wolfgang Goethe University
Frankfurt, Germany

Anthony J. Hickey
University of North Carolina School of
Pharmacy
Chapel Hill, North Carolina

Jeffrey A. Hughes
University of Florida College of
Pharmacy
Gainesville, Florida

Ajaz Hussain
Sandoz
Princeton, New Jersey

Trevor M. Jones
The Association of the
British Pharmaceutical Industry
London, United Kingdom

Hans E. Junginger
Leiden/Amsterdam Center
for Drug Research
Leiden, The Netherlands

Vincent H. L. Lee
University of Southern California
Los Angeles, California

Stephen G. Schulman
University of Florida
Gainesville, Florida

Jerome P. Skelly
Alexandria, Virginia

Elizabeth M. Topp
University of Kansas School of
Pharmacy
Lawrence, Kansas

Geoffrey T. Tucker
University of Sheffield
Royal Hallamshire Hospital
Sheffield, United Kingdom

Peter York
University of Bradford School of
Pharmacy
Bradford, United Kingdom

Pharmaceutical Photostability and Stabilization Technology

edited by

Joseph T. Piechocki
Piechocki Associates
Westminster, Maryland, U.S.A.

Karl Thoma
Ludwig-Maximilians-Universität
Munich, Germany

CRC Press
Taylor & Francis Group
Boca Raton London New York

CRC Press is an imprint of the
Taylor & Francis Group, an **informa** business

CRC Press
Taylor & Francis Group
6000 Broken Sound Parkway NW, Suite 300
Boca Raton, FL 33487-2742

First issued in paperback 2019

© 2010 by Taylor & Francis Group, LLC
CRC Press is an imprint of Taylor & Francis Group, an Informa business

No claim to original U.S. Government works

ISBN-13: 978-0-8247-5924-7 (hbk)
ISBN-13: 978-0-367-39035-8 (pbk)

This book contains information obtained from authentic and highly regarded sources. While all reasonable efforts have been made to publish reliable data and information, neither the author[s] nor the publisher can accept any legal responsibility or liability for any errors or omissions that may be made. The publishers wish to make clear that any views or opinions expressed in this book by individual editors, authors or contributors are personal to them and do not necessarily reflect the views/opinions of the publishers. The information or guidance contained in this book is intended for use by medical, scientific or health-care professionals and is provided strictly as a supplement to the medical or other professional's own judgement, their knowledge of the patient's medical history, relevant manufacturer's instructions and the appropriate best practice guidelines. Because of the rapid advances in medical science, any information or advice on dosages, procedures or diagnoses should be independently verified. The reader is strongly urged to consult the relevant national drug formulary and the drug companies' and device or material manufacturers' printed instructions, and their websites, before administering or utilizing any of the drugs, devices or materials mentioned in this book. This book does not indicate whether a particular treatment is appropriate or suitable for a particular individual. Ultimately it is the sole responsibility of the medical professional to make his or her own professional judgements, so as to advise and treat patients appropriately. The authors and publishers have also attempted to trace the copyright holders of all material reproduced in this publication and apologize to copyright holders if permission to publish in this form has not been obtained. If any copyright material has not been acknowledged please write and let us know so we may rectify in any future reprint.

Except as permitted under U.S. Copyright Law, no part of this book may be reprinted, reproduced, transmitted, or utilized in any form by any electronic, mechanical, or other means, now known or hereafter invented, including photocopying, microfilming, and recording, or in any information storage or retrieval system, without written permission from the publishers.

For permission to photocopy or use material electronically from this work, please access www.copyright.com (http://www.copyright.com/) or contact the Copyright Clearance Center, Inc. (CCC), 222 Rosewood Drive, Danvers, MA 01923, 978-750-8400. CCC is a not-for-profit organization that provides licenses and registration for a variety of users. For organizations that have been granted a photocopy license by the CCC, a separate system of payment has been arranged.

Trademark Notice: Product or corporate names may be trademarks or registered trademarks, and are used only for identification and explanation without intent to infringe.

Visit the Taylor & Francis Web site at
http://www.taylorandfrancis.com

and the CRC Press Web site at
http://www.crcpress.com

During the production of this text our dear friend and coeditor Dr. Karl Thoma passed away. This book is dedicated to his memory and is a fitting memory of the principles he championed and the inspiration he gave us all for over four decades. Evidence of the magnitude of his contributions and the breadth of his work can be found throughout this text.

Those of us who have had the pleasure of knowing him as a friend or mentor will miss him deeply. I am honored to have had the pleasure of working on this final work with him.

Preface

For many millennia, the effects of photonic exposure on humans and materials have been known and chronicled in literature. Wurtman (1) presents a good introduction to "The Effects of Light on the Human Body."

Lachman, Schwartz, and Cooper (2) presented a good review of the field of photostability testing along with a review of the theory involved and a recommended cabinet design for use in the industry, which is still in use today in many laboratories.

For over seven decades, since the seminal works of Eisenbrand (3,4,5), Arny, Taub, and Steinberg (6,7), and Arny, Taub, and Blythe (8) no text devoted to the subject of photostability testing has been printed. Articles have appeared in several journals but no unified approach was given.

With the recognition that photostability testing was an integral part of the pharmaceutical development process, by the International Conference on Harmonization, efforts were made to develop a unified worldwide approach. Unfortunately, the process became more political than scientific, resulting in a less than clear guideline, as will be highlighted in many of the chapters of this book.

In the chapters that follow, world-renowned experts in the various subjects will present their areas of expertise in a clear and concise manner and discuss the pros and cons of the many alternatives available.

Recently, at least one manufacturer has sought to develop their own proprietary lamp and system, supposedly meeting International Conference on Harmonization requirements. Major problems of this approach are the lack of proper scientific justification and peer review, as well as the dependence on a sole source, available in only one country. The danger of this approach is well illustrated in the demise and resurrection of Vita-Lite Lamp.

The purpose of this text is to give the user an understanding of the fundamentals of the science involved so that they may make an informed choice as to which procedure is best for their purposes. After reading this text, the user should have a better understanding of the advantages and disadvantages of the various approaches advocated in the literature.

Karl Thoma
Joseph T. Piechocki

REFERENCES

1. Wurtman RJ. The effects of light on the human body. Sci Amer 1975; 233(1):68–77.
2. Lachman L, Swartz CJ, Cooper J. A comprehensive pharmaceutical stability testing laboratory III: A light stability cabinet for evaluating the photosensitivity of pharmaceuticals. JAPhA 1960; 49(4):213–218.
3. Eisenbrand VJ. Protection of light-sensitive drugs. Pharm Ztg 1927; 72(81):1275–1276. (Scientific Communication, A method for the Testing of the Light Sensitivity of Drug Substances). CA 21, 1689.
4. Eisenbrand, VJ. Protection of light-sensitive drugs by colored glass. Pharm Ztg 1927; 72:247–248. CA 21, 1927.
5. Eisenbrand, VJ. Protection of light-sensitive drugs by means of colored glassware. Pharm Ztg 1929; 74:263. CA 22, 137.
6. Arny HV, Taub A, Steinberg A. Deterioration of certain medicaments under the influence of light. JAPhA 1931; 20(10):1014–1023.
7. Arny HV, Taub A, Steinberg A. Deterioration of certain medicaments under the influence of light. JAPhA 1931; 20(10):1153–1158.
8. Arny HV, Taub A, Blythe, RH. Determination of certain medicaments under influence of light. JAPhA 1934; 23(7):672–679.

Contents

Contributors

John M. Allen   Department of Chemistry, Indiana State University, Terre Haute, Indiana, U.S.A.

Sandra K. Allen   Department of Chemistry, Indiana State University, Terre Haute, Indiana, U.S.A.

Wolfgang Aman   Research and Development, Gebro Pharma GmbH, Fieberbrunn, Austria

Robert Angelo   Gigahertz-Optik, Inc., Newburyport, Massachusetts, U.S.A.

Ahmed F. Asker   College of Pharmacy and Pharmaceutical Sciences, Florida A&M University, Tallahassee, Florida, U.S.A.

Steven W. Baertschi   Eli Lilly and Company, Analytical Sciences Research and Development, Lilly Research Laboratories, Indianapolis, Indiana, U.S.A.

William E. Bowen   Global Pharmaceutical Commercialization, Merck Manufacturing Division, Merck and Co., Inc., West Point, Pennsylvania, U.S.A.

Günter Gauglitz   Institut für Physikalische und Theoretische Chemie, Universität Tübingen, Tübingen, Germany

Hans U. Gonzenbach   Roche Vitamins Ltd., Geneva, Switzerland

Paul A. Harmon   Pharmaceutical Research and Development, Merck Research Laboratories, Merck and Co., Inc., West Point, Pennsylvania, U.S.A.

Stephan M. Hubig   Department of Chemistry, University of Houston, Houston, Texas, U.S.A.

Lee J. Klein   Pharmaceutical Research and Development, Merck Research Laboratories, Merck and Co., Inc., West Point, Pennsylvania, U.S.A.

Irene Krämer   Department of Pharmacy, Johannes Gutenberg-University Hospital, Mainz, Germany

Norbert Kübler   Temmler Pharma GmbH and Co. KG, Marburg, Germany

Yu Lu   Vertex Pharmaceuticals, Cambridge, Massachusetts, U.S.A.

Douglas Moore   Department of Pharmacy, The University of Sydney, Sydney, Australia

Joseph T. Piechocki   Piechocki Associates, Westminster, Maryland, U.S.A.

Robert A. Reed   Xenoport, Inc., Santa Clara, California, U.S.A.

Heiko Spilgies   GlaxoSmithKline, Ware, Herts, U.K.

Allen C. Templeton   Pharmaceutical Research and Development, Merck Research Laboratories, Merck and Co., Inc., West Point, Pennsylvania, U.S.A.

Scott R. Thatcher   Analytical Services, The Chao Center for Industrial Pharmacy & Contract Manufacturing, West Lafayette, Indiana, U.S.A.

Karl Thoma   Institut für Pharmazie—Zentrum für Pharmaforschung, Lehrstuhl für Pharmazeutische Technologie, Ludwig-Maximilians-Universität, Munich, Germany

Hanne Hjorth Tønnesen   Institute of Pharmacy, University of Oslo, Blidern, Oslo, Norway

1 The History of Pharmaceutical Photostability Development

Joseph T. Piechocki

Piechocki Associates, Westminster, Maryland, U.S.A.

INTRODUCTION

The objective of this chapter is to give the reader an appreciation of the history of the science as it developed, who the major contributors were and who today's major researchers are. This history is abbreviated because it is impossible to include all researchers.

Pre-1900

The ability of ultraviolet (UV) light to interact with chemicals was known and reported by Scheele in 1777 (1). Ritter in 1801 (2), Davy in 1812 (3), and Becquerel 1868 (4) reported similar results.

It is reported that in 1836 Theodor von of Lemberg, Germany suggested that black bottles, prescribed in the old apothecaries, be replaced with yellow glass, which he found to be more practical (5). This change allowed for the easy inspection of the clarity of liquids stored therein.

The first edition of the Dutch Pharmacopoeia (6) published in 1851 prescribed that iodine and silver nitrate is to be stored in closed black bottles. The warning that "…medicinal preparations have to be stored in such a fashion that any changes in light should be avoided as much as possible" was extended to other substances in later editions (1871 and 1889).

Much of the history of pharmaceutical photostability-testing—pre-1900—has been reviewed and reported in the doctoral thesis of Jacobus Bakduinus Meinardus Coebergh, published in 1920 (5).

Schoorl

Nicholass Schoorl and Van Den LM, at the University of Utrecht Berg, studied the photostability of a number of chemicals in the early 1900s. Two of their papers, published in 1905, dealt with the photodecomposition of bromoform (7) and chloroform (8). Other studies by Schoorl (9–11) and De Graff et al. (12) remain unknown, even today, due to the language barrier (Old Dutch) and the fact of their publication in local journals.

Bordier

In 1916, Henri Bordier reported on the action of light on iodine and starch iodide in water. He observed that dilute solutions of 10% tincture of iodine (pale yellow) and three drops of 10% tincture of iodine plus a little starch paste (blue) were both decolorized after a few hours of exposure to light (13) or a few minutes exposure to X rays (14) and that yellow glass had no effect on the reaction. Both solutions were faded by exposure to light and the protective ability of any colored glass could be judged using these solutions by its intermediate color. Thus, we have one of the first reported methods for judging the photo protective ability of glass using a simple chemical solution.

Eisenbrand

In 1927, Johannes Eisenbrand, at the University of Frankfurt, Germany published three papers dealing with light-sensitive pharmaceutical products (15–17), which were reviewed by Arny (18).

In the first paper, Eisenbrand (15) discussed "…the well-known action of light on certain chemicals; the fact that red light has little action while blue light has distinct action; the modern views of this phenomena in terms of wavelength; the fact that the shorter the wavelength, the greater the action on chemicals; the time element involved and the fact that glass lets through a smaller amount of ultraviolet rays." He considered UV rays of lesser importance in the study of the preservation of chemicals in glass containers." This last statement has since been shown to be an erroneous conclusion on his part.

In his second paper (16) Eisenbrand "…discusses the classification of light-sensitive chemicals and means of protecting them from harmful light rays." He suggested that "…they should be grouped into three classes: (i) extremely sensitive to light, (ii) sensitive to light, and (iii) slightly sensitive to light." He also noted an anomaly in the German Pharmacopoeia regarding the listing of light-sensitive benzaldehyde.

Eisenbrand recommended the use of a quartz lamp for testing pharmaceuticals because of its ability to provide "…a uniform quantity of light, whereas sunlight is, of course, variable in quality." His recommended test procedure included treatment (i) with direct quartz lamp rays, and (ii) with these rays passes through a thick plate of glass; as passed through the customary light filters.

An important observation by Eisenbrand is that no single method of analysis would suffice to determine the amount of deterioration of a light-sensitive chemical. He rightly noted, "…in some cases biological assays will be required."

For the protection of light-sensitive chemicals he recommended a violet-manganese glass as the best for all cases because it has "…a very low transmission value for all light rays…," but acknowledged that the best "…bottle glass is somewhere between the amber and the violet." For most purposes, he concluded that amber glass was sufficient to protect light-sensitive materials.

His third paper (17) is devoted to a study of the permeability of various glasses by UV radiation. Obviously, he considered this band of radiation (though low in percentage when compared to the visible band) very important for photostability.

In his correspondence to Arny (18) and his Committee, a response to the survey they had issued, he stated that, "…based on his work with von Halben in revising Landolt's Physikalisch–Chemischen Tabellen, he studied the light absorption by various glasses and this work showed him that colored glass containers for medicines are, in a large degree, of little value, and our knowledge of the protective value of amber glass leaves much to be desired."

While he did include a bibliographic list on the subject, this list is limited as to the years it covers, 1916–1924. Two references are to actual drug product research and the other seven are glass related papers.

Coebergh

Jacobus Balduinus Meinardus Coebergh published his doctoral thesis, "Protection of medicines against the action of light," at Utrecht University in 1928 (19). This thesis was entirely devoted to the problems of photochemistry and photo protection. The objectives of this work were the following:

> To expose test chemicals to sunlight and light of known frequency and bandwidths to determine which kind of rays were responsible for their deterioration. To determine the light-protecting value of various colored glasses.

In this first comprehensive work on the subject of pharmaceutical photostability-testing, he subjected compounds listed as "should be protected from light" in the fourth edition of the Dutch Pharmacopoeia, to various bands of light and measured their effect on the chosen test substances.

He accomplished this task by using various glasses and filters as well as a spectrograph. With the spectrograph, he was able to more accurately identify the wavelength characteristics of the source and filter combinations used. After much work, he discovered that certain chemical solutions were capable of acting as filters for sunlight, passing selected bands of UV and/or visible radiation. The solutions used in this study and the wavelength bands transmitted by them are presented in Table 1.

From his studies, he was able to classify compounds according to the wavelength bands in which they were most sensitive. Table 2 lists the various solutions he used as filters in his studies. Table 2 also gives the compounds studied according and their wavelengths of highest sensitivity and Table 3 lists his findings for the different glasses studied. Table 4 presents the bandwidth protections offered by the glasses employed in this study.

Based on the data listed in Tables 3 and 4, Coebergh was able to make storage recommendations for the materials studied. A summary of this work was also published in the scientific literature (21).

One objective of Coebergh's work was "…to find a more simple practical way for judging colored flasks on their light-protecting qualities…(19,21)." He realized that access to a spectrograph was not possible for most practitioners at that time and a more practical method was needed. Based upon his work and that of others, he was able to make recommendations for certain test solutions. To test light-amber or light-green glasses for their ability to protect substances/products from radiation

Table 1   Filter Solutions Used by Coebergh to Isolate Specific Bandwidths for Testing Which Wavelengths Were Most Deleterious to the Substances Tested

Solution	Wavelength range transmitted (mμ)
Quinine bisulphate 0.8 solution	408 to beyond 720
Copper sulfate 2% behind	
dark blue–cobalt glass	350–480
Copper chloride 75 g to 200 cc	450–560
Copper chloride 10 g + calcium chloride	
10 g diluted to 200 cc	495–610
Copper sulphate 8.8 g potassium	
dichromate 9.4 diluted to 200 cc	550–645
Ruby glass	600–?
Potassium dichromate 0.125% solution	
behind dark blue–cobalt glass	700–?

Source: From Ref. 20.

Table 2 Coebergh's List of Wavelengths of Light-Decomposing Chemical

Group	Wavelength range (mµ[a])	Compounds/products
a	Up to 408	HBr, HNO_3, monobromocamphor, quinine and its salts, dimethylaminoantipyrine, CHI_3, physistygmine-H_2SO_4 and mercurous tannate
b	408–450	Et_2O, HgCl, santonin
c	450–500	Cl, water
d	500–550	$Fe_4(P_2O_7)_3$ with $(NH_4)_2$ citrate
e	550–600	BzOH, Ag proteinate, HgI_2, resorcin, α-naphthol, pyrogallol
f	600–700	Apomorphine-HCl, HgI, PhOH
g	Below 400 and above 600	$CHBr_3$, sweet spirit of niter
h	Below 500 and above 600	H_2O_2 solution
i	Above 600	HCN, tincture of Iodine
j	All wavelengths	HgO, P, liquified PhOH, adrenaline HCl solution

Source: From Ref. 20.

below 408 mm, he recommended the test container be filled with a 1% solution of benzidene in chloroform and exposed to bright sunlight for 5.5 hours. After exposure, the solution in an acceptable container should show no turbidity. Similarly, to test brown, green and red glasses for their ability to protect substances/products from radiation above 408 mm, he recommended that a piece of celloidin (silver chloride photographic paper) paper be placed in the test container and then exposed to light for 80 minutes. The paper in an acceptable glass container should show no discoloration. The attachment of a strip of dark blue–cobalt glass to the paper would allow one to determine if any of the exposure noted is caused by radiation below 480 mµ.

In 1927, Coebergh, at the urging of Kolthoff, contacted Arny (20) and made him aware of his research and dissertation. Kolthoff had received a copy of a survey then being circulated by the American Pharmaceutical Association's Section on Dispensing, "Committee on Colored Glass Containers" (16). Coebergh furnished the Committee with a copy and synopsis of his thesis.

Table 3 Various Glasses Used by Coebergh and Their Transmission Wavelengths

Color (color agent)	Transmission wavelengths (mµ)
Dark red (Cu_2O)	Above 700
Dark brown (C and S)	Above 500 (light brown transmits even the ultraviolet rays)
Thick dark–green (FeO)	Above 450
Light yellow–green (Cr_2O_3–CuO)	Above 450
Dark yellow–green	470–660
Blue	Red, ultraviolet and blue rays
Black (not opaque)	Red, ultraviolet and violet

Source: From Ref. 19.

Table 4 Protective Actions of the Glasses Employed by Coebergh

Glass type	Below	Above
	(in millimicrons)	
Dark-green glass	480	620
Dark-amber glass	500	750
Light-amber glass	430	800
Light-green glass	430	800
Ruby glass	600	800

Source: From Ref. 20.

Coebergh's work stands as the first known, detailed, spectroscopic study of the photostability of pharmaceutical substances and products and shows how the glass color can be used to protect photosensitive materials. This work might have been more well known if the author had published it in an English or German journal or if the translation, produced by Krak had ever been published (18).

Arny
Henri V in 1926, the American Pharmaceutical Industry recognized light as an important factor in the deterioration of chemicals and pharmaceuticals along with the important role glass played in protecting them from this problem. While individual companies may have had specific problems, this was the first time it was finally recognized as an industry problem and formally discussed by members of the American Pharmaceutical Association. To this end, on November 1, 1926, the Association's Section on Dispensing, "Committee on Colored Glass Containers" was formed (18). The committee, composed of Arny, Becker, and Dunning was charged with the task of reviewing all available information found in the chemical and pharmaceutical literature on this topic. The committee decided that the following general principles should apply to their work (18):

1. While it is generally accepted that amber glass is the proper type of glass for delicate chemicals, we take the protecting value of such glass too much for granted.
2. The mere fact that certain glass is colored yellow does not necessarily indicate that such glass will give adequate protection against the actinic action daylight and other sources of light.
3. Our committee should therefore investigate the light-protecting properties of various types of glass containers; not merely those made of amber glass but also those of other colors.
4. The study should also include the amount of protection afforded or deterioration produced by light rays from UV and infrared regions of the spectrum.
5. The field is too vast to be satisfactorily solved by this Committee of three. That we must obtain the cooperation of other research agencies goes without saying.
6. One of our first tasks will be to get some information as to work already performed or being now carried out by these other research agencies. In short, the preparation of a bibliography of work of this character is our first obligation.
7. As to outside agencies to which we will turn for information and aid the following have been suggested.

   a. U.S. Bureau of Standards
   b. Mellon Institute

c. Chemical Manufacturers
d. Pharmaceutical Manufacturers
e. Corning Glass Works
f. Manufacturers of glass bottled for chemicals and pharmaceuticals
g. Bausch and Lomb Optical Company
h. Eastman Kodak Research Laboratories

Supplementing this list, we should get in touch with all individual research workers in the field of Pharmacy (or outside of it) who are engaged in this sort of research work. In short, we should bring our problem to the 400 (or more) persons now on the mailing list of the National Council of Pharmaceutical Research.

To this end, the committee members drew up a letter and questionnaire soliciting the help of all of the above cited agencies in addition to the National Research Council physicists and chemists, members of the American Drug Manufacturers' Association, United States Pharmacopoeia, National Formulary, Bureau of Chemistry of the Association of Official Agricultural Chemists, teachers in the leading college of pharmacy, etc. In all, 700 questionnaires were sent out. Of these 700, only 52 answers were received while 13 stated they "…were not interested in the problem."

In their initial work, the committee identified two other important factors from the works of Anton Hogstad of the St. Louis College of Pharmacy (22) and, Tapley and Giesy of E.R. Squibb & Sons (23) which could influence their studies. Hogstad had studied the deterioration of cresol and noted that cresol kept in colorless, sealed tubes, did not darken when exposed to sunlight while those exposed in open tubes did.

Tapley and Giesy (23) had studied the photostability of bismuth subnitrate and concluded that only amber bottles protected the chemical from deterioration by light. Flint glass and blue bottles were found to have little photo protective ability.

The committee concluded from these two works that their planned study should also include the final conclusions on the following factors:

1. If air rather than light is the probable cause of deterioration.
2. If certain constituents of the glass-mix produce the deterioration we ascribe to light.

At the time of these studies, the Committee found that there were a total of 423 chemicals listed in the U.S. Pharmacopoeia and the National Formulary bearing the warning "…that experience has shown are sensitive to light."

The committee's 1928 report (20) also included a synopsis of the work of the three of Eisenbrand's papers (15–17) and Coebergh thesis (19). These papers have already been discussed, in this chapter.

In 1929, experimental work was begun on the "Deterioration of Certain Medicaments Under the Influence of Light" by Arny et al. at the College of Pharmacy of Columbia University (20). The work was sponsored by a number of Pharmaceutical Companies along with the Proprietary Association and the American Pharmaceutical Association Research Fund. The sponsoring companies and organizations are listed in Table 5.

Research was divided into three parts:

1. A study of the light transmission of commercial glass containers currently employed.
2. Examination of each chemical studied as to its pharmacopoeial quality.

Table 5 Companies Who Participated in Arny et al. Study of Glasses

Participating Companies	
Dow chemical company	Drug products company
Hynson, Wescott and Dunning	Lehn & Fink
Eli Lilly & Co.	William S. Merrell
Merck & Co.	H.K. Mulford Co.
Parke, Davie & Co.	Charles Pfizer & Co.
Perdue Fredrick Co.	Sharpe & Dohme
Dr. William J. Schieffelin	E.R/Squibb & Sons
Frederick Stearns & Co.	The Upjohn Co.
W.R. Warner & Co.	Proprietary Association
American Pharmaceutical Association Research Fund	

Source: From Ref. 24.

3. Exposure of each medicament under observation to varying degrees of light with subsequent examination of each sample as to the degree of deterioration.

Taub and Steinberg reported on the "Spectral Transmission of Colored Glass" (24). Instead of using isolated bands of the spectrum for their study, they employed a Bausch and Lomb spectrophotometer equipped with a white light source for work in the visible, 400 to 800 m:, range. For work in the UV (230–500 m) a high voltage iron arc, Hilger quartz spectrograph, with a sector photometer was employed. Both units were capable of providing emission spectra of the sources and absorption spectra of the glasses studied.

As opposed to Coebergh's studies, all glasses used came from commercially available bottles. The thickness of the glasses used to obtain transmission spectra in these studies were controlled. Representative pieces were cut out of the sides of the bottles and ground flat when necessary.

Pyrex® glass was included in the study because it was used as the inner container for all of the chemicals. This avoided any possibility of glass acidity differences that might have influenced the results. Window glass was used in all of the cabinets because all light passes through it first before reaching the shelves of a normal pharmacy. The transmission spectra of the glass containers, Corning color filters and special glass plates used in the study are reproduced in Figure 1.

These authors were quick to note from their study results that, although red glass gave the best overall photo protection of those glasses studied, "...its cost prohibits its use for commercial purposes."

Arny and Steinberg (24) reported in their paper titled "Deterioration of Chemicals," the photostability of products stored in commercially available glass bottles. These researchers eliminated the possibility of chemical differences that might be due to batch-to-batch variations in the glass bottles by using sealed Pyrex tubes as their sample containers. These sealed samples were then put into the various colored glass bottles to be tested. All samples were then stored in a "dark room" until used. When it became necessary to open these samples, a dark room, illuminated by an electric lamp of low intensity was used.

The Pyrex sample tubes were in most cases sealed with a cork stopper unless there was danger of chemical attack; in which case, rubber stoppers were then used. In some cases, when necessary, the stopper was covered with tin foil prior to

(A)

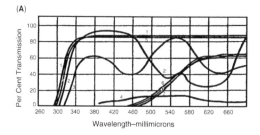

(B)

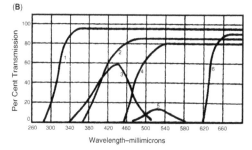

(C)

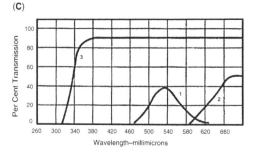

Figure 1 Transmission Spectra of the various containers and plates used by Taub and Steinberg for their studies. (**A**) Colored Glass Containers: 1 Flint A, (2.6 mm), 2 Blue A (2.5 mm), 3 Green A (2.8 mm), 4 Green B (2.5 mm), 5 Green (2.5 mm), 6 Amber A (2.4 mm), 7 Amber B (2.5 mm), 8 Amber C (2.4 mm). (**B**) Corning™ Color-Filters: 1 Pyrex (2.0 mm), Nultra (7.4 mm), Lantern (4.2 mm), 4 Noviol (6.0 mm), 5 Sextant Green (4.9 mm), Heat Resisting Red (3.3 mm). (**C**) Special Glass Plates: 1 Coeburgh Green 9 (2.3 mm), 2 Red Flash (3.0 mm), 3 Window Pane (3.0 mm). Reproduced with permission of the American Pharmaceutical Association (APhA). *Source*: From Ref. 25.

insertion. In some special cases, it was necessary to coat the stopper with melted paraffin before insertion. Lastly, in very special cases the tube might be sealed in an ampoule, either with a pocket of air above the chemical or the air displaced with an inert gas (nitrogen or carbon dioxide). Bottles containing the sealed Pyrex tube samples were sealed with cork or cotton plugs.

To simulate drug store (in use) conditions, the bottles containing the tubes were placed in specially constructed bookcases equipped with panes of window glass (Figs. 2 and 3). These bookcases were exposed on the roof to direct sunlight or daylight during the daytime hours for period ranging from 1 to 12 months. The authors called this condition "direct daylight."

A second, duplicate set of chemicals was arranged in an orderly manner on the shelves of a bookcase provided with glass doors that was situated in the office of the senior author (Fig. 1). This latter condition was designed to simulate lighting conditions in an average drug store. The authors referred to this condition as "diffused light." A third batch of samples was stored in the "dark room" and served as controls for the experiments.

These experiments employed 35 of the 400 chemicals then classified as "light-sensitive" by the U.S. Pharmacopoeia and the National Formulary. The chemicals

Table 6 Transmission Spectra of the Various Containers Used by Taub and Steinberg for Their Studies

			Colored glass containers		
Number	Type	Thickness (mm)	Number	Type	Thickness (mm)
1	Flint A	2.8	5	Green	2.5
2	Blue A	2.5	6	Amber A	2.4
3	Green A	2.8	7	Amber B	2.5
4	Green B	2.5	8	Amber C	2.4

Source: From Ref. 25.

selected were frequently cited in the literature as to their light sensitivity, a fact not necessarily always borne out by subsequent testing. The degree of deterioration of the chemicals studied was determined using the official assay whenever possible. In other instances, assays for specific impurities or color changes were used.

One important factor noted by the authors was that "…there is a 3 to 12 times as much deterioration in direct sunlight as is noted in samples kept in diffused light." They also noted that "…other factors such as evaporation, freezing in winter and explosions in summer rendered our analytical figures as to deterioration, in direct daylight, less reliable than those obtained from samples maintained under the normal condition of diffused light."

Analytical figures for each sample exposed to direct daylight can be found in Steinberg's Ph.D. dissertation (25). Results obtained on some of these same chemicals are reported in Table IV of their report as to their "General Findings as to Deterioration" (24). Other tables (V, VI, VII and VIII) report on 33 of the 35 selected products, their exposure to direct daylight and diffused light, whether color change played a role in the assay and the speed of deterioration by special testing.

This report was followed by a continuing report by Arny et al. in which they addressed the "Peculiarities of the Individual Chemicals" (26). As part of the study, these authors reviewed the literature and evaluated several photostabilizers suggested in the literature. All but one of these stabilizers gave satisfactory results in their hands. Tin compounds were found useful for stabilizing phenol and peroxides. Acetanilide was found to protect hydrogen peroxide.

Table 7 Transmission Spectra of Corning Color-Filters Used by Taub and Steinberg in Their Studies

			Corning color-filters™		
Number	Color	Thickness (mm)	Number	Color	Thickness (mm)
1	Pyrex	2.0	4	Noviol	6.0
2	Nultra	7.4	5	Sextant green	4.9
3	Lantern blue	4.2	6	Heat-resisting red	3.3

Source: From Ref. 26.

Table 8 Special Glass Plates Used by Taub and Steinberg
in Their Studies

Number	Color	Thickness (mm)
1	Coeburgh green	2.3
2	Red flash	3.0
3	Windowpane	3.0

Source: From Ref. 26.

Conclusions drawn from these studies were as follows:

Amber and green glass containers were the best for protecting light-sensitive chemicals.

Of the 35 chemicals studied, not all deteriorated because of the action of light. Other factors such as oxidation, evaporation, and atmospheric impurities (such as chlorine) also played a role.

Direct sunlight is the more deleterious of the two conditions (direct and diffuse) studied.

Not all literature recommended stabilizers are effective.

In 1934, Arny et al. reported on the third and fourth years of this on-going study, basically, a condensation of the Ph.D. thesis of Blythe (28). This paper summarized the findings of Steinberg and Blythe and identified other areas, which they recommended for further study.

Unfortunately, at this point Arny handed over the responsibility for the study to Husa of the University of Florida and further activity in this area was not found. Also, since the Columbia University school of Pharmacy ceased existence about 30 years ago and all records had been transferred elsewhere, it was impossible to find further information related to this study.

Brunner

In 1936 K. Brunner reported on the "Influence of Light on Pharmaceuticals." His studies included many acids, solvents, and drugs of the time, including chloroform. He reported this latter compound photo decomposed into phosgene and hydrogen chloride (30). This was an important finding in as much as this solvent was used for many extractions and crystallizations and influenced the results obtained.

Blythe

In 1954, R.H. Blythe presented his findings from a survey regarding the then current status of stability testing at the February 2, 1954 meeting of the Scientific Section of the American Pharmaceutical Manufacturers Association in New York City. From his survey, "...of 40 pharmaceutical companies. He found that the greater number of companies use exaggerated conditions of temperature, humidity, and light to test the stability of new formulations. However, from the results presented in his report, it is clearly evident that correlations between accelerated data to shelf storage conditions varied considerably due to different interpretations by the manufacturers (31)." This paper is important because it drew attention to the fact that stability testing, in general, at that time, was not a standardized procedure.

Figure 2 Arny showing the bookcases used for simulating the "diffused light effect" in their studies. *Source*: From Ref. 28.

Lachman et al.

In 1958, Leon Lachman and Jack Cooper of Ciba Pharmaceutical Products, spurred on by the work of Blythe (31) and others like Arny et al. (28), Garrett and Carper (32), and Garrett (33,34), and their own desire to speed up the drug development stability evaluation process, published the first in a group of papers titled, "A Comprehensive Pharmaceutical Stability Testing Laboratory." Part I, which concerned itself with the "Physical Layout of Laboratory and Facilities Available For Stability Testing (34)." This was the start of their effort to promote the standardization of normal stability and photostability testing in the pharmaceutical industry. In this paper, the authors presented details of their own laboratory, its design, and control systems.

Their major objective in writing this paper was to develop an accelerated stability-testing system that would mimic as closely as possible, the results obtained in a normal pharmacy. The parameters they identified as most important to control were temperature, humidity, and light exposure.

The design, discussed in detail in this paper, is that of a photostability cabinet capable of carrying out accelerated studies under controlled conditions of light and heat. A new feature of this cabinet was its incorporation of the major new light source of the day, the "cool-white" fluorescent lamp. This lamp was chosen because it was used at that time in most areas where pharmaceutical raw materials, in-process and final products, were to be found. This also included storage areas, pharmacies and homes.

This lamp produced the best replication of sunlight, to date, had the greatest amount of light output (foot-candles) per watt, produced the least amount of heat and was more efficient than the current filament lamps. A blower was incorporated in the cabinet design to remove the heat generated by the lamps and their ballasts, which was still substantial.

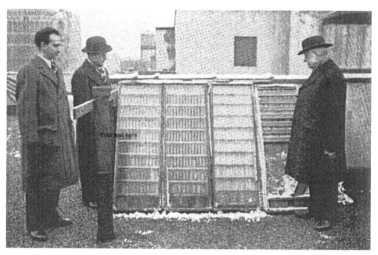

Figure 3   The "roof test" used by Arny et al. for their work. *Source*: From Ref. 27.

The design intensity for accelerated studies was 20 times that of ordinary room light (then considered to be 50 ft-candles) and only for visible light. The authors did not consider UV rays of any consequence because "...the amount of ultraviolet rays entering from the sunlight coming through the glass store front or window panes of a home or a warehouse is less than 5%. Therefore, the sun ultraviolet rays would seemingly contribute very little..." Hence, no UV source was used in this cabinet. This assumption proved to be false and eventually UV-A sources were also eventually added. Figure 4 presents a diagram of this cabinet.

The first reported application of this new photostability chamber was to the study the fading of 10 water-soluble, certified dyes by Kuramoto et al. (36). They found that sugars such as dextrose, lactose, and sucrose increased the rate of fading of these dyes whereas sugar alcohols such as mannitol and sorbitol did not appreciably affect the rate. They also discovered that if trace amounts of strong reducing catalysts remained in the alcohols used, they could significantly affect color stability. In general, antioxidants were not found effective in reducing fading. They did, however, find that nordihydroguaiaretic acid (NDGA) did have a substantial affect on reducing the fading due to the presence of dextrose.

In the second paper, Part II of the previously cited series, Lachman et al. (37) discussed many of the problems encountered when dealing with solid materials, particularly the fact that the photodegradation of solids was a surface phenomenon. They noted that different rates of fading were found, ranging from 1.5 to 240 and they could observe no particular pattern related to the phenomena.

In Part III of this series titled, "A Light Stability Cabinet for Evaluating the Photosensitivity of Pharmaceuticals," authored by Lachman et al. (38) they discussed the history of photostability testing in allied fields, the different sources that had been used along with their benefits and short-comings, and the rationale for the source they selected. The authors also presented some of the basic theory involved in this field, e.g., the relationship between wavelength and energy.

These authors referred to the work of Esselen and Barnaby (35) who reported the light levels found on shelves in typical retail locations. Their results are reproduced in Table 9.

In Part III the authors, Lachman et al., also give a more comprehensive description of their photostability cabinet, along with a plot of the spectral power distribution (SPD) of the lamp used. For "ordinary illumination" (room illumination) studies, the authors constructed their own light stability cabinet of smaller dimensions and used lamps of smaller size and wattage.

As part of their testing program, the authors exposed sulfonamide tablets and tablets containing the dye D&C Orange No, 3 and a benzothiazine derivative, in an open dish and packaged in amber glass containers. The samples stored in the amber bottles showed little photo deterioration with time. This confirmed Arny et al. conclusion regarding the use of amber bottles for photo protection (24).

Based on the results obtained, the authors concluded that the following items were responsible for the differences noted in their studies:

1. Wavelengths of the spectrum absorbed.
2. Amount of electromagnetic radiation of a suitable wavelength required to cause the molecules of a chemical to reach its excited state.
3. Amount of radiation absorbed.
4. Amber glass was successful in preventing the photodeterioration of the two tablet preparations studied.

Many U.S. firms through the years have replicated this cabinet and some units are still in use today. The only modification made to many over time has been the addition of UV-A fluorescent lamps to better replicate the UV portion of the solar spectrum and/or window-glass filtered daylight.

Reisch

One of the most prolific researchers in European pharmaceutical photostability testing was Johannes Reisch. Starting in 1966 he (39) and his associates, Abdel-Khalek (40–41), Ezik and Güneri (42), Ezik and Takacs (43) Ezik-Gücer and Güneri

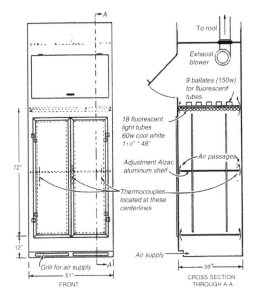

Figure 4 Comprehensive description of the lighting cabinet used for exaggerated light stability testing by Lachman and Cooper. © American Pharmacists Association (APhA), reproduced with permission. *Source*: From Ref. 35.

(44), Ezik-Gücer et al. (45) Ezik-Gücer et al. (46), Ezik-Gücer and Tewes (47), Fitzek (48–49), Henkel et al. (50), Henkel and Nolte (51), Henkel and Topaloglu (52), Iranshashi and Ezik-Gücer (53), Müller and Münster (54), Niemeyer (55), Ossenkop (56), Topaloglu (57), Weidman (58–60) Zappel (61), Zappel et al. (62), Zappel et al. (63,64), Zappel and Rao (65) produced many papers dealing with the photochemical fate of pharmaceuticals. Compounds studied include Pyrazole-(3)-on-(5) derivatives (39),[(E-1-acetyl-2-(3,3-dimethyl-1-phenyl-ureido)-3-oxo-but-1-enyl]-phenyl-carbaminsäure (40), aminophenazones (41), 4-isopropyl-2,3-dimethyl-l-phenyl-3-pyrazolin-5-on (propyphenazone)(42),testosterone and methyl testosterone (43),isopropylaminophenazones (44), testosteroneproplonate dimerization (45), azaporobe photoisomerization (46),1,4-benzodiazepines (47), amidopyrine solutions (48), 3-pyrazolonones (49), halometasone and prednicarbate (50), halometason and prednisocarbate (51), santonin (52), glucocorticoiden (53), barbiturates (54), sulfanilamide, sulfacetamide and 4-aminobenzoic acid ethyl ester (55), antipyrene (56), photodecomposition during TLC analysis (57), chloramphenicols (58), dichloracetamids (59), paranitrobenzaldehydes (60), sodium warfarin (61), ethisterone and norethisterone (62), methylprednisolone, prednisolone and triamcinolone acetonide (63), oubain, acetyldigitoxin and digoxin (64) and levonorgesterol (65).

Most of these papers concerned themselves with solid-state photochemistry and helped to show just how important photochemistry was to the understanding of modern drug development.

Thoma

In 1978 Karl Thoma, while at the Johann Wolfgang Goethe University at Frankfurt on the Main, published a book on the stability of drugs, which contained a chapter devoted to the "Photostability of Drugs" (66). This is one of the first published texts addressing this subject in scientific detail. To the best of our knowledge, this is the first textbook to appear with a chapter exclusively devoted to the principles of pharmaceutical photostability testing. Included in this chapter are a review of typical photochemical reactions, photosensitive drug substances found in the current version of the German Drug Codex and a description of the apparatus used for testing and directions for protection against photodegradation. He applied the principles taught to practical problems and was author of the first known patent issued regarding the protection of photolabile drug products, DE3136282, in 1983 (67).

Later, Thoma joined the faculty at the Ludwig-Maximillians University in Munich and continued his work, teaching and research in stability testing, and more particularly, photostability testing. The general lack of knowledge of Thoma's work,

Table 9 Light Intensities of Typical Retail Store Locations

Meter location	Average light intensity (foot-candles)	Outside reading/inside reading
Outside store windows	6500	–
Shelves, front of store	15–302	16–433
Shelves, middle of store	5–15	1300–433
Shelves, shaded parts of store	1–5	6500–1300

Source: From Ref. 38.

especially in the United States can be attributed once again to the language barrier (German–English).

This same barrier is also responsible for some of his student's dissertations, e.g., that of Kerker (68), which contains many of the details of his and Thoma's more important papers (69–71), not being more well known in photochemical circles outside of Europe. These papers showed that not all sources, UV or VIS, gave the same quantitative results. Kerker's thesis also includes the results of a survey of (free) German pharmaceutical industry photon source usage, which indicated that the most used source then in use was sunlight or a sunlight-simulating device.

Thoma deserves to be called the father of modern pharmaceutical photostability testing. Under his tutelage many students (over 60) were trained in the principles of pharmaceutical photostability testing more than any other researcher. Even though he only spent about 20% of his research time devoted to photostability testing, he was able to publish many significant papers in this area.

Practically all other studies in the literature, preceding and following these, were very limited in the number of sources studied or lacked sufficient details, to be of much use. The spectrum of the source used for testing was generally considered either equivalent or too variable. All fluorescent lamps were, and for the most part are still to this day, considered equivalent. Many lamp suppliers are unaware of the differences as noted in the works of Cole et al. (72) and Sequeira and Vozone (73).

Thoma was concerned about standardizing the photostability testing of drugs and drug products. Evidence of this is found in his papers and presentations. At the 1985 stability conference in Shiakabako, Japan, he cited the problems of photostability testing (74). He and Kerker mentioned these problems in their 1992 poster presentation, at an Arbeitsgemeinschaft für Verfahrenstechmik (APV) conference, titled "Photostability of Drugs—A Neglected Quality Criteria" (75). A third paper, circulated by Heraeus Instruments (now part of Atlas Electric Devices Company) titled "Standardized Photostability Tests—a Neglected Field of Drug Safety (personal communication)" again reiterated these points.

Thoma's interests included methods of testing (76–78), the influence of excipients and formulations on the photostability of products (71,79–83) and methods of photoprotection via packaging (67,71,84–88) or formulation (67,89–91). Several of his papers dealt with specific photostability problems such as antibiotics (92) antimitotics (93–96), adrenaline (97), molsidomin (98), phenothiazines (99), quinolines (100) and nifedipine (101–107).

Thoma published over 86 papers on photostability testing during his lifetime.

Kerker

Reinhard Kerker, a graduate student of Thoma's, in his dissertation "Untersuchungen zur Photostabilität von Nifedipin, Glucocorticoiden, Molsidomin,und ihren Zubereitungen" presented in Munich 1.7.91. (68) presents the most recent thesis devoted entirely to the problems and resolution of the problems of pharmaceutical photostability testing.

His thesis, for example, gives the results of a survey of sources then in current use in the German, Federal Republic, the spectral power distributions of these sources, their effect on various selected test substances as a function of time of exposure, total dose exposed to and absorbed. It is apparent from this data that the sources studied are not equivalent.

His study of blister packaged Nifedipin tablets is most revealing. He found that there was little consistency between the photostability of either unpackaged or packaged tablets. Some manufacturers used no protection; others protected the dosage form, some used only the package for protection and still others used both methods of protection. His work clearly illustrates a problem for generic drug manufacturers and acceptance of the position that the "use of the same container closure system as the manufacturer" as an acceptance criteria for generic drug products.

One section of his thesis gives examples of the use of pigments, UV-absorbers and suspended pigments as photo protective agents.

Stahl

P. Heinrich on June 5, 1991, Stahl of the Ciba-Geigy AG presented a paper at the Second International Symposium on Stability Testing held in Darmstadt, Germany, describing a powder sample holder, designed to expose a defined and reproducible thickness of a drug–drug product (108). While specifically designed to fit into an Atlas Weatherometer, it is easily adaptable to use in other instruments and test situations. Figure 12 in Chapter 10 of this book is a photograph of this holder and a detailed drawing is presented in Figure 5 of this chapter.

Others have made reference to the fact that photosensitivity is a surface phenomena but no one up to this time had offered a practical solution. Using this device, one can conveniently expose a precise depth of powder, simply and reproducibly.

Gauglitz

Gunter Gauglitz of the Eberhard–Karls University, Tübingen, Germany and his associates have done much research in the field of photochemistry and in particular chemical actinometry. His contributions include the development of new actinometers as well as practical guides to their usage and other articles. Four good

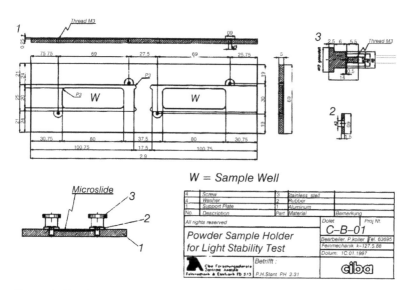

Figure 5 Detailed drawing to Stahl's powder sample holder for light stability testing. Reproduced with permission PH Stahl and Novartis International AG.

examples of his works are the "Modern Chemical Actinometry" (107), "Actinometry" (108), "Measurement of Radiation Intensities by Chemical Actinometers" (109) and the "Photokinetic Basis of Modern Actinometry" (110).

Much of his work, discussed in Chapter 8 of this book, is important because chemical actinometry is the only way of truly measuring the absorbed dosage of the incident radiation. A chemical actinometer corrects for container problems such as reflection, refraction, geometry, absorption, internal reflections, etc.

Brauer et al. developed an actinometer for the visible wavelength ranges (475–610 nm) (111). Brauer and Schmidt supplied the material and Gauglitz and Hubig did the development work necessary to commercialize it as Actinochrome N. Gauglitz and Hubig also developed a UV actinometer for the range, 245 to 440 nm, commercialized as Actinochrome 2R. This work preceeded the International Conference on Harmonization (ICH) effort by about a decade.

Both of these actinometers were produced by Gauglitz and Hubig as ready to use solutions in sealed UV cuvettes and once distributed commercially by Photon Technology International (PTI), Princeton, New Jersey, U.S., as Actinochromes. They had limited success in the marketplace.

A third PTI actinometer, Actinochrome 1R (248–334 nm), synthesized by Brauer and Schmidt (112) and was sold as a powder. Because of the limited market for these actinometers, they are no longer commercially available.

Kuhn

Hans Jochen Kuhn and Silvia Braslavsky, headed the International Union of Pure and Applied Chemistry (IUPAC), Organic Chemistry Division Commission on Photochemistry Committee on Chemical Actinometry. The duty of this group was to provide a list of chemical systems that have been found suitable for the integration of incident radiation by chemical conversion. This list was published in 1989 in the IUPAC Journal, Pure and Applied Chemistry (113).

In the paper are included solid phase, gas and liquid chemical systems, a listing of electronic actinometers and recommended actinometric procedures. Also included in this paper are "Recommended Actinometric Procedures," and "General Considerations on Chemical Actinometry." This latter topic contains items such as the pros and cons of using chemical actinometry, quality marks of a good actinometer, fields of application, as well as potential errors such a refractive index, temperature, absorption by photoproducts and the degree of absorption by the chemical actinometer itself.

The authors included in this paper, useful comments regarding the listed chemical actinometers. They graded the various systems as "well established," "still under discussion" or placed on a separate list of "disproved" systems.

Braslavsky and Kuhn also published a list of recommended actinometers titled "Provisional List of Actinometers," for the European Photochemical Association in their newsletter (114).

Tønnesen

Hanne-Hjorth of the University of Oslo, Norway is a well-respected educator and researcher in pharmacy, particularly pharmaceutical photochemistry. She is founder of the Photostability of Drugs Interest Group (PDIG). This international group of concerned scientists from academia, industry and government, formed in 1994 and with a common interest in all aspects of pharmaceutical photostability testing and its standardization. She has served on the Organizing Committee of each of the PDIG's international meetings. She has also served on the European Pharmacopoeia, Expert

Group XII, within the field of Photostability as well as the IUPAC Committee on Photostability.

Tønnesen chaired the First International Meeting on the Photostability of Drugs and Drug Products in Oslo, Norway in 1995 and has been a member of the PDIG Steering Committee since it's founding. She has graduated many students trained in pharmaceutical photostability testing and is editor of the first collection of papers devoted to this topic, "Photostability of Drugs and Drug Formulations."

Her research interests include antimalarials [Tonnesen HH, Karlsen J. Proposed Monograph on Light Stability Testing of Drugs and Drug Products Proposed IUPAC Monograph (personal communication)] (115–130), curcumin (131–143), cancer drugs (144,145) and epinepherine (146) as well as testing standardization (147–150) and source selection (151,152).

Tønnesen was the first researcher to bring to everyone's attention, the "gap," referred to by this writer as the "Tønnesen gap" to make it more easily identifiable, which exists when using as combinations of cool-white and UV-A phosphors or more commonly referred to as daylight fluorescent lamps (153). This "gap," low radiation in the 380 to 430 nm range, could effect the results obtained when irradiating compounds having their maximum absorption in this spectral region, such as curcumin and nifedipine. Also, depending on which UV-A phosphor is used, it is possible to influence the results of UV-sensitive materials.

A study by Tønnesen and Karlsen reported on the amount of window-glass filtered sunlight to fluorescent light exposure found in 20 Norwegian Hospitals (154). Based on these findings, they concluded, "From the European viewpoint, it is therefore, essential that filtered sunlight is included in the stability of drug products."

Asker

Ahmad Asker, Professor of Pharmaceutics and Industrial Pharmacy (retired) at Florida A&M University, U.S., was and educator and the leading U.S. academic researcher in the field of pharmaceutical photo protection of liquid samples. He has published many papers and educated several students in this field.

His papers and those of his associates dealt with the photo protection of colchine (155), dacarbazine (156), doxorubicin (157,158), dyes (159–162), metronidazole (163), nitroprusside (164,165), phenobarbital (166), physostigmine (167), reserpine (168), sulfathiazole (169), tetracycline (170) and vitamins (171–173). Most of his work centered upon the use of additives for the photoprotection of liquid drug products.

Moore

Douglas E. Moore, of the University of Sydney, Australia, has worked to establish testing standards in pharmaceutical photochemistry for many years. He has published several papers on this topic. His focus has generally been on the kinetics and mechanisms of photochemical reactions, particularly photosensitizers. He is the author of Chapter 11 of this book.

Moore began his work in photochemistry studying the "Antioxidation Efficiency of Polyhydric Phenols in Photooxidation of Benzaldehyde" in 1976 (174) and has made several significant contributions to the science of photochemistry, since then. He then went on to study the "Photosensitization by Drugs" (175) by examining "the effects of UV irradiation on compounds with light absorption in the 300 to 400 nm region." He studied 20 different drugs including phenothiazines, thiazides, diazepam, chlordiazepoxide, pyrilamine maleate, hexachlorophene, quinine hydrochloride, tetracycline hydrochloride, demeclocycline hydrochloride, anthracene, and L-tryptophan. He demonstrated in this work that quinine and anthracene were very efficient photosensitizers.

Moore's study of quinine (176) included the photosensitizing characteristics of both its hydrochloride and sulfate salts in water and the presence and absence of 2,5-dimethyfuran (an efficient acceptor for singlet oxygen) as well as neat aqueous, highly acidic, and buffer solutions. Both salts yielded identical reaction rates. "Over the range of 0.25 to 2.5×10^{-5} M, their rates were linearly dependent on the quinine concentration…" He also noted that, "In the absence of dimethylfuran, the oxygen consumption was very slight…" He found that "typically, at pH 6.0, the absorbance at 334 nm increased about by 5% while the smaller 280 nm peak increase by 10% when dimethylfuran was absent. In the same time period (10 minutes), when dimethylfuran was present 80% of the oxygen in the solution was consumed and the absorbance at 280 nm increased fourfold, corresponding to the oxidation product of dimethylfuran." Moore also found that xanthine, theophylline, uric acid reacted similarly to dimethylfuran and tryptophan. He also noted that the UV spectrum, λ_{max} of quinine changes from 334 to 348 nm when the pH is changed from 6.0 to 2.0 and the absorption band becomes broader resulting in a more significant absorbance at 365 nm.

Guilbault (177), Moore notes, reported that maximum fluorescence intensity of quinine occurred in 0.1 N sulfate–sulfuric acid solutions and was quenched by the presence of halide ions. Moore noted a similar quenching of the photosensitizing ability of quinine.

Moore's paper established the fact that aqueous, unbuffered, air saturated solutions of quinine are not true actinometers. Their UV spectrum varies with pH, and are subject to interference from oxidizable organic contaminants and sensitive to oxygen concentrations. As such, its adoption as a reference actinometer for photostability testing is very questionable.

Other drugs studied by Moore et al. include chlorine containing photosensitizers (178), the effect of surfactants on photosensitizers (179), photosensitization by malarial drugs (180), nalidic and oxolinic acids (181), 6-mercaptopurine (182), azathioprine and nitroimidazole (183), 7-methylbenz[c]acridine and related products (184), naproxen, benoxaprofen and indomethacin (185), mefloquine (186), sulfamethoxazole and trimethoprim (187), benzydamine (188), photodecomposition of hydrochlorthiazide (189), tetracycline (190), frusemide (191), 6-mercaptopurine (192), 7-methylbenz[c]acridine (193), metronidazole, misonidazole and azathioprine (194), misonidazole and metronidazole (195), benzydamine (196), components in drug formulations (197) and sulfamethoxazole (198,199).

Moore has written several informative book chapters pertinent to this field. The subject matters covered are the photochemistry of diuretic drugs (200), drug-induced cutaneous photosensitivity (201), principles and practices of photodegradation studies (202), techniques for the analysis of the photodegradation pathways of medicinal compounds (203), photochemistry of photosensitizing drugs (204), photochemistry and photophysical aspects of drug stability (205,206), standardization of photodegradation studies and kinetic treatment of photochemical reactions (207) and the standardization of kinetic studies (208). He and his colleagues also studied and reported on the photolytic rearrangement of metronidazole (209), mechanisms of photosensitization by phototoxic drugs (210), enhancement of the reactivity of nitroimidazoles with UV-induced free radicals (211), photolytic rearrangement of metronidazole (212) and its common UV and γ-radiolysis photolysis products (213), reaction of non-steroidal anti-inflammatory drugs (NSAIDs) with polyacrylamide free radicals (214), oxidation photosensitized by tetracyclines (215), promotion of

UV-induced skin carcinogenesis by azthioprine (216), photochemical interaction between triamterene and hydrochlorthiazide (217) photochemistry of diclofenac (218), kinetic treatment of photochemical reactions (219) and a direct electron paramagnetic resonance (EPR) and spin-trapping study of light-induced free radicals from 6-mercaptopurine and its oxidation products (220).

Anderson

Nick Anderson of the Sanofi-Winthrop (retired), was the European industry representative to the ICH and one of the researchers who brought to the world's attention, the many different photon sources being used in the industry in Europe, prior to the ICH effort (221). He presented an invited paper titled "Light Stability Studies II" (222) at the Eighth Annual Meeting and Exposition of the American Association of Pharmaceutical Scientists in 1993. This paper, Part 2 of a joint presentation with Piechocki of the U.S. Food and Drug Administration (FDA), was designed to give the industry an idea of what the ICH was doing, what FDA was thinking about with regards to photostability testing and how the industry might apply the new guideline to their problems.

Anderson was very active in European activities related to the topic of pharmaceutical photostability testing. This activity included problems with "full-spectrum lamps" (223) as well as the quinine actinometer (224). He is also the author of a book chapter titled "Photostability Testing: design and interpretation of tests on drug substances and drug forms" (225).

Beaumont

Terry Beaumont of Glaxo Smith Kline (GSK), U.K. (retired), was the Rapporteur for the ICH Quality Group on Photostability from its outset until the guidelines final acceptance. During this time, his service was exemplary and very difficult given the blending of different cultures, philosophies and the politics involved (226). The fact that any document evolved from the process without changing his great wit and humor is a sign of the mettle of the man.

A good presentation of just what the working group's philosophies and objectives were can be found in a chapter he wrote on photostability testing (227) for a recent book chapter on the topic of photostability testing. There are several startling revelations to be found there, namely:

1. No reference was made to the German surveys by Kerker (65) or Thoma (71).
2. Few instances of companies in the United States using xenon light sources were reported.
3. In the United States most evaluations were aimed at mimicking pharmacy indoor lighting with white fluorescent sources at the 500 to 1500 ft-candle level (500–1500 lux) for between one and four weeks.
4. Daylight radiation varied during the course of a day, throughout the year, and by geographical location.
5. Different exposure levels might be encountered in the three regions identified by the ICH Stability Expert Working Group.
6. Different light sources within the industry are used already to successfully evaluate photostability.
7. There are different ways of measuring light intensity.

As to item 1, the surveys of Thoma and Kerker are important because they show a great preponderance for xenon lamps use in Europe. These surveys taken together with those of the United Kingdom would show xenon to be the preferred source for that region.

For item 2, the American data used for the ICH survey was obtained from a survey taken of stability, not development labs. Most major companies do have xenon lamp based units in their development labs (personal communication).

Item 3 reflects the work of Lachman et al., previously cited (35,38). The maximum level of illumination cited 1500 ft-candles is more than what anyone might find in a pharmacy. Very bright workplaces have 500 ft-candle levels which many people find irritating.

Items 4 and 5 are correct and have been taken in to account by the Commission Internationale de L'Eclairage [CIE, International Commission on Illumination] in their development of the D_{65} standard, the basis for other national and international standards.

Item 6 assumes that the finding of photoinstability at any level is the equivalent of quantitative reproducibility. The concepts of "significant change" and "equivalence" have quantitative limits. How can one have a reproducible procedure if the major source of the problem is not well defined and controlled? This is the reason why much of the published literature is not in agreement.

Item 7 shows that the group recognized that many different measuring systems for intensity were known to exist and were known to be nonequivalent. The group chose not to use a well-developed and documented actinometer and selected an actinometer that could not qualify for listing in the IUPAC Actinometer list of actinometers. A number of articles (176,224,228) and presentations (229,230) also questioned its acceptance, but were ignored. The only paper in support of the use of this substance as an actinometer was published by Drew et al. (231) only served to document the shortcomings of this system for actinometry. This fact was also acknowledged by Drew at Photostability 97, 2nd International Meeting, Photostability of Drugs held in Pavia, IT, September 14–17, 1997.

The fact that the Expert Working Group agreed that, "It is important to note that the spectral power distributions for each option are substantially different" questions if they were developing a political or a scientific document. These items bring great concern as to just what exactly was "the purpose of the Expert Working Group." How were they going to correct the current situation of the nonstandardization of test sources, procedures and interpretations that have led to differences, even in the Pharmacopoeias?

No better example of their confusion can be given than the statement, "The case for a single reference light source that could be used legally in the event of disputes was extensively debated but was extensively rejected, as there was concern that this reference would become the only light source used and accepted." If this were truly the case, there would be no need for any of the Pharmacopoeias and their reference standards, for the same recently given reason—which in reality is not the case.

While Beaumont does report that the Working Group thought that a single defined CIE source spectrum was a better selection than an actual source, they failed to recognize the fact that such a standard is based upon a defined source, daylight. They wrongly note in their discussions on Option 1, the CIE ID_{65} is a standard; ID_{65} is not an official CIE standard nor it's equivalent. With all of the just presented problems, it appears incongruous that they should have decided "There is clearly an opportunity for specialist manufacturers to provide equipment with defined spectral distributions together with information on absolute outputs so that the equipment can be used with the minimum of subsequent validation and calibration." How is this possible given the lack of a standard on which the industries can base such a development program?

Piechocki

Joseph Piechocki of the FDA (retired) was the drafter of FDA's first position paper on pharmaceutical photostability testing, a member of FDA's Expert Working Group on Photostability Testing, a founding member of the PDIG, Chairman of the CIE Technical Committee-50: Photodegradation of Pharmaceuticals 1998–2003 and Chair of Photostability 99 and 01, the Third and the Fourth International Meetings on the Photostability of Drugs. He is author of several papers and book chapters on the topic, a consultant, and frequent lecturer on the topic and a founding member of the PDIG. He developed the first formal course on the "Principles of Pharmaceutical Photostability," a course given at the PDIG's PPS '99, PPS '01 and PPS '04 meetings.

Piechocki was the author of FDA's first position paper on photostability testing. As a member FDA's Expert Working Group on Photostability Testing, he presented a paper titled "Light Stability Studies—A Misnomer" (232) at ICH 2 pointing out problems with the documents terminology, particularly light, and other items to then current Six-Party Drafting Groups Committee. The Committee voted to immediately change the term "light" to "photostability" in the items title but failed to correct other wrong uses in the document. This change in the topic's title was immediately accepted by the ICH. This paper was later published (233).

Other items brought to the Committee's attention at this meeting, in this presentation (234), were the need for "experts in the field of pharmaceutical photostability testing" to assist in the drafting this document, the need to reduce the obfuscation in the document as it presently existed. Piechocki suggested the group should consider adopting IUPAC terminology, actinometry, and dosage measurement standards as well as select a single photon reference source and a ranking system as to a samples photosensitivity. Of all of these items, the need for expert assistance was considered the most urgent.

Piechocki, representing FDA, presented a paper titled "Light Stability Testing—Theory of Chemical Reactions" (235) in conjunction with Anderson who dealt with the applications of this science to industrial problems.

At a joint meeting of FDA and industry representatives in Rockville, Maryland, at FDA, Piechocki presented a paper titled "Photon Sources in Photostability Studies" (236). The purpose of this meeting was to plan a joint industry FDA photostability collaborative study. At the meeting, there was unanimous support for the use of a "full-spectrum" lamp, i.e., one containing both a UV-A and visible phosphor for such a study. Unfortunately, FDA overrode this decision just as the study was about to begin. They decided, based on an industry survey and the fact that no single lamp source could meet their and the ICH draft documents selected exposures limits, that two separate lamps should be used. This decision was challenged by Piechocki (237) and by Searle (238) an expert on the field of photostability testing as well as by Wozniak (239) from the drug industry.

Piechocki presented a paper at the 1999 Pittsburgh Conference on Analytical Chemistry titled "Pharmaceutical Photostability Testing—Black Box Chemistry" (240) to illustrate the problems with both the industry and instrument suppliers and highlight areas where opportunities for the future exist. Without a clear standard from the regulatory authorities, the instrument industry had little incentive to invest in this field.

As a frequent lecturer on photostability and its problems, Piechocki has covered such topics as photostability equipment, it's design and availability (241), designing studies within FDA and ICH guidelines (242), selecting the right lamp option (243–245), how the basic principles of pharmaceutical photostability testing apply to the ICH guidelines (246), basics and potential uses and abuses (247), the basic regulations and their application to product development (248), study design(249–254) and the facts and fiction surrounding pharmaceutical photochemistry (255).

Piechocki is author of a book chapter entitled "Selecting the right source for pharmaceutical photostability testing" in Photostability of Drugs and Drug Formulations (256) as well as several chapters in the current book that address the historical development of this field (Chapter 1), Action and Activation spectra (Chapter 5), Sources (Chapter 6), Filters, Glasses and Containers (Chapter 7), and Testing Chambers (Chapter 13) as well as being coeditor of this book.

Baertschi

Steve Baertschi of the Eli Lilly Company, Indianapolis, Indiana, U.S., is, in our estimation, the primary modern American researcher in the field of pharmaceutical photostability testing and is a founding member of the PDIG. He alone and with collaborators have numerous publications and presentations which include such topics as quinine and nitrobenzene actinometry (257–259), designing and implementing studies as per the ICH guideline (260–268), the practical aspects of pharmaceutical photostability testing (269), issues in evaluating the "in-use" photostability of transdermal patches (270,271), sample presentation (272) and source selection problems including the effect of heat on the results obtained (273).

He is the author of this volume's Chapter 9 on the "Photostability Testing Sample Presentation." As well as two others in other texts addressing "Sample Presentation for Photostability Studies: Problems and Solutions" (274) and "Photostability Testing: the Questions Most Frequently Asked" (275).

Important facts Baertschi et al. have presented include the importance of container shape, the effect of sample closure on testing, the effect of source selection on kinetics, chamber mapping and their effect the results obtained (272).

Matsuda

Yoshioko Matsuda of the Kobe Women's College in Osaka, Japan is the leading Japanese educator and publisher in pharmaceutical photostability testing. Much of his work has been published in English language journals.

One of Matsuda's papers dealing with the testing of pharmaceuticals (276) gives a very good example of his ability to clearly present an idea. The figure presented in Figure 6, taken from this paper, clearly illustrates the effect of particle size upon the result obtained.

Matsuda reported with Minimada on the coloration and photolytic degradation of sulfisomidine tablets (277), Inouye and Nakanishi on the use of film coating containing a UV absorber to stabilize sulfisidomine tablets (278), Mihara on the coloration and photolytic degradation of some sulfonamide tablets (279), Kouzukio et al. on the photostability of gelatin capsules of indomethacin (280), Ito on the action spectra for the coloration and photolytic degradation of sulphisomidine tablets (281), Itooka, and Mitsushashi on the effect of film thickness and TiO_2 concentration on indomethacin gelatin capsules (282), Mashara on a comparative

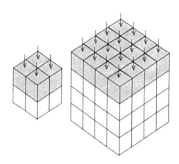

Surface area   :   4

Volume   1   :   8

Conversion (%)   50   :   25

Figure 6   Illustration of the effect of sample area on the results obtained in photostability tests *Source*: From Ref. 281.

evaluation of the coloration of photosensitive drugs under various light sources (283), Mashara on the photostability of solid-state ubidecarenone (284), Teraoka on the improved photostability of ubidecarone microcapsules achieved by incorporating fat soluble vitamins (285), Teraoka and Sugimoto on the photostability of niphedipine under normal and intensive radiation conditions (286) and Akazawa et al. on the comparative study of carbamazepine polymorphs using Fourier-transformed reflection–absorption infrared spectroscopy and colorimetric measurement (287).

In many of Matsuda's works, there is reference to the principle of "action spectra," an important procedure used to determine the region of the spectra most responsible for photodegradation. He is the only pharmaceutical photostability scientist to apply this very important procedure to actual product development work (276,281,285,286). The apparatus he used for his studies is described in one of his papers on the photostability of ubidecarenone (284).

Matsuo

Masaaki Matsuo, Deputy General Manager of the Tanabe Seiyaku Company, Ltd., Analytical Chemistry Research Laboratory (retired) was active in the development of the Japanese in country and ICH proposed photostability standard and served as the Japanese industry representative on the ICH Photostability Expert Working Group until his retirement.

Matsuo was the principal compiler of information for the Research and Technology Committees for Drug Standards, the Osaka Pharmaceutical Manufacturers Association and the Pharmaceutical Manufacturers Association of Tokyo's Basic Study on Establishing a Light Stability Testing Method both basic study and draft Light Stability Test. This information was published by Yatani et al. (288).

At ICH 2, Matsuo presented data he had obtained showing the effect of pH on the spectrum of quinine. This presentation was important because it demonstrated the need for pH control of the media.

A good presentation of the various ICH sources cited in the guideline was published by Matsuo et al. (289) and with Takeda at the First International Meeting on the Photostability of Drugs (290). This is the first published figure to compare

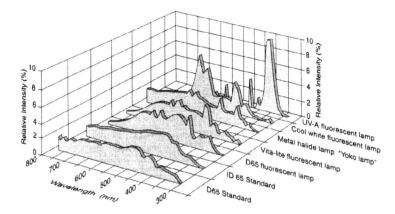

Figure 7 Three-dimensional plot of the relative spectral power distributions of the D_{65} standard, ID_{65} standard and other ICH lamps. *Source*: From Ref. 288.

directly all of spectral power distributions for all lamps specified in the ICH document. Figure 7 is a three-dimensional representation of this data.

At ICH 2, Matsuo presented data he had obtained showing the effect of pH on the spectrum of quinine. This presentation was important because it demonstrated the need for pH control of the media if accurate results were to be obtained. Figure 8 is a copy of this figure (personal communication from M. Matsuo).

Matsuo's work to help establish an international standard for pharmaceutical photostability testing, both on a national and international scale, cannot be exaggerated.

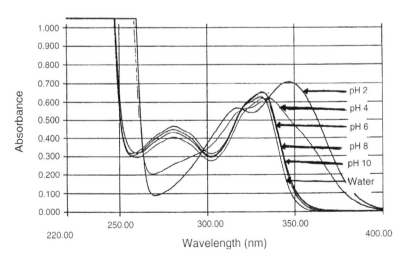

Figure 8 Illustration of the effect of various pH values on the absorption spectra of quinine.

Yoshioka

Sumie Yoshioka of the Japanese National Institutes of Health Sciences, Tokyo, Japan, was a member of the Japanese Expert Working Group and worked very closely with their industry to bring the problem of a need for a standardized testing method to everyone's attention. Her paper with Ishihara, Terazono, Tsunakawa, Murai, Yasuda, Kitamura, Kunihiro, Sakai, Hirose, Tonooka, Takayama, Imai, Godo, Matsuo, Nakamura, Aso, Kojima, Takeda and Terao titled "Quinine Actinometry as a Method for Calibrating Ultraviolet Radiation Intensity in Light Stability Testing of Pharmaceuticals" (291) attempted to develop correction factors based upon quinine actinometry for the various sources specified in the ICH guideline.

Yosihioka also served as one of the Editors of Drug Stability magazine during its short existence.

SURVEYS

There have been many surveys of industry practices related to pharmaceutical photostability testing through the years. All identify two common problems. First, there is no uniformity of practices and second, quite often, the -essential details are not reported. These problems are due to the lack of textbooks on this subject as well as the absence of regulatory standards.

The intentions of all of the researchers surveyed were of the highest order but the basic understanding as to what to report was lacking.

As previously noted, the avowed role of the ICH to relieve this situation has not been achieved. A brief summary of past surveys can possibly guide future researchers.

American

Arny

Arny (18) reported the first survey ever done to determine what the common pharmaceutical photostability testing practices of his day were. At the time, little modern equipment (radiometers, spectrophotometers, etc.) was available and window glass filtered daylight was the primary source of radiation used. All of the exposures used at the time were time based and any variations in intensities were not possible to compensate for, at that time. The results of his survey have already been discussed earlier in this chapter.

Blythe

Blythe (32) made a survey of the stability practices of 40 pharmaceutical companies. He found that a majority used exaggerated conditions of temperature, humidity, and light to test the stability of pharmaceutical products. Correlations of the data obtained with that of shelf-life data differed considerably.

Lachman et al.

Lachman and Cooper, as part of their effort to standardize accelerated stability testing procedures (36) performed surveys to determine what current practices at their time were. To get the best information, they and their associate Swartz (292) even surveyed allied fields. Based on this information, they developed what they called a "Comprehensive Pharmaceutical Stability Testing Laboratory." The basic design of this laboratory and the logic, which went into its design, are discussed in these articles. This cabinet's basic design characteristics have been reproduced in many laboratories and remain in use, in some of them, to this day (personal observation). This is also the first reported use of the newly introduced fluorescent lamp, specifically, the "cool white" in this field.

International Conference on Harmonization Survey
As part of the ICH document development process, surveys of practices was taken in each geographical region by their respective Expert Working Groups. The American Survey was performed by the Pharmaceutical Stability Discussion Group (PSDG) with the assistance of its Facilitator, O'Neill.

The survey questions were designed by the U.S. ICH Expert Working Group and sent to members without input from the PSDG. Because most members of the PSDG are concerned with quality control as opposed to research and development, their response may be slightly biased.

Results of the American survey were reported at the ICH 2 Quality Six-Party Drafting Group on Light Stability Testing, Kyoto, Japan and showed that the cool-white fluorescent lamp and a UV-A lamp were the most commonly used sources in the United States. Because the only literature source for pharmaceutical photostability testing information in the United States at this time was that of Lachman et al. (292) and they highly praised the newly available, less costly, more energy efficient and best readily available solar reproducing lamp of their time, its widespread use over the years and these results are of no surprise.

Personal communications with one of the then major xenon lamp unit manufacturers indicated that most of the major American pharmaceutical manufacturers, even at that time, had purchased them for their research and/or packaging-testing units.

Alsante et al.
Alsante et al. (262) recently reported the results a survey of American pharmaceutical companies, sponsored by Eli Lilly & Company and Pfizer Inc., U.S.

KMR, a management consulting firm (293) worked with the sponsoring companies and these researchers to assure the confidentiality of the survey. Twenty companies agreed to participate in this confidential survey, which addressed stress-testing procedures, including photostability-testing procedures.

Interesting findings of this post-ICH survey were that one company did no testing at all and 18 of 19 performed the testing on both drug substance and product. Roughly, 16% of the companies performed testing on intermediates and one only tested the drug substance.

A majority of companies (63%) use the ICH recommended over-all VIS light dosage amount while 37% used a higher dosage. Most (67%) used the ICH recommended UV dosage.

Half of the reporting companies used Option 1 for their source selection while 30% used Option 2. Two companies used both Options. Of those who did use Option 1, 86% used a xenon lamp source.

Three of the companies reported using home-built models for their exposure chambers. No indication was given as to what the design of these units might have been. For example, whether they based their design on those proposed by Lachman (36,292), a modified stability oven or the free-form design used by Drew et al. for their collaborative study was not reported.

Radiometers and photometers were used for dosage measurement. Twelve companies used an external model while seven used built-in units supplied by the manufacturer.

The difference noted between the original U.S. EWG survey and that of these authors can be attributed to several factors. Primary among these is the recognition of the time saving achieved when a high-intensity lamp is used; 10-fold or greater. When one also adds-in the savings in exposure time because only one exposure is required, only a single set of samples need be analyzed, more accurate results are obtained (free from dichromism errors) and the better agreement with actual in-use results, it is no surprise that the xenon lamp is now preferred.

European

German

Thoma and Kerker
Reinhard Kerker as part of his doctoral thesis (69), did a survey of sources used in the then (1990) German Republic. His survey, the results of which were presented by Thoma and Kerker (76), showed that the xenon lamps were the most popular, whether alone or when in combination with daylight.

English

Greenhill and McLelland
In 1990, John V. Greenhill and Margaret A. McLelland published a review chapter titled the "Photodecomposition of Drugs" (294). "This review concentrated on those drugs where one or more photoderivatives have been positively identified. Also included are photosensitizers which, although unchanged themselves, photocatalyze degradations of biomolecules." Their searches included the 1980 and 1988 editions of the British Pharmacopoeia (BP) "for all drugs which carry a 'protection from light' warning and those programmed into the Chemical Abstract Services (C.A.S) on-line search, which covered Chemical Abstracts from 1967 to the end of 1988."

These authors also noted "A cursory glance in the BP 1988 edition revealed several drugs which did not carry the 'protection from light' warning although the available evidence suggests that such would be merited." This discrepancy, observed in the United States, also still exists in other jurisdictions and internationally.

Two other important factors noted by them, which still persisted 30 years after the works of Lachman et al. are that the number of different sources that were still used for photostability testing and most importantly, "It is not always clear in published material what wavelengths and intensity light have been used…"

Anderson et al.
Nick H. Anderson et al. followed up on this information by doing a then current United Kingdom industry practices survey in 1991 (222). This survey proved that the diversity of practices in sample presentation, source selection and exposure times and the need for regulatory guidelines.

Japanese
In 1992, the pharmaceutical manufacturers of Tokyo and Osaka, the Japanese Pharmaceutical Manufacturer's Association (JPMA) sent a questionnaire to all of

their members to determine what current pharmaceutical photostability-testing practices were prevalent in their industry. This action was prompted by their recognition that no standard method existed even though their pharmacopoeia, as well those of other major countries contained substances and products which were to be labeled "unstable to light" or some variation thereof.

Based on the results obtained, a testing protocol was developed, encompassing as many of the practices as possible. Fluorescent, xenon and metal-halide lamps were all deemed to be equivalent for photostability testing. The protocol developed was used for a collaborative study by 49 member companies of 175 items described in the Japanese Pharmacopoeia as being "unstable to light." Based on the results obtained, a proposed monograph General Test was sent to the Japanese Pharmacopoeia.

The Japanese protocol was submitted to the ICH and with a few minor modifications, to become, eventually, the current ICH document.

INTERNATIONAL CONFERENCE ON HARMONIZATION GUIDELINE DEVELOPMENT

ICH Steering Committee was established and had its first meeting in April 1989. The main purpose of this organization was "...to seek ways to eliminate redundant and duplicative technical requirements for the registering new medicinal substances and products. The main objective of this understanding is to expedite global development and availability of new medicines without sacrificing safeguards on quality, safety or efficacy."

ICH 1, the first international ICH meeting, was held in Brussels in November of 1991. This meeting, sponsored by the Commission of the European Communities and the European Federation of Pharmaceutical Industries' Associations (EFPIA), was composed of three parallel workshops on the topics of Quality, Safety, and Efficacy and identified 11 topics for harmonization.

The topic of Pharmaceutical Stability Testing actually started on June 5–7, 1991 during discussions held in parallel with a conference organized by the APV in Darmstadt, Germany on the topic of "Stability Testing, New Trends and Requirements: An International Symposium for the EC, Japan and the USA."

All drafts of the ICH Guideline state in their preamble that "The guideline seeks to exemplify the core data required for light stability testing of new drug substances and products. It is not always necessary to follow this when there are scientifically proven justifiable reasons for using alternative approaches." Unfortunately, little scientific justification is given for the various options and selections listed. Details of the procedures used to obtain the data are lacking. The Expert Committees failed to seek input from experts in the field, as was brought to their attention at ICH 2 (232), before the proceedings began.

Draft 1

On September 15–16, 1992, the first draft guideline of what was then called "Light Stability" was distributed to all ICH participants for submission of comments by the end of October and subsequent discussion at the forthcoming Brussels conference in March 1993. This document was based on the results of a Japanese study of pharmaceutical photostability practices in their country.

The first draft, resulting from the initial circulation, was dated October 22, 1992 and contained two options, 1 and 2. Option 1 specified use of a white fluorescent lamp with a bandwidth from 450 to 650 nm, a Thorn Pluslux 3500 or equivalent. Option 2 was to be a near UV lamp with a bandwidth from 300 to 400 nm and a peak maximum between 355 to 370 nm. A Philips black light, Toshiba FL20L-BL chemical lamp or artificial daylight lamp were all considered acceptable. Users of the UV-A lamps were advised to burn them in for 100 to 500 hours.

The exposure time for Option 1 was stated to be 30 days, for a total illumination of not less than (NLT) 10^4 lux hours. For Option 2, the exposure time was to be for a period of NLT 24 hours and no irradiance was specified.

No references were made to types of sources used to develop the proposed guideline information or the instruments to be used for dosage measurement. The quinine actinometer was not mentioned in this draft.

Draft 2

The second draft was issued in June of 1993. This draft added statements such as, "With liquid products it may be appropriate to only test in the original container" and "Certain products such as infusion liquids may require specific testing to underwrite their light stability in-use."

Option 1 specifications remained the same, except that references to any particular lamp were removed. The original Options 1 and 2 were now combined and Option 1 became a two-part specification. Part 1 for the VIS range and 2 for the UV range. Option 2 now became a specification for "Use (of, authors insert) a xenon lamp fitted with a window glass filter to eliminate radiation below 310 nm." A new Option 3 was introduced. Option 3 covered the "Use (of, authors insert) a metal halide lamp fitted with a window glass filter to eliminate radiation below 310 nm." The particular metal-halide lamp (s) that were acceptable were not specified.

"Recipients were requested to propose the level of exposure in terms of energy output. Inclusion of an approximate relationship between this and real exposures would be welcomed" for all three of these options. The JPMA proposed a 24-hour exposure time for part 2 of Option 1. The EFPIA suggested an exposure time of 24 hours for Option 2 (no irradiance/illumination level given) while the JPMA suggested and overall illumination of 10^6 lux hours, considered equivalent "to 90 days average exposure." The maximum energy emission of the UV-A lamp was widened in this draft to now include 350 nm, for a new specification of "between 350 and 370 nm".

Also added in this draft was a reference on how the VIS exposure should be measured. For Option 1, it was suggested that "White fluorescent lamp (Option 1)—measure light intensity with a lux meter."

Quinine was introduced as a "…primary calibration standard for light testing." In this draft, a "…5% solution of quinine monochloride dihydrate in water…" Approximately 20 mL of this solution, hermetically sealed in two colorless ampoules (Japanese Industrial Standards No. 5) colorless ampoules or equivalent were to be used. One of these ampoules was to be protected from exposure by being wrapped in aluminum foil, the other exposed at a distance of about 30 cm for three hours. It was specified that "the transmittance of the exposed solution in 1 cm cells should lie between 55% and 60%, and that the control should show no significant change (about 70%)." Radiometers were permitted as "…secondary standard UV measuring devices."

Two literature references were added to this draft. One reference is to an article by Thoma and Kerker (69) which was misidentified and the other was to the original Japanese study results published in Iyakuhin-Kenkyu, which did not identify the journal in sufficient detail.

Draft 3

This draft, issued in November 1993, included for the first time a decision tree detailing how a product should be treated. This scheme proposed that if the directly exposed product was stable, the testing should end and no labeling statement would be required. This discounted any possible contribution of packaging leachables to the photoinstability of drug substances and products.

A statement regarding temperature control was added in this draft. "To eliminate the effect of localized temperature changes it is necessary either to maintain an appropriate control of the temperature or to include a dark control in the same environment."

Option 1, part 2 was changed to eliminate the use of a daylight fluorescent lamp, but the use of this lamp as a possible substitute for both parts of Option 1 was suggested, "…provided the source meets the appropriate calibration criteria," which were not specified.

Options 2 and 3 remained the same. A statement that "The power input to the lamps or the distance of the light source from the samples may be adjusted to provide the appropriate intensity," was added to this draft. For stress testing, light with an overall illumination of 1.2 million lux hours and an integrated UV energy of NLT 200 W hr/m^2 were proposed. This draft also added "Samples may be exposed side-by-side with the appropriate actinometric system to ensure the specified light exposure is obtained or for the proper duration of time with validated and calibrated light sources."

Comments were requested "…on the suitability of the relative spectral distribution of the range of light sources that could be encompassed within Option 3."

The concentration of the quinine actinometric solution was changed to 2% because of solubility problems with the 5% solution and the absorbance change specification changed to at least 0.6 at 400 nm. An editorial note allowed for the possible future revision of this value to 0.8.

Dimensional specifications, devoid of thickness values, for the ampoule to be used were presented along with a note stating that transmission characteristics would be provided in due course.

Maintenance of the physical characteristics of samples under test was to be maximized. The methods should be chosen to prevent this and should "…provide the minimal interference with the irradiation of samples under test." The particle size of powders was to be "…with a particle size distribution representative of material as released for use…" and "…a thickness of not more than (NMT) 5 mm."

Draft 4

This draft was issued in September 1994 and included a revised decision-tree. The tree was now divided into two parts, drug substance and drug product.

The establishment of a reference method was suggested and reference was made to the distinct difference between the United States and EC, with regard to test sources. The former preferring as a reference source UV-A plus a fluorescence source while the latter favored the xenon source.

The specification for the white fluorescent lamp for Option 1 was proposed to be that of ISO 10977, with a wavelength range of 300 to 800 nm. It was also proposed that a light source should be comparable to the spectral distribution of the standard and that said distribution be divided into intervals according to DIN 53387 and the limits of each interval be those specified in said standard.

The cut-off frequency for the window-glass filter used for Options 2 and 3 was raised to 320 nm, even though the standardization was specified to be for the 300 to 800 nm range for all three options.

Under the "Choice of Light Sources" Option 1 was listed as the xenon lamp. Option 2 as the metal-halide and Option 3 as the "Fluorescent lamp matching the spectral distribution of the UV and visible components of indoor indirect daylight between 300 and 800 nm."

It was noted in this draft, "The Committee for Proprietary Medicinal Products (CPMP) is keen to have the information on the quinine actinometer formally published and accepted as an internationally recognized standard.

Testing of a single batch during the development phase and another single definitive pilot scale or larger was proposed as sufficient to establish the photostability of a product, FDA noted its preference for three batches.

The CPMP proposed that for accelerated testing window glass filtered daylight be used or alternatively "...6000 W hr/m^2 with a spectral distribution defined according to the ISO 10977 (DIN 53387), including approximately 9% UV light, equating to 540 W hr/m^2 and 1.1 M lux." These figures were based upon an average global radiation of 572 W/m^2 (defined by the CIE), 6000 W hr/m^2 corresponds approximately to 10.5 hours (direct daylight without protection)."

The volume of the 2% quinine monochloride dihydrate in water to be used in a 20 mL ampoule was reduced from 20 to 10 mL. The change in absorbance was changed from at least 0.6 to 0.8 absorbance units at 400 nm.

The two literature references added in the third draft were removed.

Final Document

The final document was "signed off" by the six cosponsors at Step 2 in the ICH process November 29, 1995 at the ICH 3 meeting in Yokohama, Japan and transmitted to the three regional regulatory agencies for formal consultation in accordance with their normal internal and/or external consultation process.

In this document, the decision-tree was again revised to cover only drug products. No reference was made the drug substance in this tree.

The number of ICH source options was reduced to two, with Option 1 listing 5 separate standards one actual, i.e., D65 and ID65 (ISO 10977) a synthetic window-glass filtered daylight a daylight fluorescent lamp, xenon or metal halide lamp. "For a light source emitting significant radiation below 320, a window glass filter may be outfitted to eliminate such radiation."

Option 2 is defined as, "A cool-white fluorescent lamp as defined in ISO 10977 (1993)" and "A near ultraviolet fluorescent lamp having a spectral power distribution from 320 to 400 nm with a maximum transmission energy emission between 350 and 370 nm; a significant proportion of UV should be in both bands of 320 to 360 nm and 360 to 400 nm."

The sample depth for a powdered drug substance was reduced to "...no more than 3 mm." This, in contrast to the previously stated NMT 5 mm. "Where solid drug substance samples are involved, sampling should ensure that a representative portion is used in individual tests. Similar sampling considerations, such as homogenization of the entire sample, apply to other materials that may not be homogeneous after exposure."

A reference to the use of "Quinine actinometry as a method for calibrating ultraviolet radiation intensity in light stability testing of pharmaceuticals was added."

REFERENCES

1. Scheele. Aris atque ignis examen chemicum. Upsala Leipzig 1777.
2. Ritter (1801). Coebergh JBM. Beschutting Van Geneesmiddelen Tegen den den Invloed Van Licht. Ph.D. dissertation, Utrecht University, The Netherlands 1920.
3. Davy (1812). Coebergh JBM. Beschutting Van Geneesmiddelen Tegen den den Invloed Van Licht. Ph.D. dissertation, Utrecht University, The Netherlands 1920.
4. Becquerel E. La Lumire, Ses Causes et Ses Effets. 1868. Coebergh JBM. Beschutting Van Geneesmiddelen Tegen den den Invloed Van Licht. Ph.D. dissertation, Utrecht University, The Netherlands 1920.
5. Torosiewice Theodor von R, Coebergh JBM. Beschutting Van Geneesmiddelen Tegen den den Invloed Van Licht. Ph.D. dissertation, Utrecht University, The Netherlands 1920.
6. Nederlandse Farmacopee (4 ed.).
7. Schoorl N, Van Den Berg LM. De ontleding van bromoform onder invloed van licht en lucht. Pharm Wkblad 1905; 43:2–10.
8. Schoorl N, Van Den Berg LM. De Ontleding van chloroform onder invloed van licht en lucht. Pharm Wkbld 1905; 43:877–888.
9. Schoorl N. De chemicsche artikelen van de vierde uitgave der Nederlandsche Pharmacopee. 1907; (2 dl.) Holdings of Utrech University, The Netherlands.
10. Schoorl N. Dankbare proeven over pharmaceutische chemie. 1932:56. Holdings of Utrech University, The Netherlands.
11. Schoorl N. De verhouding van de pharmacie tot het hooger onderwijs. Utrecht Oosthoek 1908; 48.
12. De Graff WC, Schoorl N, der Wielen P. Commentaar op de Nederlandsche Pharmacopee 4 dl. Utrech Oostjoek, 1927–1931.
13. Bordier H. Action of light on iodine and starch iodide in water. Compt Rendu 1916; 163:205–206.
14. Bordier H. Action X-rays on iodine and starch iodide in aqueous media. Compt Rendu 1916; 163:291–293.
15. Eisenbrand J. Protection of light-sensitive drugs by colored glass. Pharm Ztg 1927; 72:247–248.
16. Eisenbrand J. Protection of the light sensitive drugs, II. Pharm Ztg 1927; 72:1275–1276.
17. Eisenbrand J Protection of light-sensitive drugs by means of colored glassware. III. Pharm Ztg 1929; 74:263.
18. Arny HV. Report of committee on colored glass containers. JAPhA 1928; XVI(No.10): 999–1003.
19. Coebergh JBM. Beschutting Van Geneesmiddelen Tegen den den Invloed Van Licht. Ph.D. dissertation, Utrecht University, The Netherlands 1928.
20. Arny HV. Report of committee on colored glass containers. JAPhA 1928; XVII(10): 1056–1060.
21. Corbergh JBM. Protection of medicines against the action of light. Pharm Wkblad 1920; 57:1452–1457.
22. Hogstad A. Effect of glass on cresol deterioration, not light. (see Arny Reference 18) 1001.
23. Tapley MW, Giesy PM. Light sensitivity of bismuth subcarbonate. JAPhA 1929; 15:46.
24. Arny HV, Taub A, Steinberg A. Deterioration of certain medicaments under the influence of light. JAPhA 1931; 20(10):1014–1023.
25. Steinberg A. Deterioration of Chemicals and Pharmaceuticals Under the Influence of Light. Ph.D. Dissertation, Columbia University, School of Pharmacy, NY, NY, June 3, 1931.
26. Arny HV, Taub A, Steinberg A. Deterioration of certain medicaments under the influence of light. JAPhA 1931; 20(10):1153–1158.
27. Arny HV. Further studies with Light Rays. The Glass Packer 1933; 12:497–499.
28. Arny HV, Taub A, Blythe RH. Determination of certain medicaments under influence of light. JAPhA 1934; 23(7):672–679.
29. Arny HV. Glass standardization JAPhA 1933; XXII(11):1176.

30. Brunner K. Einwirkung des lichtes auf arzneimittel. Pharm Zentr 1936; 77:721–723.
31. Blythe RH. Tests on shelf life and stability. Scientific Section of the Scientific Section of the American Pharmaceutical Manufacturers Association Meeting, New York City, NY, February 2, 1954.
32. Garrett ER, Carper RF. Prediction of stability in pharmaceutical preparations: color stability in a liquid multisulfa preparation. JAPhA 1955; 44:515–518.
33. Garrett ER, Prediction of stability in pharmaceutical preparations, vitamin stability in liquid multivitamin preparations. JAPhA 1956; 45:171–178.
34. Garrett ER, Prediction of stability in pharmaceutical preparations: Comparision of vitamin stabilities in different multivitamin preparations. JAPhA 1956; 45:470.
35. Lachman L, Cooper J. A comprehensive pharmaceutical stability-testing laboratory. JAPA 1959; XLVIII(4):226–233.
36. Kuramoto R, Lachman L, Cooper J. A study of certain pharmaceutical materials on color stability. JAPhA 1058; XLVII(3):175–180.
37. Lachman L, Swartz CJ, Urbanyi T, Cooper J. Color stability of tablet formulations II. JAPhA 1960; 49(3):165–169.
38. Lachman L, Swartz CJ. Cooper J. A comprehensive pharmaceutical stability testing laboratory III: A Light stability cabinet for evaluating the photosensitivity of pharmaceuticals. JAPhA 1960; 49(4):213–218.
39. Reisch J Der photochemische zerfall von pyrazolin-(3)-on-(5)-derivaten. Pharmazie 1966; 21:183.
40. Reisch J, Abdel-Khalek M. Part 66: [(E-1–Acetyl-2–(3,3–dimethyl-1–phenyl-ureido)-3–oxo-but-1–enyl]-phenyl-carbaminsure, ein neues photoprodukt desinophenazons. Acta Pharm Turcica 1993; 35:31–36.
41. Reisch J, Abdel-Khalek M. Zur fotooxidation von kristallinem aminophenazon. Pharmazie 1979; 34:408–410.
42. Reisch J, Ekiz N, Gneri T. Fotoabbau von 4-isopropyl-2,3-dimethyl-l-phenyl-3-pyrazolin-5-on (propyphenazon) 49. Mittellung: foto und strahlenchemische studien. Pharmazie 1986; 41:287–288.
43. Reisch J, Ekiz N, Takacs M. Photochemische studien, 52. Mitt. untersuchungen zur Photostabillät von testosteron und methyltestosteron im kristallinen zustand. Arch Pharm 1989; 322:173–175.
44. Reisch J, Ekiz-Gucer N, Gneri T. Photo- und strahlenchemische studien, 46, Mitt. zur photochemie des isopropylaminophenazons in kristallinen zustand und wsseriger lsung. Arch Pharm 1986; 319:973–978.
45. Reisch J, Ekiz-Gucer N, Takacs M, Henkel G. Photochemische studien, 54. Photodimerisierung von testosteronproplonat in kristallinem zustand und kristallstruktur von testosteronproplonat. Liebigs Ann Chem 1989; 595–597.
46. Reisch J, Ekiz-Gucer N, Takcs M, Gunaherath GM, Kamal B. Photochemische studien, 53. Mitt. ber die photoisomerisierung des azapropazons. Arch Pharm 1989; 322: 295–296.
47. Reisch J, Ekiz-Gucer N, Tewes G. Photochemische studien, 63. Photostabilität einiger 1,4–benzodiazepine in kristallinem zustand. Liebigs Ann Chem 1992:69–70.
48. Reisch J, Fitzek A. Über die zersetzung von wässrigen amidopyrinlsungen unter dem einfluss von licht und γ-strahlen. Dtsch Apoth Ztg 1967; 107:1358–1359.
49. Reisch J, Fitzek A. Photolyse von 3-Pyrazolinonen-(5). Arch Pharm 1974; 307:211.
50. Reisch J, Henkel G, Ekiz-Gucer N, Nolte G. Untersuchungen zur photostabilisuft von halometason und prednicarbat in kristallinem zustand sowie deren kristallstrukturen. Liebigs Ann Chem 1992; 63–67.
51. Reisch J, Henkel G, Nolte G. 62. Untersuchungen zur photostabilität von halometason und prednicarbat in kristallinem zustand sowle kristallstruktur von halometason. Liebigs Ann Chem 1992; 63–67.
52. Reisch J, Henkel G, Topaloglu Y, Simon G. Crystal structure of santonin: contribution to the lattice controlled photodimerization. Pharmazie 1988; 43:15–17.
53. Reisch J, Iranshahi L, Ekiz-Gucer N. Photochemische studien, 64. Photostabilität von glucocorticoiden in kristallen zustand. Liebigs Ann Chem 1992; 1199–2000.

54. Reisch J, Mller M, Mnster G. Ber die photostabilität offizineller arznei- und hilfsstoffe 1: Barbiturate. Pharm Acta Helv 1984; 59:56–61.
55. Reisch J, Niemeyer DH. Photochemische studien am sulfanilamide sulfacetamid und 4–aminobenzoeäsureäthylester Arch Pharm (Weinheim) 1972; 305:135–40.
56. Reisch J, Ossenkop WF. Ist bei der aufbewahrung von Phanazon Lichtschutz erforderlich? Pharm Ztg 1971; 116:1472.
57. Reisch J, Topaloglu Y. Ist die dunschichtchromatographie zur onkontrolle der lichtstanilität von arzneistoffen geeignet? Pharm Acta Helv 1986; 61(5–6):142–147.
58. Reisch J, Weidman KG. Photo- und radiolyse des chloramphenicols. Arch Pharm 1971; 304:911–919.
59. Reisch J, Weidman KG. Notiz zur photochemie des dichloracetamids. Arch Pharm 1971; 304:920–922.
60. Reisch J, Weidman KG. Photochemie des p-nitrobenzaldehyds. Arch Pharm 1971; 304: 906–910.
61. Reisch J, Zappel J. Photochemische studien, 70. Photostabilität suntersuchungen an natrium-warfarin in kristallinem zustand. Sci Pharm 1993; 61:283–286.
62. Reisch J, Zappel J, Henkel G, Ekiz-Gcer N. Photochemische studien, 65. Mitt. Untersuchungen zur photostabilität von ethisteron und norethisteron sowie deren kristallstrukturen. Monatshefte fr chemie 1993; 124:1169–1175.
63. Reisch J, Zappel J, Raghu A, Rao R. Photochemical studies part 68. Solid state photochemical investigations of methylprednisolone, prednisolone and triamcinolone acetonide. Acta Pharmaceutica Turcica 1995; XXXVII:13–17.
64. Reisch J, Zappel J, Raghu A, Rao R. Photostability studies of ouabain, (-acetyldigoxin and digoxin in the solid state. Pharm Acta Helv 1994; 69:47–50.
65. Reisch J, Zappel J, Rao AR, Henkel G. Dimerisation of levonorgestrel in solid state ultraviolet light irradiation. Pharm Acta Helv 1994; 69:97–100.
66. Thoma K. Arzneimittelstabilität. Franfurt am Main: Johann Wolfgang Goethe-University, 1978:98–116.
67. Thoma K. Stabilizing light-unstable drug materials. German Patent No. DE 3136282, issued 1983–03–23, applicant HOECHST AG (DE):pp 14.
68. Kerker R. Untersuhungen zur Photostabilität von Nifedipin, Glucocorticoiden, Molsidomin und Ihren Zubreirungen. Dissertation der Fakultt fr Chemie und Pharmazie der Ludwig Maximillians Universitt, Mnchen, Germany, 1991. (Available from Dissertations Und Fotodruck Frank GmbH, 8000 Mnchen 2, Gabelsbergerstrasse 15, Germany).
69. Thoma K, Kerker R. Photoinstabilität von arzneimittelin, 1. Mittelung ber das verhalten von nur im UV bereich absorbierenden substanzen bei der tagleslichtsimulation. Pharm Ind 1992; 54(1):169–77.
70. Thoma K, Kerker R. Photoinstabilität von arzneimittelin, 2. Mittelung ber das verhalten von im sichtbaren bereich absorbierenden substanzen by der tageslichtsimulation. Pharm Ind 1992; 54(3):287–293.
71. Thoma K, Kerker R. Photoinstabilität und stabilisierung von arzneimittelin. Moglichkeiten eines allgemein anwendbaren stabilisierungsprinzips. Pharm Ind 1991; 53(5):504.
72. Cole C, Forbes PD, Davies RD. Different biological effectiveness of blacklight fluorescent lamps available for therapy with psoralens plus ultraviolet A. J Acad Dermatol 1984; 11(1):599.
73. Sequeira F, Vozone C. Photostability studies of drug substances and products. Pharm Tech 2000:30–35.
74. Thoma K. Stability of Drugs–Current Problems in Pharmaceutical Technology. 10th Conference on Pharmaceutical Technology, Shirakabako, Japan, July 1985:1–20.
75. Thoma K, Kerker R. Photostabilität von arzneimitteln ein vernachläsigtes qualitätskriterium? APV Symposium Regensburg, 14 April, 1992.
76. Thoma K, Kübler N. Anwendung der HPLC-MS-kopplung fr die photostabilitätsprfung. Pharmazie 1996; 51(12):940–946.
77. Thoma K, Kuebler N. Photodegradation of active ingredients and current methods for its analytical investigation. Pharmazie 1996; 51(12):919–923.

78. Thoma K, Kübler N. Einflubder wellenlänge auf die photozersetzung von arzneistoffen. Pharmazie 1996; 51(9):660–664.
79. Thoma K, Kübler N. Einflub von hilfisstoffen auf die photozersetzung von arzneistoffen. Pharmazie 1997; 52(2):122–129.
80. Thoma K, Pfaff G. Use of solubilizers as adjuvants in the production of drugs. Part 5. Effect of solubilization with polyethylene glycol-glycerol fatty acid esters on the photostability of menadione. Pharmazie 1976; 31(7):477–480.
81. Thoma K, Pfaff G. Relations between the micellar bond and the photoinstability of solubilized menadione. Part 6. Use of solution aids as adjuvant substances in the preparation of drugs. Pharmazie 1976; 31(9):628–629.
82. Thoma K, Struve M. Einfluss hydrophiler osungsmittel auf die instabilität von arzneistoffen. Pharm Ind 1986; 48:179–183.
83. Aman W, Thoma K. The influence of formulation and manufacturing process on the photostability of tablets. Int J Pharmaceut 2002; 243:33–41.
84. Thoma K. Photodecomposition and stabilization of compounds in dosage forms. In: Tonnesen HH, ed. Photostability of Drugs and Drug Formulations. London: Taylor & Francis Ltd., 1996:111–140.
85. Thoma K, Klimek R. Photostabilization of drugs in dosage forms without protection from packaging materials. Int J Pharm 1991; 67(2):169–175.
86. Thoma K, Aman W. Influence of Blister Colouration on the Photostability of Nifedipine Tablets 15th Pharmaceutical Technology Conference, Oxford, U.K., March 19th-21st, 1996, Kongreband. Vol. 3:114–127.
87. Aman W, Thoma K. Photostabilization of tablets by blister covers. Pharm Ind 2002; 64(12):1287–1292.
88. Thoma K, Klimek R. The photoinstability of nifedipine. Part 3. Photostabilization of nifedipine in pharmaceutical formulations. Pharm Ind 1991; 53(4):388–396.
89. Thoma Karl, Holzmann C. Dithranol formulations. Stability in pharmaceuticals. Deutsch Apoth Ztg 1997; 137(38):3280–3286.
90. Thoma Karl, Strittmatter T, Steinbach D. Studies on the photostability of antibiotics. Act Pharm Tech 1980; 26(4):269–272.
91. Thoma K, Kbler N. Untersuchung der photostabilität von antimykotika 1. Mitt. Photostabilität von azolantimykotika. Pharmazie 1996; 51(11):885–893.
92. Thoma K, Kubler N. Photodegradation of antimyotic drugs. Part 2. Communication. photodegradation of polyene antibiotics. Pharmazie 1997; 52(4):294–302.
93. Thoma K, Kbler N, Reimann E. Untersuchung der photostabilität von antimykotika 3. Mittelung: photostabilität lokal wirksamer antimykotika. Pharmazie 1997; 52(5):362–372.
94. Thoma K, Kbler N, Reimann E. Untersuchung der photostabilität von antimykotika 4. Mittelung: photostabilität von flucytosin and griseofulvin. Pharmazie 1997; 52(6):455–302.
95. Thoma K, Struve M. Studies on the photo- and thermostability of adrenaline solutions. Part 1. Stability of adrenaline solutions. Pharm Acta Helv 1986; 61(1):2–9.
96. Thoma K, Kerker R. Photoinstabilität von arzneimitteln 6. Mitteilung: untersuchungen zur photostabilität von molsidomin. Pharm Ind 1992; 54:630–638.
97. Thoma Karl, Vasters E. Photoinstability and stabilization of phenothiazine derivatives. I. colloidal association of chlorpromazine, triflupromazine, and homophenazine. Pharm Ztg 1974; 119(37):1430–1435.
98. Thoma K, Kubler N. Photodegradation of quinolones. Pharmazie 1997; 52(7):519–529.
99. Thoma K, Klimek R. Stabilittsspezifische polarographische gehaltsbestimmung von nifedipin in arzneiformen. Dtsch Apoth Ztg 1980; 120:1967–1972.
100. Thoma K, Klimek R. Untersuchungen zur photoinstabilität von nifedipin, 1. Zersetzungkinetik und reaktionmechanismus. Pharm Ind 1985; 47(2):207–215.
101. Thoma K, Klimek R. Untersuchungen zur photoinstabilität von nifedipin, 2. Einfluss von milieubedingungen. Pharm Ind 1985; 47(3):319–326.
102. Thoma K, Klimek R. Untersuchungen zur photoinstabilität von nifedipin 3. Mitt. photoinstabilität und stabilisierung von nifedipin in arzneizubereitungen. Pharm Ind 1991 53: 388–396.

103. Thoma K, Kerker R. Photoinstabilität von arzneimittelin, 3. Mittelung: photoinstabilität und stabilisierung von nifedipin in arzneiformen. Pharm Ind 1992; 54(4):359.
104. Thoma K, Kerker R. Photoinstabilität von arzneimittelin, 4. Mittelung: untersuchung zu den zersetzungprodukten von nifedipin. Pharm Ind 1992; 54(5):465.
105. Thoma K, Aman W. Influence of Blister Colouration on the Photostability of Nifedipine Tablets. 15th Pharmaceutical Technology Conference, Oxford, U.K., March 19th–21st, 1996, Kongreband, Vol. 3, 114–127.
106. Stahl PH. I. Prformulierung: Die Voraussetzung zur Herstellung Haltbarer Arzneiformen: 15–31, 1985. Ein Seminar der APV vom 20–22 Mai, 1985, Whurzburg. APV. Vol. 15. Stuttgart: Wissenschaftliche Verlagsgesellschaft, 1986.
107. Gauglitz G. Modern chemical actinometry. Eur Photochem Assoc (EPA) Newsletter 1983; 83(Nov):49–53.
108. Gauglitz G. Actinometry. Studies in Organic Chemistry. Amsterdam: Elsevier Science Publishers, 1990:883–902.
109. Frank R, Gauglitz G. Measurement of radiation intensities by chemical actinometers. CAV 1978:19–22.
110. Gauglitz G. Photokinetische grundlagen moderner chemischer aktinometer zeitschrift fiir physikalische chemie. Neue Folge 1984; 139:237–246.
111. Brauer HD, Drews W, Schmidt R, Gauglitz G, Hubig S. Chemical actinometry in the visible (475–610 nm) range. Photochem Photobiol 1983; 37:595–598.
112. Brauer HD, Schmidt R. A new reusable chemical actinometer for UV irradiation in the 248–334 nm range. Photochem Photobiol 1983; 37(5):587–591.
113. Kuhn HJ, Braslavsky SE, Schmidt R. Chemical actinometry. Pure Appl Chem 1989; 61(2):187–210.
114. Braslavsky SE, Kuhn HJ. Provisional list of actinometers. EPA Newsletter 1987; 29:49–60.
115. Tonnesen HH. Photostability of Drugs and Drug Formulations. 1st ed. U.K.: Taylor & Francis, U.K. 1996.
116. Tonnesen HH, Kristensen S, Nord K. Photoreactivity of selected antimalarial compounds in solution and in the solid state. In: Albini A, Fasani E, eds. Drugs—Photochemistry and Photostability. Cambridge, U.K.: The Royal Society of Chemistry, 1998:87–99.
117. Tonnesen HH, Grislingaas AL. Photochemical stability of biologically active compounds. II. Photochemical stability of mefloquine in water. Int J Pharm 1990; 60(2):157–162.
118. Tonnesen HH, Moore DE. Photochemical stability of biologically active compounds. III. Mefloquine as a photosensitizer. Int J Pharm 1991; 70:95–101.
119. Tonnesen HH, Skrede G, Martinsen BK. Photoreactivity of biologically active compounds. XIII. Photostability of mefloquine hydrochloride in the solid state. Drug Stab 1997; 1:249–253.
120. Tonnesen HH. Photoreactivity of biologically active compounds. XV. Photochemical behaviour of mefloquine in aqueous solution. Pharmazie 1999; 54:590–594.
121. Tonnesen HH, Grislingaas AL, Woo SO, Karlsen J. Photochemical stability of hydroxychloroquine. Int J Pharm 1988; 43:215.
122. Kristensen S, Grislingaas AL, Greenhill JV, Skjetne T, Karlsen J, Tonnesen HH. Photochemical stability of biologically active compounds. V. Photochemical degradation of primaquine in aqueous medium. Int J Pharm 1993; 100:15–23.
123. Kristensen S, Karlsen J, Tonnesen HH. Photoreactivity of biologically active compounds. Photohemolytical properties of antimalarials in vitro. Pharm Sci Comm 1994; 4:183–191.
124. Kristensen S, Orsteen AL, Sande SA, Tonnesen HH. Photoreactivity of biologically active compounds. VII. Interaction of antimalarial drugs with melanin in vitro as a part of a phototoxicity screening. J Photochem Photobiol B Biol 1994; 26:87–95.
125. Kristensen S, Wang RH, Tonnesen HH, Dillon J, Roberts JE. Photoreactivity of biologically active compounds. VIII. Photosensitized polymerization of lens proteins by antimalarial drugs in vitro. Photochem Photobiol 1995; 61:124–130.

126. Kristensen S, Grinberg LN, Tonnesen HH. Photoreactivity of biologically active compounds. XI. Primaquine and metabolites as radical inducers. Eur J Pharm Sci 1997; 5:139–146.

127. Kristensen S, Nord K, Orsteen AL, Tonnesen HH. Photoreactivity of biologically active compounds. XIV. Influence of oxygen on light-induced reactions of primaquine. Pharmazie 1998; 53:98–103.

128. Motton AG, Martinez LJ, Holt N, et al. Photophysical studies on antimalarial drugs. Photochem Photobiol 1999; 69:282–287.

129. Navaratnam S, Hamblett I, Tonnesen HH. Photoreactivity of biologically active compounds. XVI. Formation and reactivity of free radicals in mefloquine. J Photochem Photobiol B: Biol 2000; 56:25–38.

130. Nord K, Karlsen J, Tonnesen HH. Photochemical stability of biologically active compounds. IX. Characterization of the spectroscopic properties of the 4-aminoquinolines chloroquine, hydroxychloroquine and selected metabolites by absorption, fluorescence and phosphorescence measurements. Photochem Photobiol 1994; 60:427–431.

131. Nord K, Karlsen J, Orsteen AL, Tonnesen HH. Photoreactivity of biologically active compounds. X. Photoreactivity of chloroquine in aqueous solution. Pharmazie 1997; 52:598–603.

132. Nord K, Andersen H, Tonnesen HH. Photoreactivity of biologically active compounds. XII. Photostability of polymorphic modifications of chloroquine diphosphate. Drug Stab 1997; 1:243–248.

133. Tonnesen HH, Karlsen J. Studies on curcumin and curcuminoids. VI. Kinetics of curcumin degradation in aqueous solution. Z Lebensm-Unters-Forsch 1985; 180:402.

134. Tonnesen HH, Karlsen J, van Henegouwen BG. Studies on curcumin and curcuminoids. VIII. Photochemical stability of curcumin. Z Lebensm-Unters-Forsch 1986; 183:116.

135. Tonnesen HH. Chemistry, Stability and Analysis of Curcumin—a Naturally Occurring Drug Molecule. Ph.D. Dissertation, University of Oslo, Oslo, NO, 1986.

136. Tonnesen HH, de Vries H, Karlsen J, van Henegouwen GB. Studies on curcumin and curcuminoids. IX. Investigation of the photobiological activity of curcumin using bacterial indicator systems. J Pharm Sci 1987; 76:371.

137. Tonnesen HH. Studies on curcumin and curcuminoids. XIV. Effect of curcumin on hyaluronic acid degradation in vitro. Int J Pharm 1989; 50:91.

138. Tonnesen HH. Studies on curcumin and curcuminoids. XIII. Catalytic effect of curcumin on the peroxidation of linoleic acid by 15-lipoxygenase. Int J Pharm 1989; 50:67.

139. Tonnesen HH. Studies on curcumin and curcuminoids. XVI. Effect of curcumin analogs on hyaluronic acid degradation in vitro. Int J Pharm 1989; 51:259.

140. Tonnesen HH, Greenhill JV. Studies on curcumin and curcuminoids. XXII. Curcumin as a reducing agent and as a radical scavenger. Int J Pharm 1992; 87:79.

141. Tonnesen HH, Msson M, Loftsson T. Studies on curcumin and curcuminoids. XXVII. Cyclodextrin complexation; solubility, chemical and photochemical stability. Int J Pharm 2002; 244:127–135.

142. Tonnesen HH. Studies on curcumin and curcuminoids. XXVIII. Solubility, chemical and photochemical stability of curcumin in surfactant solution. Pharmazie 2002; 57: 820–824.

143. Tonnesen HH, Karlsen J. Studies on curcumin and curcuminoids. X. The use of curcumin as a formulation aid to protect light sensitive drugs in soft gelatin capsules. Int J Pharm 1987; 38:247.

144. Tonnesen HH, Karlsen J. Studies on curcumin and curcuminoids. XI. Stabilization of photolabile drugs in serum samples by addition of curcumin. Int J Pharm 1988; 41:75.

145. Grinberg LN, Shalev O, Tonnesen HH, Rachmilewitz ER. Studies on curcumin and curcuminoids. XXVI. Antioxidant effects of curcumin in the red blood cell membrane. Int J Pharm 1996; 132:251–257.

146. Dyvik O, Grislingaas AL, Tonnesen HH, Karlsen J. Methotrexate in infusion solutions. A Stability test for the hospital pharmacy. J Clin Hosp Pharm 1986; 11:343.

147. Tonnesen HH. Fotokjemisk stabilitet av cytostatika. Samlerapport, Faggruppen for cytostatika, Bergen 1987.
148. Tonnesen HH, Greenhill JV. Photodecomposition of drugs. In: Swarbrick J, Boyland JC, eds. Encyclopedia of Pharmaceutical Technology. Vol. 12. New York: Marcel Dekker, 1995:105–135.
149. Brustugun J, Tonnesen HH, Klem W, Kjnniksen I. Photodestabilization of epinephrine by sodium metabisulfite. PDA J Pharm Sci Tech 2000; 54:136–143.
150. Tonnesen HH. Photochemical degradation of components in drug formulations. I. An approach to the standardization of degradation studies. Die Pharmazie 1991; 46:263.
151. Tonnesen HH. Formulation and stability testing of photolabile drugs. Int J Pharm 2001; 225:1–14.
152. Tonnesen HH, Karlsen J. Standardization of photochemical stability testing of drugs and drug formulations. Pharm Eur 1993; 5:1.
153. Tonnesen HH, Moore DE. Photochemical degradation of components in drug formulations. Part 2: Selection of radiation sources in light stability testing. Pharm Tech 1993; 5:27–34.
154. Tonnesen HH, Karlsen J. A comment on photostability testing according to ICH guidelines: Calibration of light sources. Pharm Europa 1997; 9:735–736.
155. Habib MJ, Asker AF. Influence of certain additives on the photostability of colchicine solutions. Drug Dev Ind Pharm 1989; 15:845–849.
156. Islam MS, Asker AF. Photostabilizatlon of dacarbazine with reduced. PDA J Pharm Sci Tech 1994; 48(1):38–40.
157. Asker AF, Habib MJ. Effect of glutathione on photolytic degradation of doxorubicin hydrochloride. J Parent Sci Tech 1988; 42(5):153–156.
158. Habib MJ, Asker AF. Photostabilization of doxorubicin hydrochloride with radioprotective and photoprotective agents: potential mechanism for enhancing chemotherapy during radiotherapy. J Parent Sci Tech 1989; 43:259.
159. Asker AF, Collier A. Influence of uric acid on photostability of FD&C Blue #2. Drug Dev Ind Pharm 1981; 7:563.
160. Asker AF, Colbert DY. Influence of certain additives on the photostabilizing effect of acid for solutions of FD&C Blue No. 2. Drug Dev Ind Pharm 1982; 8(5):759–773.
161. Asker AF, Jackson D. Photoprotective effect of dimethyl sulfoxide for FD&C Red No. 3 solutions. Drug Dev Ind Pharm 1986; 12(3):385–396.
162. Asker AF, Andrews S. Influence of thiourea on the photostability of FD&C Red No. 3 solutions. Drug Dev Ind Pharm 1987; 13:1081.
163. Habib MJ, Asker AF. Complex formation between metronidazole and sodium ureate: effect on photodegradation of metronidazole. Pharm Res 1989; 6(1):58–61.
164. Asker AF, Canady D. Influence of certain additives on the photostabilizing effect of dimethyl sulfoxide for sodium nitroprusside solutions. Drug Dev Ind Pharm 1984; 10:1025.
165. Asker AF, Gragg R. Dimethyl sulfoxide as a photoprotective agent for sodium nitroprusside solutions. Drug Dev Ind Pharm 1983; 9(5):837–848.
166. Asker AF, Mohammad Islam MS. J. Effect of sodium thiosulfate on the photolysis of phenobarbital: evidence of complex formation. J Parent Sci Tech 1993; 48(4):205–210.
167. Asker AF, Harris CW. Influence of certain additives on the photostability of physostigmine sulfate solutions. Drug Dev Ind Pharm 1988; 14(5):733–748.
168. Asker AF, Helal MA, Motawi MM. Light stability of some parenteral solutions of reserpine. Pharmazie 1970; 26:90–92.
169. Asker AF, Larose M. Influence uric acid on photostability of sulfathiozole sodium solutions. Drug Dev Ind Pharm 1987; 13:2239.
170. Asker AF, Habib MJ. Effect of certain additives on photodegradation of tetracycline hydrochloride solutions. J Parent Sci Tech 1991; 45(2):113–115.
171. Asker AF, Habib MJ. Effect of certain stabilizers on the photobleaching of riboflavin solutions. Drug Dev Ind Pharm 1990; 16(1):149–156.

172. Asker AF, Harris CW. Influence of storage under tropical conditions on the stability and dissolution of ascorbic acid tablets. Drug Dev Ind Pharm 1990; 16:165.
173. Habib MJ, Asker AF. Photostabilization of riboflavin by incorporation into liposomes. J Parent Sci Tech 1991; 45:124.
174. Moore DE. Antioxidant efficiency of some polyhydric phenols in the photooxidation of benzaldehyde. J Pharm Sci 1976; 65:1447–1451.
175. Moore DE. Photosensitization by drugs. J Pharm Sci 1997; 66:1282–1284.
176. Moore DE. Photosensitization by drugs: quinine as a photosensitizer. J Pharm Pharmacol Comm 1980; 32; 216–218.
177. Guilbault GG. Practical Fluorescence. New York: Marcel Dekker, 1973; 23.
178. Moore DE, Tamat SR. Photosensitization by drugs: photolysis of some chlorine containing drugs. J Pharm Pharmacol 1980; 32:172–177.
179. Moore DE, Burt CD. Photosensitization by drugs in surfactant solutions. Photochem Photobiol 1981; 34:431–439.
180. Moore DE, Hemmens VJ. Photosensitization by antimalarial drugs. Photochem Photobiol 1982; 36:71–77.
181. Moore DE, Hemmens VJ, Yip H. Photosensitization by drugs: nalidixic and oxolinic acids. Photochem Photobiol 1984; 39:57–61.
182. Hemmens VJ, Moore DE. Photochemical sensitization by azathioprine and its metabolites. I. 6–mercaptopurine. Photochem Photobiol 1986; 43:247–255.
183. Hemmens VJ, Moore DE. Photochemical sensitization by azathioprine and its metabolites. II. Azathioprine and nitroimidazole metabolites. Photochem Photobiol 1986; 43:257–262.
184. Burt CD, Moore DE. Photochemical sensitization by 7-methylbenz[c]acridine and related compounds. Photochem Photobiol 1987; 45:729–739.
185. Moore DE, Chappuis PP. A comparative study of photochemical sensitization by the non-steroidal anti-inflammatory drugs, naproxen, benoxaprofen and indomethacin. Photochem Photobiol 1988; 47:173–180.
186. HH Tonnesen, Moore DE. Photochemical stability of biologically active compounds. III. Mefloquine as a photosensitizer. Int J Pharm 1991; 70:95–101.
187. Zhou W, Moore DE. Photosensitizing activity of the antibacterial drugs sulfamethoxazole and trimethoprim. Photochem Photobiol 1997; 39:63–72.
188. Moore DE, Wang J. Electron transfer mechanisms in photosensitization by the anti-inflammatory drug benzydamine. J Photochem Photobiol 1998; 43:175–180.
189. Tamat SR, Moore DE. Photolytic decomposition of hydrochlorothiazide. J Pharm Sci 1983; 72:180–183.
190. Moore DE, Fallon MP, Burt CD. Photo-oxidation of tetracycline—a differential pulse polarographic study. Int J Pharm 1983; 14:133–142.
191. Moore DE, Sithipitaks V. Photolytic degradation of frusemide. J Pharm Pharmacol 1983; 35:489–493.
192. Hemmens VJ, Moore DE. Photo-oxidation of 6-mercaptopurine in aqueous solution. J Chem Soc Perkin Trans 1984; II:209–211.
193. Burt CD, Cheung HTA, Holder G, Moore DE. Photooxidation of 7-methylbenz[c]acridine in methanol—identification of two primary photoproducts. J Chem Soc Perkin Trans 1986; I:741–745.
194. Moore E, Chignell CF, Sik RH, Motten AG. Generation of radical anions from metronidazole, misonidazole and azathioprine by photoreduction in the presence of EDTA. Int J Radiat Biol 1986; 50:885–891.
195. Moore DE. Flash photolyis of misonidazole and metronidazole. Int J Radiat Biol 1987; 51:45–51.
196. Wang J, Moore DE. A study of the photodegradation of benzydamine in pharmaceutical formulations. J Pharm Biomed Anal 1992; 7:535–540.
197. Tonnesen HH, Moore DE. Photochemical degradation of components in drug formulations. Pharm Tech Int 1993; 5(2):27–33.

198. Moore DE, Zhou W, Photodegradation of sulfamethoxazole: a chemical system capable of monitoring seasonal changes in UVB intensity. Photochem Photobiol 1994; 59: 497–502.

199. Zhou W, Moore DE. Photochemical decomposition of sulfamethoxazole. Int J Pharm 1994; 110:55–63.

200. Moore DE. Photochemistry of diuretic drugs in solution. In: Albini A, Fasani E eds. Drugs—Photochemistry and Photostability. Royal Society of Chemistry, 1998:100–115.

201. Moore DE. Drug-induced cutaneous photosensitivity incidence, mechanism, prevention and management. Drug Safety 2002; 25:345–372.

202. Moore DE. Principles and practice of drug photodegradation studies. J Pharm Biomed Anal 1987; 5:441–455.

203. Moore DE. Photochemistry in medicine. Techniques for the analysis of photodegradation pathways of medicinal compounds. Trends Anal Chem 1987; 6:234–238.

204. Moore DE. Photochemistry of photosensitizing drugs. Trends Photochem Photobiol 1990; 1:13–23.

205. Moore DE. Photophysical and photochemical aspects of drug stability. In: Tonnesen HH, ed. The Photostability of Drugs and Drug Formulations. London: Taylor and Francis, 1996:9–38.

206. Moore DE. Photophysical and photochemical aspects of drug stability. In: Tonnesen HH, ed. The Photostability of Drugs and Drug Formulations. 2nd ed. Taylor and Francis, London. (in press)

207. Moore DE. Standardization of photodegradation studies and kinetic treatment of photochemical reactions. In: Tonnesen HH, ed. The Photostability of Drugs and Drug Formulations. 1st ed. London: Taylor and Francis, 1996:63–82.

208. Moore DE. Standardization of kinetic studies of photodegradation reactions. In: Tonnesen HH, ed. The Photostability of Drugs and Drug Formulations, 2nd ed. CRC Press: LL Boca Raton USA, 2004.

209. Wilkins BJ, Moore DE. Photolytic rearrangement of metronidazole to 1-hydroxyethyl-4-hydroxyimino-5-oxo-imidazole and the formation of copper complexes of these compounds. Photochem Photobiol 1988; 47:481–484.

210. Moore DE. Mechanisms of photosensitization by phototoxic drug. Mut Res 1998; 442:165–173.

211. Moore DE. Reactions of nitroimidazoles with free radicals—enhancement of reaction by u.v. radiation. Int J Radiat Biol 1985; 47:563–568.

212. Wilkins BJ, Gainsford GJ, Moore DE. Photolytic rearrangement of metronidazole to a 1,2,4-oxadiazole. Crystal structure of its 4-nitrobenzoate derivative. J Chem Soc Perkin Trans 1987; I:1817–1820.

213. Moore DE, Wilkins BJ. Common products from (-radiolysis and UV photolysis of metronidazole. Radiat Phys Chem 1990; 36:547–550.

214. Moore DE, Chen BC. Reactions of non-steroidal anti-inflammatory drugs with polyacrylamide free radicals. Redox Report 1997; 3:41–47.

215. Wiebe JA, Moore DE. Oxidation photosensitized by tetracyclines. J Pharm Sci 1977; 66:188–189.

216. Kelly GE, Meikle WD, Moore DE. Promotion of UV-induced skin carcinogenesis by azathioprine; role of photochemical sensitization. Photochem Photobiol 1989; 49:59–65.

217. Moore DE, Mallesch JL. Photochemical interaction between triamterene and hydrochlorthiazide. Int J Pharm 1991; 76:187–190.

218. Moore DE, Roberts-Thomson S, Dong Z, Duke CC. Photochemical studies on the anti-inflammatory drug diclofenac Photochem Photobiol 1990; 52:685–690.

219. Moore DE. Kinetic treatment of photochemical reactions. Int J Pharm 1990; 63:R5–R7.

220. Moore DE, Sik RH, Bilski P, Chignell CF, Reszka K. Photochemical sensitization by azathioprine and its metabolites. Part 3. A direct EPR and spin-trapping study of light-induced free radicals from 6-mercaptopurine and its oxidation products. Photochem Photobiol 1994; 60:574–581.

221. Anderson NH, Johnston D, McLelland MA, Munden P. Photostability testing of drug substances and drug products in U.K. pharmaceutical laboratories. J Pharm Biomed Anal 1991; 9(6):443.

222. Anderson NH. Light Stability Studies II. APQ Symposium American Association of Pharmaceutical Scientists Eighth Annual Meeting and Exposition, Nov 14–18, 1993.

223. Anderson NH, Jackson SL, Smail A. Comparison of a xenon source with full spectrum fluorescent tubes for pharmaceutical product photostability. Photostability 95, 1st International Meeting, Photostability of Drugs, Oslo, NO, June 8–9, 1995.

224. Anderson NH. Quinine Actinometry. IBC Meeting on Photostability, July 1, 1999 in London, U.K.

225. Anderson NH. Photostability testing: design and interpretation of tests on drug substances and dosage forms. In: HH Tonnesen, ed. The Photostability of Drugs and Drug Formulations. London: Taylor and Francis, 1996.

226. Sager N. F-D-C Reports—The Pink Sheet. August 28, 1995, T&G-5.

227. Beaumont T. Photostability testing. In: Mazzo DJ. International Photostability Testing. Interpharm Press, 1999:59–71.

228. Christensen KL, Christensen J, Frjakr S, Langballe P, Hansen LG. The influence of temperature and light on the quinine chemical actinometric system. Photostability 97, 2nd International Meeting, Photostability of Drugs Pavia, IT, Sep 14–17, 1997.

229. Kester TC, Zhan S, Bergstrom DH. Quinine chemical actinometry—studies under two light sources. ICH Guideline on Photostability Testing 1996. Annual Meeting of the American Association of Pharmaceutical Scientists.

230. Drew HD, Brower JF, Juhl WE, Thornton LK. Quinine photochemistry in aqueous solution: a chemical actinometer system for photostability studies of pharmaceutical drug substances and drug products. PF1998; 24(3):6334–6346.

231. Piechocki J, Wolters R. Light stability studies a misnomer. ICH 2 Six-Party Drafting Group on Light Stability Testing Committee Meeting Orlando, Florida, U.S.A., October 25, 1993.

232. Piechocki J, Wolters R. Light Stability Testing: A Misnomer. Pharm. Tech. 1994; January:60–655.

233. Piechocki J. Light testing I: theory of chemical reactions. APQ Symposium Eighth Annual Meeting and Exposition, of the American Association of Pharmaceutical Scientists, at Lake Buena Vista, FL, Nov. 14–18, 1993.

234. Piechocki J, Wolters R. Use of actinometry in light stability testing. Pharm Tech 1993; 6:46–52.

235. Piechocki J, Wolters R. Photon sources in photostability studies; Part I. Joint Meeting of FDA/Pharmaceutical Manufacturers Association (PMA) in Rockville, MD, December 14, 1993.

236. Piechocki J. Comments on Docket No. 96D-0010. Notice in Federal Register, Vol. 61, No. 46, Thursday, March 7, 1996:9310–9313; Docket Management Branch (HFA-305): Item code C6.

237. Searle ND. Comments on Docket No. 96D-0010, Notice in Federal Register, Vol. 61, No. 46, Thursday, March 7, 1996, Pgs. 9310–9313; Docket Management Branch (HFA-305): Item code C4.

238. Wozniak TJ. Comments on Docket No. 96D-0010, comments on the Notice in Federal Register, Vol. 61, No. 46, Thursday, March 7, 1996:9310–9313; Docket Management Branch (HFA-305): Item code, C10.

239. Piechocki J. Pharmaceutical Photostability Testing—Black Box Chemistry 1999. Pittsburgh Conference on Analytical Chemistry in Orlando, FL March 1999.

240. Piechocki J. Equipment designs and availability. IBC Photostability Conference, 1 July 1999 in London, U.K.

241. Piechocki J. Pharmaceuticals—design photostability studies within FDA and ICH guidelines. Center for Business Intelligences. 2nd Annual Regulatory Compliance and

Expedited Design and Execution of International Stability Programs, in Philadelphia, PA June 22–23, 2000.

242. Piechocki J. Pharmaceutical photostability: how to select the right option. University of Wisconsin, Pharmaceutical Stability, Current Trends and Practices, Huntington Beach, CA, October 21–22, 2000.

243. Piechocki J. How to choose the right option. University of Wisconsin, Annual Stability of Pharmaceuticals—Currents Issues, Practices and Technologies, Nov 14, 2000, in Costa Mesa, CA.

244. Piechocki J. Pharmaceutical photostability: how to select the right option. University of Wisconsin, Pharmaceutical Stability, Current Trends and Best Practices Course held in Durham, NC, Nov. 12–13, 2001.

245. Piechocki J. Basic principles of pharmaceutical photostability testing and how they apply to the ICH guidelines. Core Stability Issues and Industry Practices Update by AAI Pharmaceutical Seminars in Clearwater Beach, Fl, Feb. 26–27, 2001.

246. Piechocki J. Pharmaceutical photostability testing. A primer on regulations and product development. Center for Business Intelligence, 3rd Annual Stability Testing Forum, June 13–14, 2001 in Philadelphia, PA.

247. Piechocki J. Pharmaceutical Photostability. Basics, Abuses and Potential Uses Interphex 2000. Philadelphia, PA, March 20, 2001.

248. Piechocki J. Designing and Conducting Photostability Studies. Center for Business Intelligence CBI. 5th Annual Stability Testing Forum, in Philadelphia, PA. July 17–18, 2003.

249. Piechocki J. Designing an Effective and Compliant Photostability Study. Barnett International Conference on Stability Testing, in Philadelphia, PA, Sept. 24, 2003.

250. Piechocki J. Pharmaceutical Photostability Protocols. University of Wisconsin, Pharmaceutical Stability Currents Trends and Best Practices Course held in Las Vegas, NV October 20–21, 2003.

251. Piechocki J. Designing a Photostability Study. International Pharmaceutical Academy 2003 Conference on Design and Implementation of a Comprehensive and Compliant Stability Program for Pharmaceutical and Biopharmaceutical Industries, in Toronto, Canada, Sept 22, 2003.

252. Piechocki J. Designing an Effective and Compliant Photostability Study. Barnett International Stability Testing Conference, Philadelphia, PA June 24–25, 2004.

253. Piechocki J. Pharmaceutical Photostability Protocols. University of Wisconsin, Pharmaceutical Stability Currents Trends and Best Practices Course held in Las Vegas, NV November 5, 2004.

254. Piechocki J. Pharmaceutical Photostability Fact, Fiction and Future. AAI International Seminar, Pharmaceutical Stability: Current Regulations and the Analytical Search for Degradants held in Wilmington, NC, March 27, 2003.

255. Piechocki J. Selecting the right source for pharmaceutical photostability testing. Photostability of Drugs and Drug Formulations. London, UK: Taylor & Francis, Ltd., 1998:247–271.

256. Baertschi SW. Commentary on the quinine actinometry system described in the ICH draft guideline on photostability testing of new drug substances and products. Drug Stab 1997; 1(4):193–195.

257. Allen JA, Allen SK, Dreiman J, Baertschi SW. 2-Nitrobenzaldehyde: a convenient UV-A and UV-B chemical actinometer for drug photostability testing. Photostability 99, Joint Meeting with 27th Annual Meeting of the American Society for Photobiology, Washington, D.C., July 10–15, 1999.

258. Allen JA, Allen SK, Baertschi SW. 2-Nitrobenzaldehyde: a convenient UV-A and UV-B chemical actinometer for drug photostability testing. J Pharm Biomed Anal 2000; 24 167–178.

259. Thatcher SR, Mansfield RK, Miller RB, Davis CW, Baertschi SW. Pharmaceutical photostability: a technical and practical interpretation of the ICH guideline and its application to pharmaceutical stability: Part I. Pharm Tech 2001; 25(3):98–110.

260. Thatcher SR, Mansfield RK, Miller RB, Davis CW, Baertschi SW. Pharmaceutical photostability: a technical and practical interpretation of the ICH guideline and its application to pharmaceutical stability: Part II. Pharm Tech 2001; 25(4):50–62.

261. Alsante KM, Martin L, Baertschi SW. A Stress Testing Benchmarking Study. Pharm Techn 2003; 27(2):60–72.

262. Baertschi SW. Designing and implementing photostability studies in alignment with ICH guidelines. Barnett-Parexels Conference on Stability Testing, Washington, D.C., September 14, 1998.

263. Baertschi SW. Designing and implementing photostability studies in alignment with ICH guidelines. Preparing for Changing Paradigms in Pharmaceutical Stability Programs, Institute for International Research, Philadelphia, PA, September 16, 1998.

264. Baertschi SW. Designing and implementing ICH-compliant photostability studies. Pharmaceutical Stability: Statistical Applications, Regulatory Updates, and Current Industry Practices, AAI Seminar Series, East Brunswick, NJ, February 22–23, 1999.

265. Baertschi SW. Designing and implementing photostability studies in alignment with ICH guidelines. Statistical Methods for Determining and Justifying Your Specifications and Shelf-Life, Institute for International Research, Bethesda, MD, Sept. 22–24, 1999.

266. Baertschi SW. Practical aspects of pharmaceutical photostability testing, Photostability 99. Joint Meeting with 27th Annual Meeting of the American Society for Photobiology, Washington, D.C., July 10–15, 1999.

267. Baertschi SW. Current trends in photostability. Practical considerations in designing and implementing photostability studies. Stability of Pharmaceuticals: Current Issues, Practices, and Technologies, University of Wisconsin-Madison Seminar Series, Costa Mesa, CA, November 13–14, 2000.

268. Baertschi SW, Kinney H, Snider BG. Issues in evaluating the in-use photostability of transdermal patches. Pharm Tech 2000; 9.

269. Baertschi SW, Kinney HD, Snider BG. Issues in evaluating the in-use photostability of transdermal patches. Photostability 99, Joint Meeting with 27th Annual Meeting of the American Society for Photobiology, Washington, D.C., July 10–15, 1999.

270. Baertschi SW. Pharmaceutical photostability testing: sample presentation. Photostability 01: 4th International Conference on the Photostability of Drug Substances and Drug Products, Research Triangle Park, NC, July 16–19, 001.

271. Baertschi SW. ICH Option 1 and Option 2 sources: potential for different photodegradation pathways. Photostability 01: 4th International Conference on the Photostability of Drug Substances and Drug Products, Research Triangle Park, NC, July 16–19, 2001.

272. Baertschi SW. Photostability testing: practical issues related to sample exposure. Photostability, a 1997 Update; Scientific, Regulatory, and Practical Issues, AAI Seminar Series, Arlington, VA, February 24–25, 1997.

273. Baertschi SW, Thatcher SR. Sample presentation for photostability studies: problems and solutions. In: Piechocki J, Thoma K, eds. Pharmaceutical Photostability and Stabilization Technology New York.

274. Tonnesen HH, Baertschi SW. Photostability testing: the questions most frequently asked. In: Tonnesen HH, ed. Photostability of Drugs and Drug Formulations. 2nd ed. London: Taylor and Francis. (chapter accepted for publication, book in preparation).

275. Matsuda Y. Some aspects on the evaluation of photostability of solid-state drugs and pharmaceutical preparations. Pharm Tech Japan 1994; 10(7):7–17.

276. Matsuda Y, Minamida Y. Stability of solid dosage forms. II Coloration and photolytic degradation of sulfisomidine tablets by exaggerated ultraviolet irradiation. Chem Pharm Bull 1976; 24(9):2229–2236.

277. Matusuda Y, Inouye H, Nakanishi R. Stabilization of sulfisomidine tablets by the use of film coating containing UV absorber: protection of coloration and photolytic degradation from exaggerated lighting. J Pharm Sci 1978; 67(2):196–201.

278. Matsuda Y, Mihara M. Coloration and photolytic degradation of some sulfonamide tablets under exaggerated and ordinary ultraviolet radiation. Chem Pharm Bull 1978; 26(9): 2649–2656.

279. Matsuda Y, Kouzuki K, Tanaka M, Tanigaki J. Photostability of gelatin capsules; effect of ultraviolet irradiation on the vapor transmission properties and dissolution rates of indomethacin. Yakugaku Zassi 1979; 99(9):907–913.

280. Matsuda Y, Ito M. Photostability of sulphisomidine tablets: action spectra for colouration and photolytic degradation. Asian J Ph Sci 1979; 1:107–118.

281. Matsuda Y, Itooka T, Mitsuhashi Y. Photostability of indomethacin in model gelatin capsules: effects of film thickness and concentration of titanium dioxide on the coloration and photolytic degradation. Chem Pharm Bull 1980; 28(9):2665–2671.

282. Matsuda Y, Mashara R. Comparative evaluation of coloration of photosensitive solid drugs under various light sources. Yakugaku Zasshi 1980; 100:953–957.

283. Matsuda Y, Mashara R. Photostability of solid-state ubidecarenone at ordinary and elevated temperatures under exaggerated UV irradiation. J Pharm Sci 1983;72(10): 1198–1203.

284. Matsuda Y, Teraoka R. Improvement of the photostability of ubidecarenone microcapsules by incorporating fat-soluble vitamins. Intern J Pharmaceut 1985; 26:289–301.

285. Matsuda Y, Teraoka R, Sugimoto I. Comparative evaluation of photostability of solid state nifedipine under ordinary and intensive light irradiation conditions. Intern J Pharmaceut 1989; 54:211–221.

286. Matsuda Y, Akazawa R, Teraoka R, Otsuka M. Pharmaceutical evaluation of carbamazepine modifications: comparative study for photostability of carbamazepine polymorphs by using Fourier-transformed reflection-absorption infrared spectroscopy and colorimetric measurement. J Pharm Pharmacol 1994; 46:162–167.

287. Yatani K, Ueno M, Tsunakawa N, Shimizu R, Matsuo M, Murayama S. Investigation to establish photstability testing. Iyakuhin Kenkyu 1988; 19(6):1028–1052.

288. Matsuo M, Machida Y, Furuichi H, Nakamura K, Takeda Y. Suitability of photon sources for photostability testing of pharmaceutical products. Drug Stab 1996; 1:179–187.

289. Matsuo M, Takeda Y. Investigation of photon sources for photostability testing requirement for registration of pharmaceutical products. Photostability 95, 1st International Meeting, Photostability of Drugs, Oslo, NO, June 8–9, 1995.

290. Yoshioka S, Ishihara Y, Terazono T, et al. Quinine actinometry as a method for calibrating ultraviolet radiation intensity in light stability testing of pharmaceuticals. Drug Dev Ind Pharm 1994; 20(13):2049–2062.

291. KMR Group, Inc., Chicago, IL, U.S.

292. Esselen WB, Barnaby HA. Modern Packaging 1939.

293. Drew HD, Thornton LK, Juhl WE, Brower JF. An FDA/PhRMA Interlaboratory Study of the International Conference on Harmonization–proposed photostability testing procedures and guidelines. PF 1998; 24(3):6334–.

294. Greenhill JV, McLelland MA. Photodecomposition of Drugs. In: Ellis GP, West GB, eds. Progress in Medicinal Chemistry. Vol. 27. Amsterdam: Elsevier, 1990:51–121.

2

The International Conference on Harmonization Photostability Guideline

A Discussion of Experimental Conditions

Hanne Hjorth Tønnesen
Institute of Pharmacy, University of Oslo, Blidern, Oslo, Norway

INTRODUCTION

Daylight has destructive effects on many natural and manufactured products in common use. The photochemical reactions these products undergo are complex and depend on the chemical and physical structure of the materials exposed to irradiation. Many drug substances are white in color, and hence their degradation depends mostly on the amount of ultraviolet (UV) radiation absorbed by the material. For colored products, radiation of any wavelength that is absorbed may be effective. Sometimes photochemical reactions are difficult to identify or prevent. The reactions take place from the electronic excited state of a molecule and lead in many cases to high-energy products, such as radical ions and free radicals. These may eventually react to form final, stable products through thermal reactions. In some cases, the photochemical degradation products may be the same products formed in the thermal reaction of the ground state (dark reaction); this similarity is not generally the rule but coincidental. Some drug substances or excipients have a sensitizing effect and may initiate reactions of products in which they are present.

ICH GUIDELINE

Stability testing aims to document how environmental factors, such as humidity, temperature, and UV-visible (VIS) radiation may alter the quality of a drug substance or a product. The resulting data establishes recommended storage conditions, retest periods, and shelf life. Common standards for stability testing in the United States, Europe, and Japan have been developed. The Guideline for the Photostability Testing of New Drug Substances and Products, prepared under the auspices of the International Conference on Harmonization (ICH), describes a useful basic protocol for testing of new drug substances and products for first submissions. This guideline recommends radiation sources and the amount of exposure needed to assess the photostability of drug substances and drug products.

Forced Degradation Studies

The guideline calls for "forced degradation" studies and a "confirmatory" study. Forced degradation studies should be carried out early in the formulation development process to evaluate overall photosensitivity.

Confirmatory Studies

The purpose of the confirmatory study is to estimate the photostability characteristics under standardized conditions, and this will normally be carried out as late as possible in the development process. The guideline also allows for the possibility of applying alternative approaches, if properly justified. The aim of the photostability testing is to determine photosensitivity and the appropriate package or labeling combination needed during synthesis, manufacturing, handling, packaging, and storage to ensure satisfactory product quality. In most cases, the photostability problems are preventable and it will be thermal processes that determine the product shelf life. The results obtained using the photostability guideline should be regarded as essentially qualitative rather than quantitative, i.e., this is equivalent to a limit test.

GUIDELINE PROBLEMS

A discussion of photostability is not as straightforward as that of thermal stability due to the complexity of photoexposure. Correlation between devices of the same type, either in the same laboratory or at different locations, can be very good if they are operated under the same test conditions. Test results between devices with different sources of radiation (e.g., xenon arc and fluorescent tube) will vary according to the spectral absorption characteristics of the product being tested. Thus it is possible that some products may decompose at an equal rate in systems having different types of irradiation sources; however, other products may react quite differently. Although the proposed test is reasonably simple to conduct, there are some practical problems not definitively resolved (e.g., selection of adequate irradiation source and calibration). This chapter will focus on practical problems related to the use of the current guideline.

Irradiation Sources

Many different irradiation sources can be used in the stability studies of drugs and drug products. The source(s) used should be comparable in spectral power distribution to those to which products are exposed in practical use. It is however, difficult to predict the actual exposure of a pharmaceutical product during practical usage.

A drug substance and its products can be exposed to any one of a number of artificial sources (e.g., full spectrum fluorescent tubes and mercury vapor) during synthesis, manufacturing, or storage in the warehouse. It may also be exposed to a combination of indoor lighting (e.g., white fluorescent tubes or tungsten–halogen) and window-glass–filtered sunlight during storage by the consumer.

Routines for the handling of drug products seem to vary from continent to continent. In Europe, it is recognized that products may be removed from their secondary (outer) carton and thereby be exposed to window-glass–filtered daylight for longer periods in hospital wards or in the pharmacy, whereas in the United States and Japan, product exposure to window-glass–filtered daylight is believed to occur rarely (1). In the United States, buildings utilize less natural lighting for illumination and are therefore constructed with less window glass.

In some cases, a drug product can even be exposed to direct sunlight. Clearly, it is difficult to predict the amount of UV and VIS irradiation to which the product is exposed during the shelf life. The spectral power distribution and the overall illuminance used in a photostability study should therefore provide a "worst-case" exposure.

Source Selection

The ICH guideline gives two options for the selection of irradiation source. Option 1 addresses exposure to outdoor daylight or window-glass–filtered daylight. According to Option 1, the lamp should produce an output similar to the internationally recognized standards for outdoor daylight (D65) or indoor indirect daylight (ID65). D65 represents an artificial reproduction of normal daylight. ID65 is D65 radiation filtered through window glass. In most cases, it is unlikely that the product will be exposed to direct sunlight for any length of time and a source providing glass-filtered daylight (corresponding to the ID65 standard) should therefore be appropriate.

Option 1 can be achieved by use of a fluorescent lamp combining UV and VIS outputs or by use of a xenon or metal halide lamp. In actuality, Option 1 offers a choice between three different types of sources. A window-glass filter should be used in combination with sources producing significant radiation below 320 nm (e.g., xenon- and metal halide lamps, near-UV fluorescent tubes). It has also been questioned whether filters should be recommended for fluorescent lamps (Option 2) (2).

Although the artificial daylight and the full spectrum fluorescent tubes, like Duro-test Vitalite/Truelite®, emit in the UV region, a window-glass filter is usually not required for these sources. This is because only a small fraction (approximately 0.5%) of the irradiance in the UV region is below 320 nm.

Some lamp suppliers provide filters according to the standard for window glass and ID65 (International Organization for Standardization 10977). It is important to be aware of the fact that high-intensity UV energy causes aging of the glass used in lamps and filters. This leads to development of color or a significant increase in the absorption of UV radiation, e.g., a change in the spectral power distribution, which may influence the rate of product degradation.

Changes in the spectral power distribution of the lamp caused by glass aging can be an important cause of the variability in results between different devices running identical test cycles.

The effect of changing lamps in a Suntest cabinet is demonstrated in Figure 1.

Replacement of old lamps did lead to a 20% increase in the product degradation rate. This very well illustrates the fact that lamps should be changed at defined

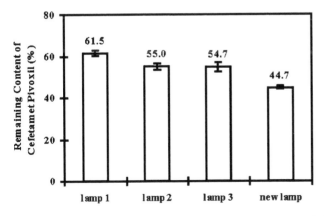

Figure 1 The influence of xenon lamp aging on the photodegradation rate of drug substance. Lamps 1 to 3 had not been changed according to the manufacturer's recommendation. Lamp 4 is new. *Source*: Courtesy of H. Spilgies, SmithKline Beecham, Essex, U.K.

intervals in accordance with the manufacturers specifications. This same procedure should also be applied to filters.

Many lamps do also have a typical "burn in" period before they can be used in stability testing. Preaged filters have been introduced recently as a convenience to the user.

Exposure Level

Any source simulating window-glass–filtered daylight will provide a relatively greater UV exposure than most practical situations. Inside a room, the UV level decreases rapidly outside the region of direct window-glass–filtered daylight. During its shelf-life, a product is likely to be exposed to a mixture of natural radiation filtered through window glass and artificial light (e.g., fluorescent tube). Although most lamps, even incandescent ones, emit some radiation in the UV region of the spectrum, the ratio UV/VIS under "real" storage conditions is probably less than under "standard" indoor indirect daylight (ID65) conditions. Please note that this is not true for tungsten–halogen lamps, which have very high UV levels.

The ICH guideline recommends a total exposure of 200 Whm^{-2} in the UV range 320–400 nm. This represents the equivalent of approximately one to two days exposure to window-glass–filtered daylight on a sunny day (Fig. 2).

For a lamp with a spectral output similar to the ID65 standard (i.e., Option 1), a total irradiance of 200 Whm^{-2} in the UV region corresponds to approximately 0.45 million lux hours in the VIS region. The exposure in the VIS region recommended in the ICH guideline is 1.2 million lux hours. A test run with the end criterion "1.2 million lux hours" will therefore exceed the requirement "200 Whm^{-2}" by a factor of 2.5–3 when a lamp according to Option 1 is used (with the exception of certain full

Visible light

~ 3 days on
windowsill

~ 3 months in
pharmacy

~ 6 - 12 months at home

UV Irradiation

1 - 2 days on
sunny windowsill

25-50 days
diffuse daylight

Figure 2 Comparison of exposure times needed to reach the International Conference on Harmonization recommended exposure of 200 Whm^{-2} (UV irradiation) and 1.2 million lux hours (visible light).

spectrum fluorescent sources where the UV exposure may be less than 200 Whm^{-2} after 1.2 million lux hours (3). A single source cannot succeed in providing the desired exposure levels.

There are several possible ways to reach the minimum exposure values suggested in the ICH guideline.

- One test can be run covering both test end criteria.
  - o Removing 50% of the samples after having reached the radiant exposure for the UV range and the other 50% after having reached the one for the VIS range
  - o Excluding UV irradiation by use of filters after having reached the radiant exposure for the UV range
- Two separate tests can be run; one for each criterion, with identical samples.

Option 2 in the ICH guideline simulates exposure to indoor fluorescent light (cool white). A UV source is added to cover the spectral region of natural light through window glass. The test according to Option 2 is more severe than indoor fluorescent light alone but does not adequately simulate daylight radiation through window glass.

A combination of a UV source and a fluorescent source makes it easier to reach the minimum exposure value in the VIS range without overexposure in the UV region because the UV lamp can be turned off when the desired UV exposure is obtained. It is thereby possible to deconvolute the effect of UV on the overall degradation.

Tests using two lamps in combination can, however, be more complicated to conduct than tests applying a single irradiation source. The sample can be exposed to both lamps simultaneously, or the irradiation can be performed sequentially. It is essential that the same sample be exposed to both sources. When sequential exposure is selected, the sequence of irradiation (UV followed by VIS or VIS followed by UV) may be of importance if the substance forms colored degradation products that can act as sensitizers or have a filter effect on the degradation process (1). It is, therefore, recommended that samples be exposed simultaneously to the UV and cool white fluorescent tubes. It has been noted that a combination of sources according to Option 2 may produce little or no output between 380 nm and 430 nm (2). It is important to check that the product does not absorb primarily in this region. No guidance is given in the ICH guideline for the choice of Option 1 or 2.

Equivalent stability evaluations have not been documented for the various alternatives under the two options or between the Option 1 and 2. Thoma and Kerker have compared several of these sources and obtained nonequivalent results (4,5). From a scientific viewpoint, the various alternatives are not equivalent assuming that experimental conditions are carefully controlled. Both the irradiance levels and the spectral power distributionsare different. For the purpose of a confirmatory study, the options could however, possibly be regarded as equivalent. The results are most likely to be combined with knowledge about the compound or product from previous tests; e.g., by the time the ICH confirmatory test is carried out, product photostability is already known from the forced degradation studies.

Could a truly unstable product be labeled stable as a result of testing using an inappropriate source? In theory this is possible, especially with Option 2, if the compound or degradation products that sensitize the reaction absorb mainly in the "gap" (380–420 nm), e.g., curcumin or bilirubin. As mentioned above, it is important to combine the results with knowledge already obtained from other tests before labeling and packaging decisions are made.

Table 1 Advantages (+) and drawbacks (−) of International Conference on Harmonization irradiation sources. ? = likely to vary between the sources

	Xenon	Metal halide	White	Fluorescent Full spectrum	Near ultraviolet
High output	+	+	−	−	+
Solar spectral distribution	+	+	−	+/−	−
Low heat output	−	−	+	+	+
Large area	−	−	+	+	+
Consistency	+	+	−	?	?
Low cost	+	+	+	+	+
Window-glass filter needed	+	+	?	−	+

Because the confirmatory study also represents "worst-case" conditions, there should be few chances that an unstable product is not discovered. Examples of sources currently used by the pharmaceutical industry are xenon- and metal halide lamps (Option 1), artificial- and full spectrum daylight fluorescent tubes (Option 1), white fluorescent- and near UV-fluorescent tubes (Option 2). In (Northern) Europe (e.g., Scandinavia, England, and Germany), it seems that Option 1 with the xenon lamp is the preferred source.

The ideal source would provide a close simulation of window-glass–filtered daylight with even illumination (±10%) over a large area. The spectral power distribution and intensity would be constant, have a low-heat output, and would be inexpensive to purchase and run. At present, the ideal source does not exist (Table 1).

It has, however, been demonstrated that the choice of source can affect the extent of degradation (3,4,6–9). An inappropriate spectral power distribution could lead to a different energy absorption by the samples relative to window-glass–filtered daylight, resulting in less or greater degradation than under "standardized" window-glass–filtered daylight conditions.

Many daylight sources are so named because they have the same color temperature although they do not simulate daylight in a spectral power distribution sense. Several papers have been published recommending the proper source to use for pharmaceutical photostability testing (6,10–16).

Instrumentation that fulfills the requirements specified in Option 1 or 2 has been developed and is commercially available.

Irradiance Level and Temperature Effects
The ICH guideline does not specify an irradiance level, only the overall illumination. The user may therefore adjust the irradiance level according to individual requirements.

One must be aware that irradiance and temperature are dependent variables in photostability testing.

Temperature will increase as the irradiance is increased. The maximum allowable irradiance is limited by the temperature threshold of the specimens. It is important to find the best balance between high irradiance levels and realistic sample temperatures, e.g., an irradiance level that is high enough to accelerate the test without causing unwanted temperature effects.

The use of a very high irradiance level compared to the "real" conditions tends to decrease the correlation of test results. At a high irradiance level, the mechanisms of sample degradation may be changed even when the spectral distribution of the lamp is kept constant. Tests conducted at significantly different irradiance levels should not be compared unless a correlation has been established. Test conditions corresponding to the maximum output of the lamp will often be the first choice because the exposure time can thereby be reduced. If any changes in product quality are observed, it is recommended that the analyst repeat the test at a lower irradiance level.

The output from the lamps can vary between the tubes and also along the length of the tubes. The UV and VIS levels should therefore be mapped across the test chamber to ensure that the samples are placed at points of equal irradiance, ±10%. The specimens should be placed only in the exposure area where the irradiance is at least 90% of the maximum irradiance unless the exposure time is increased. It can also be useful to reposition the samples during the exposure period to ensure that each specimen receives an equal amount of radiant exposure. If a combination of UV and VIS is used (i.e., Option 2, also allowed in Option 1), it is important to make sure that the sample is exposed to both radiation components independent of its location in the chamber.

A difference in irradiation uniformity can be observed between higher and lower settings of lamp power (Fig. 3A and B) (17) or when the lamp or filters are changed. The result of the irradiation intensity mapping will to a certain extent also be dependent on the spectral sensitivity of the measuring device due to wavelength-dependent reflection or scattering of the incident radiation.

Sample Cooling

Commercially available equipment for photostability testing often includes a cooling system. The real cooling effect, however, depends on whether the exposed surface is directly affected by the cooling air stream or refrigeration system. It is difficult to determine the exact surface temperature of individually exposed samples.

The maximum surface temperature that may occur under total incident radiation can be measured by use of a black standard thermometer. Alternatively, a white standard thermometer can be used to predict the surface temperature of a reflecting surface, e.g., white tablets. In our laboratory, we have found the use of a temperature level strip (Thermax, Thermographic Measurements Ltd., South Wirral, U.K.) quite convenient to predict the temperature of the exposed sample. The strip can be placed on the exposed surface (e.g., surface of an infusion bag) or inside a container (e.g., inside a petri dish). A color change indicates that the rating quoted on the indicator has been reached. The maximum temperature reached during the exposure is thereby registered with good accuracy (±1%).

The ICH guideline recommends that dark controls be placed alongside authentic samples. The dark control samples may be used to compensate for temperature and humidity effects but it should be kept in mind that this will only be valid for "ordinary" thermal reactions (i.e., thermal reactions from the ground state of the molecules). Reactions initiated by UV–VIS incident radiation (i.e., thermal reactions from the excited state of the molecules) can also be accelerated by any increase in temperature. This effect cannot be compensated for by use of a dark control.

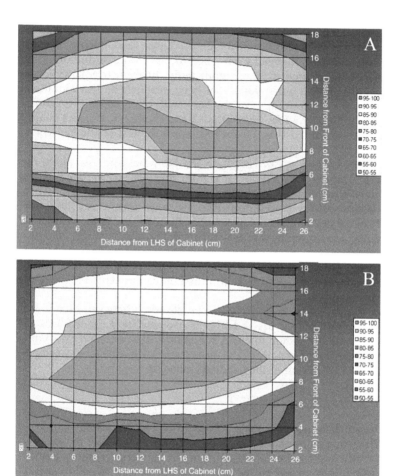

Figure 3   Percentage distribution of power within a Heraeus Suntest CPS Cabinet (xenon arc lamp). (**A**) Low power-power level 1-maximum power $=22.3\,mWcm^{-2}$ (approximately 48 kLux), (**B**) high power-power level 7-maximum power $=52\,mWcm^{-2}$ (approximately 113 kLux).

Calibration

As described above, most lamps age as a function of time. This can result in not only a change of spectral power distribution but also in a lower output. Lamps should, therefore, be changed at defined time intervals (e.g., xenon lamp after 1000–2000 hrs, fluorescent tubes after 5000–10,000 hrs, and metal halide after 3000 hrs) as specified by the producer. Care should be taken to ascertain that the producers and the user's rating criteria are the same, which may not be the case.

Some tubes need a sufficient burning period to provide a stabilized output. As an example, a fluorescent tube will change spectral intensity output when it is new and typically need to be "burned in" for 100–200 hrs before being used in a photostability study. Although the user can rely on the spectral power distribution specification provided by the manufacturer (assuming that the lamps are changed at regular usage intervals according to the specifications), it is essential to calibrate the source and periodically monitor its irradiance in order to obtain the predetermined exposure value.

Some testing cabinets suitable for photostability testing will have a built-in irradiance sensor for set-up and intensity monitoring. If the irradiance intensity drops, the lamp power will be adjusted accordingly, to keep the irradiance level constant.

For instrumentation without a built-in sensor, calibration can be performed manually using a UV filter radiometer, luxmeter, thermopile or chemical actinometer (18–20).

With the exception of the thermopile, all of these devices suffer from the same drawbacks. They do not have a constant spectral responsivity over the entire UV or VIS wavelength range (i.e., the relative weighting given to the different wavelengths is not constant) and they are only sensitive over a narrow wavelength region.

Measuring Devices

Radiometers

The ICH guideline recommends the use of a calibrated radiometer or a validated actinometric system to monitor the exposure in the UV region and a calibrated luxmeter to determine the overall illumination in the VIS range.

A UV filter radiometer provides a convenient measure of the total UV output of a photon source, if properly designed for a specific source. They are broadband meters designed to measure incident radiation in the entire selected, UV region. An optical filter is used to limit the spectral responsivity to a certain band (e.g., UVB, near-UVA, or far-UVA). Ideally, the spectral responsivity is constant within the band and zero outside, but this is not the case in practice. Also, the characteristics of the filters may change with time and the device should be calibrated at least once per year.

Unfortunately, radiometers from the same and different manufacturers can measure different fractions of radiant energy, because there are no international standards for the filters.

Luxmeters (Photometers)

A luxmeter has a photopic response curve, i.e., it will measure all incident radiation as it is perceived by the human eye. For that reason, this device will mainly detect irradiation only in that part of the spectrum. "Light" (e.g., irradiation in the VIS part of the spectrum) is defined scientifically as from 400–700 nm. The limits of the photosensitivity range of the human eye lie around 380 to –780 nm, with the highest sensitivity being at 555 nm. A typical luxmeter will therefore not detect irradiation in the upper part of the UVA and the lower part of the infrared range in addition to not detecting all wavelengths in the VIS region with equal sensitivity.

Meter Calibration

It is important to differentiate between calibration of the UV filter radiometer or luxmeter as prescribed in the ICH guideline and calibration of these devices for a specific lamp. In the first case, the calibration is carried out by the manufacturer to establish the response curve of the device. If the meters are calibrated against a standard lamp and used as received, they are well suited for measuring any variation of irradiance across the sample area and changes in total output with time. But they cannot be used to give an absolute measurement of irradiance or to compare irradiance between sources unless they are calibrated specifically for each source (21).

Table 2 Calibration for a specific source

Calibration for a specific photon source
Measure spectral irradiance from the photon source using a spectroradiometer
Determine the irradiance within the waveband of interest
Place the narrow band radiometer at the same point in space as the spectroradiometer and note the reading
Determine the multiplicative correction factor by dividing the correct irradiance by the displayed irradiance

A procedure for the calibration of radiometers against a specific light source is suggested in Table 2. It is important to remember that neither the UV filter radiometer nor the luxmeter provides information on the spectral power distribution of sources. A detailed plot of the irradiation as a function of wavelength is only obtainable by use of a spectroradiometer. At present, such equipment is not used on a regular basis.

Spectroradiometric data should be provided by the lamp manufacturer upon request. In cases where the spectral power distribution data are obtained separately from the calibration of the narrowband radiometer, it can be difficult to place the two devices exactly at the same point in space (Table 2). It is then of great importance to at least place radiometers at the same distance used for calibrating the lamp if accurate calibration measurements are to be achieved.

Chemical Actinometry

The total irradiance received by a sample can be determined by chemical actinometry, which uses a reaction of known photochemical efficiency. The ICH guideline proposes quinine actinometry as a standard based on the assumption that its increase in absorbance is proportional to the integrated UVA irradiance for a given lamp (22). The suitability of quinine as actinometer is, however, questioned in several different reports (10,23,24). The photolysis of quinine is sensitive to temperature, pH, and dissolved oxygen content. The physical characteristics of the quartz cell (e.g., cuvette dimension) are further shown to influence the result (24).

Additionally, the measurements are carried out on a steep slope of the absorption curve. It is therefore of importance that the wavelength be set accurately (e.g., by use of a photodiode array spectrophotometer).

Complex secondary reactions can follow the primary photochemical process of quinine leading to dark reactions, which occur after exposure of the quinine solution has ceased. For this reason, the absorbance should be measured immediately after exposure. The user must carefully control the variables that can affect the quinine actinometry system in order to obtain reproducible results. One should take into account that the calibration factor stated in the ICH guideline may only be valid at 25°C.

Presentation of Samples

Drug substances and drug products should be presented in a way to provide maximum area of exposure to the source. Photodegradation occurs only on the surface of solid samples (25). The surface area or weight ratio will therefore affect the extent of observed degradation. It is recommended in the ICH guideline that the sample thickness should not exceed 3 mm for solid drug substances. Preparations like tablets or capsules should be spread in a single layer.

As previously mentioned, some test cabinets may be temperature regulated by means of a cooling air stream. This air stream will cause turbulence in the test chamber, and in some cases a protective container is needed to keep the powder samples, tablets, or capsules in a fixed position. Whatever container that is used and whatever the orientation is will influence the results. It is therefore important to specify the presentation of the samples.

Containers
The containers used to hold the sample should be specified in terms of their transmittance characteristics. A glass dish protected with a transparent cover (e.g., window glass or plastic wrap cover) is convenient for most purposes. For hygroscopic substances, low-melting or highly toxic compounds, specially designed containers might be necessary.

Liquids are exposed in clear glass vials or in quartz cuvettes. A slight reduction in exposure intensity on the sample surface caused by the protective cover or container can be compensated for by a corresponding increase in lamp intensity or longer exposure time, assuming that the protective cover does not change the ratio of UV/VIS reaching the sample.

The main disadvantage of using a protective container is that a significant increase in temperature can be expected. The sample is no longer directly affected by the cooling air stream and the temperature inside a covered glass dish can easily reach 40°C or more, even though the temperature in the chamber outside the dish is kept close to room temperature. Temperatures up to approximately 70°C have been reported for capsule samples stored under a polyethylene protective cover, leading to melting of hard gelatin capsules (26). A loose fitting or perforated plastic film can reduce the problem.

A water cooling system can also be used to keep the temperature within the test chamber close to room temperature, but in some cases this makes the interior surface of the test chamber too cold, leading to the condensation of water droplets inside the protective container. To minimize this problem, the container can be placed on a thin layer of isolating material, e.g., a thin rubber ring. Condensation can further occur if the light source is automatically switched off, but the cooling system continues to operate when the desired exposure level is reached. The temperature in the chamber can easily drop 20°C within a few hours (26). This should be taken into account if the lamp is automatically switched off during the night.

The dark controls should be placed alongside the authentic sample. In the case where the sample is placed in a protective container, it is recommended that the tester wrap the individual dark control sample in a protective material (e.g., aluminium foil) before placing it in the container rather than wrapping the protective container in aluminium foil, as the latter method can change the temperature inside the container.

Stirring may influence the photochemical degradation rate of drugs in solution. In order to obtain reproducible results, stirring must be standardized. For the purpose of photostability testing according to the ICH guideline, stirring or shaking of the samples during exposure is not recommended.

Interpretation of Data
Photostability testing according to the ICH guideline will give an indication as to whether photochemical degradation of the drug substance or drug product is

likely to occur during its synthesis, production, packaging, or shelf life. The results obtained are used to make packaging and labeling decisions as well as patient use decisions (labeling directions for use).

The guideline does not include the design of in-use test, nor does it cover abridged applications (in all regions), clinical trial materials, or generic products. In-use considerations can be complicated and need to be addressed in the future.

The aim of the forced degradation studies is to evaluate the overall photo-sensitivity, as well as provide material for analytical method development and degradation pathway elucidation. The experimental design is left to the applicant's discretion, and exposure levels are not defined. It is, however, recommended that one choose a stress level significantly higher than the conditions used in the con-firmatory study (e.g., 3–10 times higher) to ensure that any degradants formed can be detected. Excessive degradation (>15%) should however be avoided in order to minimize the influence of the degradation products on the process. Forced degra-dation studies on the product are not required by the guideline, but may be used to control the validity of methods developed for the drug substance. Excipients may influence the stability or even cause a photostable active to be photodegraded via photosensitization, etc.

In principle, a product can be less stable than the drug substance, especially in solution. In some cases, different degradants will be observed for drug substance and drug product. Photodegradants are treated as other degradants or impurities, e.g., there is no need to identify them unless they are present in concentrations above the ICH threshold levels for identification (27). The identification of deg-radation products may, however, help in understanding the mechanism(s) of photodecomposition.

The testing of a drug substance in solution can give useful information about the intrinsic stability of a molecule, and the degradation rates in solution can be used as a guide for the solid state.

Experiments with solid-state samples are important, because their degra-dation mechanisms can be different from those of a solution. In the solid state, degradation is limited to the surface layer and therefore parameters like polymorphic form, particle size, and layer thickness will affect the result. It is important to remem-ber that excipients may influence the stability. The confirmatory test is carried out to provide the information necessary for handling, packaging, and labeling of the drug. The study is carried out in a sequential manner. The decision tree describes how to interpret the results. The testing should progress until the product can be demonstrated to be adequately protected.

How is one to proceed if the sample shows no change after exposure to the 200 Wh/m² UV dose but is decomposed after the 1.2 million lux hours in a test chamber designed according to Option 1 (e.g., xenon lamp)? One possibility is to rerun the test using a filter to eliminate the UV irradiation in excess. Another possi-bility is to rerun the exposure to 1.2 million lux hours using a cool white fluorescent tube. If the sample is stable to this VIS exposure, it meets the ICH guideline.

If the sample is only unstable to UV, it can be labeled "protect from daylight." If there is any risk at all of the product undergoing photodegradation it should be labeled "protect from light," indicating that the product should be stored in its sec-ondary container—normally a cardboard or a card box. In the case where a product fails to meet the requirement even when stored in the outer container, the producer should consider to choose an alternative package that offers a full protection of its content. Alternatively, reformulation of the product could be tried.

The term "significant change" is not defined in the guideline but in the main documents of which it is a subpart. Justification of impurity limits should be based on the ICH Drug Product Impurity Guideline (27). The possible effect of photodegradation on the shelf life should be taken into account to assure that the product will be within specifications for its shelf life.

CONCLUSION

The proposed ICH test for the photostability of drugs and drug products is relatively simple to conduct. Careful control of the experimental conditions is, however, essential in order to obtain reproducible results. The selection and calibration of the irradiation source used is critical. Additionally, the irradiance level and temperature effects have to be considered.

Radiation intensity mapping of the test chamber is essential to ensure that samples are placed at points of equal irradiance. Sample presentation is of great importance and the containers employed should be of known transmittance.

It is important to be cognizant of humidity changes and possible sample surface heating problems. Identification of all photodegradants is rarely necessary, but can be useful in understanding the mechanism(s) of photodecomposition and aid in the development of protective measures.

ACKNOWLEDGMENTS

The author would like to thank Dr. N. Anderson, Dr. J.T. Piechocki, Dr. C. Willwoldt, and Dr. Heiko Spilgies for useful discussions.

REFERENCES

1. Tønnesen HH, Karlsen J. Photochemical degradation of components in drug formulations. III. A discussion of experimental conditions. Pharmeuropa 1995; 7:137–141.
2. Tønnesen HH, Moore DE. Photochemical degradation of components in drug formulations. Pharm Tech Eur 1993; 5:27–33.
3. Anderson N. Forced degradation studies. Presented at Photostability, IBC Global Conference, London, July 1, 1999.
4. Thoma K, Kerker R. Photoinstabilität von Arzneimitteln 1. Mitteilung über das Verhalten von nur im UV-Bereich absorbierenden Substanzen bei der Tageslichtsimulation. Pharm Ind 1992; 54:169–177.
5. Thoma K, Kerker R. Photoinstabilität von Arzneimitteln 2. Mitteilung über das Verhalten von im sichtbaren Bereich absorbierenden Substanzen bei der Tageslichtsimulation. Pharm Ind 1992; 54:287–293.
6. Matsuo M, Machida Y, Furuichi H, Nakamura K, Takeda Y. Suitability of photon sources for photostability testing of pharmaceutical products. Drug Stab 1996; 1:179–187.
7. Kerker R. Untersuchungen zur Photostabilität von Nifedipin, Glucocorticoiden, Molsidomin und ihren Zubereitungen. Ph.D. Dissertation, Ludwig-Maximilians-Universität, München, Germany, 1991.
8. Thoma K, Kübler N. Einfluss der Wellenlänge auf die Photozersetzung von Arzneistoffen Pharmazie 1996; 51:660–664.
9. Drew HD. Photostability of drug substances and drug products: a validated reference method for implementing the ICH photostability guidelines. In: Albini A, Fasani E, eds. Drugs: Photochemistry and Photostability. Cambridge: The Royal Society of Chemistry, 1998:227–242.

10. Piechocki JT. Selecting the right source for pharmaceutical photostability testing. In: Albini A, Fasani E, eds. Drugs: Photochemistry and Photostability. Cambridge: The Royal Society of Chemistry, 1998:247–271.

11. Boxhammer J, Willwoldt C. Design and validation characteristics of environmental chambers for photostability testing. In: Albini A, Fasani E, eds. Drugs: Photochemistry and Photostability. Cambridge: The Royal Society of Chemistry, 1998:272–287.

12. Forbes DP. Design limits and qualification issues for room-size solar simulators in a GLP environment. In: Albini A, Fasani E, eds. Drugs: Photochemistry and Photostability. Cambridge: The Royal Society of Chemistry, 1998:288–294.

13. Boxhammer J. Technical requirements and equipment for photostability testing. In: Tønnesen HH, ed. Photostability of Drugs and Drug Formulations. London: Taylor & Francis, 1996:39–62.

14. Anderson N. Photostability testing: design and interpretation of tests on drug substances and dosage forms. In: Tønnesen HH, ed. Photostability of Drugs and Drug Formulations. London: Taylor & Francis, 1996:305–322.

15. Nema S, Washkuhn RJ, Beussink DR. Photostability testing: an overview. Pharm Tech 1995; 19:170–185.

16. Sayre RM, Cole C, Billhimer W, Stanfield J, Ley RD. Spectral comparison of solar simulators and sunlight. Photoderm Photoimmun Photomed 1990; 7:159–165.

17. Clapham D, Sanderson, D. The Importance of sample positioning and sample presentation within a light cabinet. 3rd International Meeting on the Photostability of Drugs and Drug Products, Washington, DC, July 10–14, 1999.

18. Moore, DE. Standardization of photodegradation studies and kinetic treatment of photochemical reactions. In: Tønnesen HH, ed. Photostability of Drugs and Drug Formulations. London: Taylor & Francis, 1996:63–82.

19. Favaro G. Actinometry: concepts and experiments. In: Albini A, Fasani E, eds. Drugs: Photochemistry and Photostability. Cambridge: The Royal Society of Chemistry, 1998:295–304.

20. Piechocki JT, Wolters RJ. Use of actinometry in light-stability studies. Pharm Tech 1993; 17:46–52.

21. Tønnesen HH, Karlsen J. A comment on photostability testing according to ICH guidelines: calibration of light sources. Pharmeuropa 1997; 9:735–736.

22. Yoshioka S, Ishihara Y, Terazono T, et al. Quinine actinometry as a method for calibrating ultraviolet radiation intensity in light stability testing of pharmaceuticals. Drug Dev Ind Pharm 1994; 20:2049–2062.

23. Bovina E, De Filippis P, Cavrini V, Ballardini R. Trans-2-Nitrocinnamaldehyde as chemical actinometer for the UV-A range in photostability testing of pharmaceuticals. In: Albini A, Fasani E, eds. Drugs: Photochemistry and Photostability. Cambridge: The Royal Society of Chemistry, 1998:305–316.

24. Baertschi SW. Commentary on the quinine actinometry system described in the ICH draft guideline on photostability testing of new drug substances and products. Drug Stab 1997; 1:193–195.

25. Sande SA. Mathematical models for studies of photochemical reactions. In: Tønnesen HH, ed. Photostability of Drugs and Drug Formulations. London: Taylor & Francis, 1996:323–339.

26. Baertshi SW. Pharmaceutical photostability testing: sample presentation. Presented at Photostability '99, Washington DC, July 10–14, 1999.

27. ICH guideline Q3A. Impurities in new drug substances. International Conference on Harmonization of technical requirements for registration of pharmaceuticals for human use, 1995, Geneva, Switzerland, International Federation of Pharmaceutical Manufacturers Associations (IFPMA).

3 Terminology

Joseph T. Piechocki
Piechocki Associates, Westminster, Maryland, U.S.A.

INTRODUCTION

In this chapter, we will not attempt to repeat the excellent work already done by others but rather highlight terminology problems specific to and which continue to cause problems in the application of the International Conference on Harmonization (ICH), Q1B, Guideline for the Photostability Testing of New Drug Substances. For general terms the reader should consult other established references such as "The Vocabulary of Photochemistry" by Pitts et al. (1), "Glossary of Terms Used in Photochemistry" by Braslavsky and Houk (2), and the "Glossary of Terms in Photocatalysis and Radiocatalysis" by Parmon et al. (3). Other useful references include the Illuminating Engineering Society of North America (IESNA) Lighting Handbooks' "Dictionary of Lighting Technology" (4), "Terminology Related to Natural and Artificial Weathering Tests for Nonmetallic Materials" by American Society for Testing Materials (ASTM) G 113 (5), Commission Internationale de L'Eclairage (CIE) Pub. Nos. 20 (6) and 85 (7), the International Standards Organizations (ISO) ISO 4892-1 thru 3 (8–10), and the new Joint ISO/CIE Standard ISO 10526:1999/CIE S005/E-1998 CIE Standard Illuminants for Colorimetry (11). Other excellent glossaries can also be found in some manufacturers' catalogs such as those of Gigahertz-Optik Inc. (12), International Lights' "Light Measurement Handbook" (13), and Newport Corporation "The Book of Photon Tools" (14).

Most of the current problems that have found their way into the ICH document are the result of the number of different fields covered by the document and the experts on the committee not truly understanding the terminology they were using. Frequently they have repeated terms found in the commercial literature, which while perfectly appropriate for color matching and lighting purposes, are inappropriate for pharmaceutical photostability testing purposes. We hope that this chapter will aid in correcting these errors.

GENERAL TERMINOLOGY

Light

The term light is perhaps one of the most misused terms in photochemistry. This problem has also plagued pharmaceutical photochemistry and chemistry throughout the last century. Many of today's scientists were taught about UV light.

"Light" is defined in the dictionary as radiation from the 390 to 800 nm region of the electromagnetic spectrum (15). Unfortunately, we also employ light meters, called photometers, to measure light. Because the objective of lighting is to provide proper illumination and not distort colors, these instruments are designed to mimic as closely as possible the normal photopic response of the human eye as indicated in Figure 1 and Table 1.

Two things that are readily obvious from Figure 1 and Table 1 are that, first, its response does not cover the entire VIS spectrum given by the definition of light and,

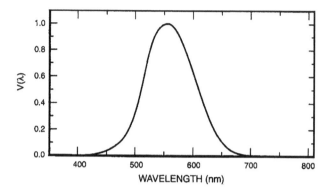

Figure 1 Photopic response of the human eye. *Source*: Courtesy of the Newport Corporation.

second, that the response of the photometer for the wavelengths actually measured is only accurate at 555 nm. All other wavelengths are measured at fractions of their true value. This topic will be discussed later in this chapter, under the topic of Photometers.

The first draft of the ICH document, from ICH 1, used the term "light stability" in its title. At ICH 2 (16), the inappropriateness of this term was brought to the attention

Table 1 Photopic Response Correction Values

Wavelength (nm)	Photopic Luminous Efficiency V(λ)	Wavelength (nm)	Photopic Luminous Efficiency V(λ)
380	0.00004	580	0.870
390	0.00012	590	0.757
400	0.0004	600	0.631
410	0.0012	610	0.503
420	0.0040	620	0.381
430	0.0116	630	0.265
440	0.023	640	0.175
450	0.038	650	0.107
460	0.060	660	0.061
470	0.091	670	0.032
480	0.139	680	0.017
490	0.208	690	0.0082
500	0.323	700	0.0041
510	0.503	710	0.0021
520	0.710	720	0.00105
530	0.862	730	0.00052
540	0.954	740	0.00025
550	0.995	750	0.00012
555	1.000	760	0.00006
560	0.995	770	0.00003
570	0.952		

Source: From Ref. 14.

of the Expert Working Group (EWG). A vote was immediately taken and a change in the name recommended and adopted. The EWG did however resist correcting all other misuses of the term in the document from that point onward, despite protests (17).

It should also be noted that acceptance of the definition for visible light given in practically all books and standards excludes UV radiation. It is therefore recommended that this term only be used when referring to the 390 to 800 nm region of the electromagnetic spectrum.

Visible Radiation

Light is defined by IESNA as "radiant energy that is capable of exciting the retina and producing visual sensation" (4). There is some variation in the literature of the upper and lower limits because of the normal variation in this subjective response. They also define the visible portion of the electromagnetic spectrum as extending from 380 to 770 nm. If strictly applied, according to this definition, light stability testing then should mean only testing within this limited range and no UV radiation need be included in the testing.

UV Radiation

The UV region is generally defined as electromagnetic radiation from 10 to 380 nm. This bandwidth is further divided into three regions of interest to pharmaceutical photo chemists, UV-A (315–400 nm), UV-B (280–315 nm), and UV-C (100–280 nm) (4). The UV-A and UV-B regions are of special interest to sunscreen developers and their possible role in immunomodulation. The UV-C region is of interest because of the very strong 280 nm mercury line from artificial sources based on mercury arcs, e.g., fluorescent, metal-halide, mercury, and mercury-doped xenon lamps. There is also confusion regarding just what the subdivisions of electromagnetic radiation in the range of 180 to 800 nm, UV-A, UV-B, UV-C, and VIS, should be preferred the band assignments are given in Table 2.

It is well to note that not all UV-A, UV-B and UV-C, and VIS lamps are equivalent. This subject will be discussed in further detail in Chapter 6 of this book.

Einstein

The term "einstein," one mole of photons, is not supported by the International Union of Pure and Applied Chemistry (IUPAC) (2). Part of the lack of support may be the fact that because photons have different energies, depending upon

Table 2 Recommended Ranges for UV and VIS Radiation

Name	Wavelength range (nm)
UV-A	315–380
UV-B	280–315
UV-C	100–280
VIS	380–770

their wavelength, the energy available per mole is not a constant, except for a particular wavelength.

If, however, one is calculating the quantum efficiency of a reaction, Φ, one must know the number of moles of photons used in the reaction because we are dealing with molecular reactions. Photochemical reactions occur on a molecular basis. In this case the current practice is to use a well-characterized chemical actinometer to measure the dosage. This is discussed in more detail in Chapter 7 of this book.

Equivalency

The dictionary (15) defines this term as "1. A. Equal in value, force, or meaning. B. Having identical or similar effects. 2. Corresponding or practically equal in effect."

In talking about sources for chemical purposes, it is wise to realize that much of the equivalency used in the color comparison, photo testing, and light illumination fields may not be directly applicable to our work. This results from the fact that we require a good reproduction of the selected test sources wavelengths and relative intensities.

Also, any discussion of equivalency must start with a reference standard and definition of just what are the limits of equivalency. For a lamp to be equivalent, it should contain the same wavelengths and intensities within a designated percentage of the reference source. Because the wavelength determines the energy of the incident radiation, which in turn determines whether a reaction with a bond will or will not take place, all wavelengths should be present. Likewise, because the intensity of each wavelength determines just how much energy at a particular wavelength will be transferred to the test substance, it should be reproduced to within certain defined limits.

As previously noted in Chapter 1 of this book, the ICH's EWGs chose not to select any reference source. "...as there was concern that this reference would become the only light source used and accepted." But they did agree that a reference spectrum should be used, D65 (17). Because there is a lamp and filter associated with the D65 CIE standard, this selection automatically establishes a single reference source. The EWGs agreed on this spectrum but not necessarily the allowable deviations from that spectrum.

Their selection of all of the other lamps in Options 1 and 2, nonequivalent to the D65 standard, is not rational from a scientific standpoint. Good examples of how a lamp should be specified can be found in ASTM standard G-155-98 (18) and ISO 4892-2 "Plastics—Methods of Exposure to Laboratory Light Sources. Part 2. Xenon-Arc Sources" (9), which have tables, such as that found in Tables 1 and 2 of ASTM G-155-98, which specify the acceptable upper and lower limits for specific bandwidths of the UV and VIS spectrum for sunlight and xenon arc with daylight filters, and xenon arc with window-glass filters and estimated window-glass filtered sunlight. Figure 2 shows the difference between the spectral power distributions (SPDs) of D65 and ID65.

The absence of radiation at the lower, more energetic region of the UV spectrum of the ID65 SPD is readily evident. Because more bonds are capable of interacting with incident photons at these lower wavelengths, this difference can lead to actual differences in the results obtained.

The CIE has worked very hard to develop alternate, equivalent sources for use in color testing. Some of this early work is illustrated in the paper of Wyszecki (19). In this paper he proposed the use of a "Goodness-of-Fit" calculation to determine "whether the spectral irradiance distribution S(λ) of a given source is a

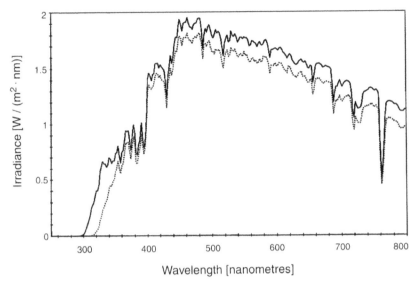

Figure 2 D65 (—) and ID65 (...) spectral power distributions. *Source*: Reproduced with permission of Atlas Electric Devices Company.

sufficiently close fit to a given CIE standard illuminant with distribution SD(λ) so that visual colorimetric tasks can be performed." Unfortunately, this once again is in support of color matching, not photochemistry.

Window Glass

While this topic will be discussed in depth in Chapter 7 of this book, it should be noted that the term "window glass" is not a scientific term. The composition of glass can differ from region to region due to impurities, and this can materially affect its transmission properties.

Along with these basic considerations one must also know if the particular window glass is single or double paned, what the fill gas might be for double paned windows, and if the glass surface has been altered for thermal reflectivity, UV transmission, or "self-cleaning." All of these factors have generally been ignored in photostability reports.

One important thing to remember is that "window glass" is defined not as an absorption spectra of its own but rather by the effect it has on the D65 daylight spectrum.

Inner Filter Effect

In emission spectroscopy, this refers to the decrease in emission and distortion of the emission band shape by preabsorption of the emitted radiation. In an absorption experiment, this refers to the absorption of incident radiation by the same or a species other than the intended primary absorber. At high concentrations the primary absorber, even if in solution and not stirred, along with any photodegradation products may act as is inner filters.

With high intensity sources, this problem manifests itself by distorting the emission line, producing a u-shaped instead of a Gaussian shaped peak.

Correlated Color Temperature, Chromaticity

The actual color appearance of light that comes from a "black body" source is called its color temperature or chromaticity. Correlated color temperature (CCT) is the terminology used when referring to the color of an artificial source. Because the color temperature of an artificial source will most likely not fall on the normal black body curve, it is the accepted practice to refer to their temperature as the temperature of the closest CIE daylight (D) temperature value. Hence, a CCT of 6500 K means that its color temperature is closest to that of CIE D6500 daylight standard.

Figure 3 shows two "cool white" lamps of the same CCT but definitely not of the same spectral power distribution. It is evident that this single criterion is not sufficient to designate a lamp for our uses.

Color Rendering Index

The ability of a light source to correctly represent colors of objects is its color rendering index (CRI), and with the CCT it suffices to help colorists select an equivalent source. Established by the CIE this standard rates lamps as to their color matching ability. A CRI value of 100 indicates that a source truly represents all colors of the CIE standard. Sources with CRIs of 70 to 80 are considered good and those with values of 80+ are rated as excellent.

Hunt (20) and Macbeth (21) give good discussions of this subject in their books.

Room Light

Room light is one of the most variable quantities. In the words of Hunt (20) "Indoor daylight varies according to the nature of the sky (and sunlight, if any), the geometry

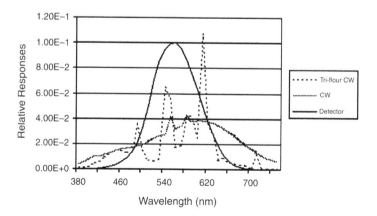

Figure 3 Spectral power distributions of the two "cool white" lamps available overlaid with detector response curve. *Source*: Courtesy of Dr. R. Levin, Osram Sylvania.

of the windows relative to the sky, the spectral transmission of the windows, and the nature of the interior of the room." Indoor daylight can therefore be even more variable than outdoor daylight. To this one must also add the effects of shelving, position relative to the windows, and artificial sources, etc. There are a number of different artificial sources to be found in any one environment whether it is home or manufacturing place. For example, a home might contain tungsten filament, tungsten halide, and fluorescent, which can change with time.

Recent legislation in the United States (22) has led to some lamps being discontinued and others being changed. For example, the standard monofluorophosphate "cool white" lamp is now manufactured in a lower wattage that gives the lamp a visible reddish color and lower intensity. This change has also led to the elimination of diffusers and their replacement by what is termed "egg crate" diffusers that transmit more light but no longer provide any filtering of the incident radiation.

Riehl and Maupin (23) define indoor lighting as artificial and natural light, neither a well-defined term. There are many different types of artificial radiation sources, and natural light is an undefined term. More correctly, their original statement should state it is a combination of "window-glass filtered" daylight and that from an artificial source.

Room light has varied and will continue to vary as time progresses. For home and business lighting, we have historically seen changes from daylight to fire, incandescent, fluorescent, and most recently metal-halide lamps. Even window glass has varied, thereby affecting the "window-glass filtered daylight" that one finds in different types of buildings. These problems will be discussed further in Chapter 7 of this book.

Similar

One of the most often misused terms is similar. Webster (15) defines this word as meaning "having characteristics in common" and "alike in substance or essentials." Either of these definitions if applied to the ICH photostability document Options section would eliminate the use of many sources they currently allow.

It should be basic in the comparison of any two or more items that their degree of similarity is given. For lamps and measuring devices, it is imperative that their degree of similarity be numerically given.

Solarization

Solarization is defined as the change in glass optical properties following exposure to high intensity radiation of short wave-length. This change is generally manifested by a darkening of the glass and can generally be reversed by exposing the glass to long wavelength radiation.

A good introduction to this topic can be found in an article by Rindone (24) who points out that one should be careful in interpreting solarization data because of the different test sources used. In this particular instance he is referring to the use of the mercury lamp versus a solar simulator.

Lux (lx)

A lux is the standard of illuminance equivalent to one lumen per square meter (lm/m^2). This is a lighting term and not very applicable to photochemical uses.

Table 3 Some Sources That Have Been Used for Pharmaceutical
Photostability Testing and Their Respective Color Rendering Indexes and
Correlated Color Temperatures

Lamp	CRI	CCT (K)
Tungsten (100W)	100	2850
Tungsten halide	100	3000
Cool white (f)	67	4100
Cool white (f) tri-fluor	85	4000
Daylight D65	100	6500
Artificial daylight (f)	92	6500
Xenon	100	6500
Metal halide (Hg)	88	6430

Abbreviations: f, fluorescent; CRI, color rendering index; CCT, correlated color
temperature.

The terms "lumens" and "lux" were introduced into pharmaceutical photostability
testing because they are units found on commercial light meters. No one ever
questioned the appropriateness of the term to this field of study. It has no application in the field of pharmaceutical photostability testing.

As was pointed out in our prior discussion of "light" and demonstrated in
Figure 1 and Table 1, if the measurement system we adopt does not cover the wavelength range of interest it is flawed. Pharmaceutical photostability testing should not
be designed to determine the effect of an inaccurately measured quantity of radiant
energy.

All references in allied photochemical standards, to the intensity of the incident
radiation, use radiometric instruments and the more accurate terms "watts/m^2" or
"microwatts/cm^2" for both the UV and VIS regions of the electromagnetic spectrum.
Using the same term for both regions of the spectrum facilitates the comparison of
alternate sources.

Photochromism

As defined by the IUPAC "Photochromism is a reversible transformation of a chemical species induced in one or both directions by the absorption of electromagnetic
radiation between two forms A and B, having different absorption spectra" (2).
Articles by Exelby and Grinter (25), Dessauer and Paris (26), and Bouas-Laurent
and Dürer (27) give a good introduction to the phenomena.

This phenomenon is important to pharmaceutical photostability testing because
it represents an equilibrium situation where the action of one wavelength is reversed
by that of another. Essentially this means that the use of single sources sequentially,
as indicated in Option 2 of the ICH guideline, can lead to results that are not related
to reality. Manufacturers who advocate using both "cool white" and UV lamps
concomitantly for part of the testing time and then shutting off the UV lamps when
they have reached the calculated UV dosage are giving their customers advice that
could lead to less than valid results.

Sanvordeker (28) reported on photochromism in a potent anticonvulsant
veterinary drug.

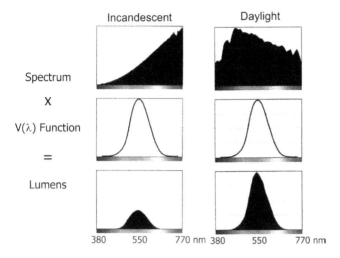

Figure 4 Resultant response of photopic detector with incandescent and daylight lamps. *Source:* From Ref. (29).

Photometer

An electronic measuring device is used for the measurement of quantities such as illuminance. It is composed of an amplifier, sensing device, filters, and a readout device. The ICH document does not specify which of the several different detector heads and filters available should be used. Photometric measurements are all weighed for human vision and as such do not measure the true amount of incident radiation. Table 1 listed the correction factors that are applied for a photometric detector and Figures 4 to 6 show how these corrections distort the actual data obtained (29).

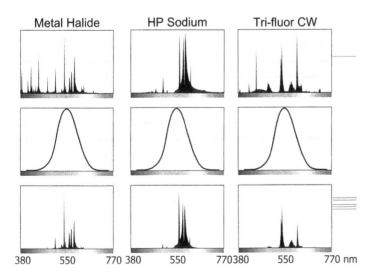

Figure 5 Resultant response of photopic detector with metal-halide, high-pressure sodium and tri-fluor cool-white lamps. *Source:* From Ref. (29).

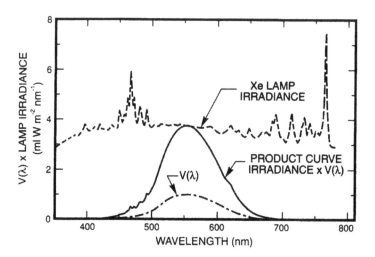

Figure 6   Resultant response of photopic detector with a xenon lamp. *Source*: From Ref. (14).

Surveys, presented at one of the ICH EWG meetings (30), support the fact that the VIS detector used by most laboratories is the one shown in Figure 1, the photometric one. Reference to detector manufacturer's catalogs will also support this statement, if they have identified some of their products as applicable to pharmaceutical photostability testing.

In Figure 6, we can see just how much radiation is not measured by this instrument when it is used with a xenon or metal-halide lamp. Upwards of 60% of the incident radiation is not measured. This leads to the "overexposure" of samples and less than accurate correlations between potentially similar/equivalent sources and significantly increases analysis times.

Quantum Detectors

The name quantum detector is given to a radiometric detector having equal or near equal response over the entire range to be measured, UV or VIS and calibrated in quantum units. As is the case for all meters, they may be calibrated for any appropriate unit.

The ideal quantum detector response curve is a square wave, i.e., equally sensitive over the entire wavelength range of measurement. These types of detectors should be used for all measurements of incident intensity, UV and VIS, if a spectroradiometer is not available. While spectrally blind they will at least give a truer reading of the total amount of radiation actually being received by the samples. Figure 7 shows the response curves of two of these detectors.

Radiometer

A radiometer is an electronic measuring device which when coupled to a proper sensing device, e.g., flat response radiometric detectors (quantum) and calibrated, measures most accurately the quantity of incident radiation falling on its sensor. These instruments are available for both the UV or VIS ranges.

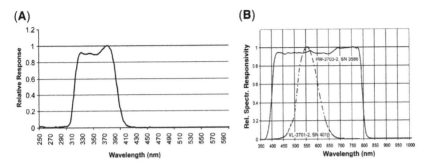

Figure 7 Response curves of (**A**) UV and (**B**) VIS photopic (VL 3701-2) and VIS (RW-3703-2) irradiance detectors. *Source*: From Ref. (12).

Measurements are generally made in terms of W/m^2 or mW/cm^2. A radiometer spectral characteristics are determined by the detector head chosen and any filters used. These instruments may be frequency sensitive.

These units when coupled with near linear (quantum) detector sensors are useful for monitoring processes for total incident intensity on a sample. One should remember, however, that because they are spectrally blind, if change is noted it will be difficult to know just where in the spectrum band this is being measured, if it occurred.

Spectroradiometer

A spectroradiometer is a spectrophotometric instrument used for the measurement of radiant flux as a function of wavelength. The output generated is referred to as the "SPD" of the source being measured. When coupled with an integrating sphere a dual monochromator unit produces data considered the gold standard of source spectral measurements.

For photostability testing purpose the small, single monochromator computer driven units are more than adequate. Slightly more expensive than a photometer/radiometer combination, these units can measure the complete spectrum on a real-time basis and output the results in several different formats. When coupled with the computer software available as part of most packages it should be the measurement system of choice. Chapter 13 of this book discusses these instruments in more detail.

ICH SOURCES

Option 1 Sources

D65

D65, daylight, with a CCT of 6500 K is defined by the CIE. The International Standards Organization Standard ISO 10977 (1993) refers to this fact. D65 is also known as D6500 or Standard Illuminant D by the CIE, represents daylight over the spectral range 300 to 830 nm was first adopted in 1966. This standard is not a particular lamp but an internationally agreed to spectral power distribution for solar radiation, issued by the CIE as "Technical Report, Solar Irradiance," first edition

1989, Publication No. 20 (10). This standard has been updated and reissued as Publication No. 85 (7). The term "6500" designates the CCT of the source, 6500 K.

This document, Publication No. 85, has recently been reissued as a combined Joint ISO and CIE Standard, ISO 10526:1999/CIE S005/E-1998 (11). The authors of these documents recognized the fact that the ultraviolet portion on the spectrum varied with many factors including geographical location, altitude, the time of year, etc.; so they agreed to one standard set of values for reference purposes. This standard has been adopted worldwide.

Reference to the previously cited CIE (7), ISO (9), and ASTM (18) standards will reveal that the xenon-arc lamps with appropriate filters are widely used to simulate this spectrum. This is because the properly filtered xenon-arc lamp is the best representation of solar light of any lamp produced.

ID65

ID65 is also referred to as simulated indoor/indirect daylight by the ISO (31) and window-glass filtered daylight by other standards such as the ASTM (18), in the context of the ICH document means ISO 10977:1993(E). It is an artificial mixture of the old CIE Publication No. 20 solar spectrum (a high-pressure xenon-arc lamp with a quartz burner tube assembly giving an illuminance of 6 klx, whose radiation is filtered through an approximately 6.5 ± 0.5 mm float, soda-lime glass, used as window-glass, and whose specifications conform closely to those given in Table 3 of the ISO 10977 standard).

It is well to note that the currently used spectrum, agreed to by the CIE, ISO, and ASTM, has been updated and will not be in true agreement with the old spectrum. The newer spectra have been adjusted for changes in the atmosphere and other climatic conditions. Figure 2, a plot of the D65 and ID65 SPDs, shows their nonequivalency in the lower UV-A and upper UV-B regions.

To equate these two standards as the ICH document does is not scientifically correct. In this spectral re gion, most compounds have some absorbance so that differences in the results obtained are possible.

Artificial Daylight

The D in all CIE standards means daylight. To say artificial daylight without a number to reflect which standard we are talking about is meaningless. The CIE has several daylight standards. They differ from one another in their CCT and CRI.

In discussing lamp equivalency, it should be noted that not all lamps labeled "daylight" are equivalent for chemical purposes. Their spectral power distributions, as demonstrated in Figure 8, can be different.

The differences noted in these figures can be attributed to the number of phosphors used, their ratios, and other factors that effect the SPDs. We must always remember that most of the lamps we have been talking about are commodity items to be produced as cheaply as possible, and they have two main objectives, first to provide illumination and second to reproduce colors as much as possible. They are not scientific standards and never claimed to be such.

Xenon Lamps

At first glance it would appear that such a simple lamp composed of only four basic parts (two electrodes, a glass envelope, and xenon gas) should not be that complicated.

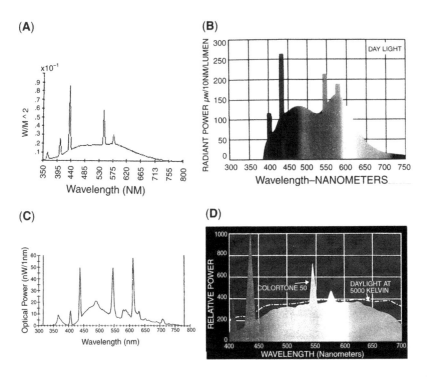

Figure 8 Spectral power distributions of various daylight lamps. (**A**) Hytron, (**B**) General Electric, (**C**) Bahren Lichttechnik, and (**D**) Philips.

If only this were true. There are long- and short-arc lamps and high, medium, and low pressure, and mercury-free and mercury-doped lamps in use (Fig. 9).

The short-arc lamps are small, generate a high internal pressure, and are generally air-cooled and produce a relatively small area of high illumination being point sources. Long-arc lamps, those having a 5-cm or larger electrode gap, are air- and/or water-cooled, depending on wattage. These lamps have a larger area of high illumination and act as bar sources.

Mercury-doped xenon lamps have a higher total emissivity but suffer from the fact that results obtained using them do not correlate as well to actual daylight samples as do the undoped lamps. These lamps are frequently encountered in Sun Protection Factor testing and evaluation. The ICH document fails to adequately describe this lamp and the filters to be used with it.

Metal-Halide Lamps

These lamps are short-arc mercury lamps containing a metal halide surrounded by a glass envelope. The metal used is generally one with a very rich line spectrum, low ionization potential, and low corrosivity. They are available as both air- and water-cooled varieties.

The frequently used halide in these lamps is iodine. The reason for adding the halide is for it to react with any vaporized or deposited metal and form a volatile halide salt, which will eventually come in contact with the hot filament,

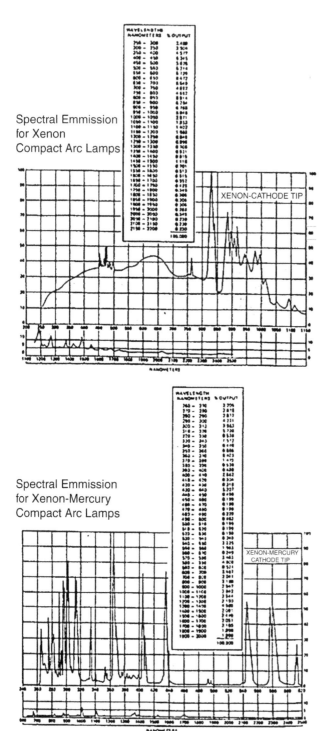

Spectral Emmission
for Xenon
Compact Arc Lamps

Spectral Emmission
for Xenon-Mercury
Compact Arc Lamps

Figure 9 Effect on the
SPD of adding mercury to a
xenon-arc lamp.

decompose to the base metal, and free up the metal and halide for further reaction. In essence, the halide helps to keep the lamp's internal wall surfaces clean by eliminating the darkening of the inner surface of the lamp, helping to maintain its emission intensity.

There are many varieties of these lamps. Two of the most commonly used, in the industry, are the mercury and sodium vapor lamps used for street lighting. These same lamps can frequently be found in recreation, manufacturing, and packaging areas. Mercury lamps are typically used for normal nonphotosensitive materials/ products and sodium vapor lamps for photosensitive materials.

For photostability testing purposes, the two most often used lamps have been the thorium-doped dysprosium and iron lamps. The SPDs for these lamps can be found in Chapter 6 of this book.

Once again the ICH document fails to correctly identify which of the over 40 varieties are acceptable.

Option 2

Cool White

"Cool white" is a term used by lamp manufacturers and to describe the sensation that people experience in rooms lighted with these fluorescent lamps. These lamps have a CCT of approximately 4100 K. This term was used originally to describe a fluorescent lamp, first produced with a monofluorophosphate phosphor. This lamp was more energy efficient and brighter than the then standard, incandescent lamp and had a good color rendition. It was also the first phosphor used for the mass production of fluorescent lamps.

As time progressed, the lamp coating composition changed to the present three rare earth phosphors (tri-fluor) of higher efficiency and brightness and superior color rendition than its predecessor. These lamps are not photochemically (spectrally) equivalent.

This nonequivalence is no trivial matter. One major chamber manufacturer made the mistake of supplying the triphosphor lamp in its equipment (personal communication), and there is at least one reported instance of this problem being reported in the literature (32). In both instances the individuals/companies had checked with their local suppliers and received assurance that these were receiving the right lamps, which turned out not to be the case.

When ordering a "cool white" lamp for pharmaceutical photostability testing, three parameters should be specified, and they are

1. the spectral power distribution desired,
2. the CCT desired (e.g., 4100 K),
3. the CRI desired.

All of these criteria should be available in manufacturers' catalogs. The most important of these three criteria is the SPD.

It is important to also remember that fluorescent lamps with the same common names, e.g., "cool white" are made in many different CCTs and intensities and may not be equivalent. If one is fortunate enough to obtain the SPD for a lamp from their supplier it will hardly ever be for the lamp they purchase or even the batch the lamp came from.

It should also be noted that the spectral power distribution for the "cool white" lamp given in ISO 10977:1993(E) is not that of "cool white" fluorescent lamp but that of a the visible "cool white" spectrum to which the mercury lines have been added (30). This lamp also has a CCT of approximately 4100 K. There is one known notable exception to this last statement and that is Luzchem Research Inc. (33), which does supply a spectrum for each lamp that they sell if it is to be used for ICH photostability testing.

The mercury line intensities vary from lamp to lamp due to the difficulty of delivering a microdrop of the liquid reproducibly into each lamp. These lines are frequently eliminated from published SPDs.

UV Lamps

A similar situation exists for UV lamps. As previously noted, there are three distinct recognized regions of the ultraviolet, so any lamp selection criteria should start with this designation, i.e., UV-A, UV-B, or UV-C. Within each region, there are several lamps available with their own specified minimum, half-maximum, maximum bandwidths, and peak maxima wavelengths. These items further define the lamp.

The information cited could be found in manufacturers' catalogs, which unfortunately are not too easily available. Most distributors do not have these catalogs and are not always aware of their value. They supply commodity lamps for lighting not scientific testing.

Two good examples of the importance of knowing exactly what type of UV lamp is being used can be found in the literature. Cole et al. reported on the "different biological effectiveness of black light lamps available for therapy with psoralens plus ultraviolet A" (34). In this instance, both the acceptable lamp and the unacceptable lamp bore the same designation on the lamp body. The lamps were not equivalent, and problems arose with the therapy received by patients.

More recently, Sequiera and Volzone encountered a similar experience (32). They and an associate both ordered UV and VIS lamps for use in their pharmaceutical photostability testing on the same product. They gave their supplier the ICH specifications and depended on him to supply them with the correct lamps, as stated in the ICH document. In fact, they both received different UV-A and VIS lamps and could not reproduce one another's work.

There is some confusion as to which of several available lamps should be used. The publication of the SPD for the UV-A lamp to be used would alleviate this problem.

CONCLUSION

Proper terminology and clear definitions are necessary in science if one has to have international standards. Without them confusion reigns. This chapter pointed out some of the present problems with the current photostability guideline. It is our hope that this work will stimulate revision of the document and put the ICH effort on a more scientific basis.

REFERENCES

1. Pitts JN Jr., Wilkinson F, Hammond GS. The vocabulary of photochemistry. Advances in Photochemistry. New York: Wiley Interscience, 1963.
2. Braslavsky SE, Houk KN. Glossary of terms used in actinometry. Pure Appl Chem 1988; 60(7):1055–1106.

3. Parmon V, Emeline AV, Serpone N. Glossary of terms in photocatalysis and radiocatalysis. Int J Photoenerg 2002; 4:91–131.
4. Rea MS. Lighting Handbook, Reference and Application. 9th ed. New York: Illuminating Engineering Society of North America, 1993.
5. ASTM Method G 113. Terminology Related to Natural and Artificial Lighting for Nonmetallic Materials. Philadelphia, PA: American Society for Testing and Materials.
6. Commission Internationale de L'Eclairage (CIE). International Commission on Illumination. Recommendation for the integrated irradiance and spectral distribution of simulated solar radiation for testing purposes. CIE, No. 20 (TC 2.2). Paris, France: Bureau Central de la CIE, 1972.
7. Commission Internationale de L'Eclairage (CIE). International Commission on Illumination. Technical Report, Solar Irradiance, 1st ed. CIE, No. 85. Paris, France: Bureau Central de la CIE, 1989.
8. International Organization for Standardization. 1st ed 1994 01. Plastics Methods of exposure to laboratory light sources, Part 1: General guidance. ISO 4892-1:1994(E). Genève, Switzerland: International Organization for Standardization
9. International Organization for Standardization. 1st ed 1994 01. Plastics Methods of exposure to laboratory light sources, Part 2: Xenon arc sources. ISO 4892-2. Genève, Switzerland: International Organization for Standardization.
10. International Organization for Standardization. Plastics methods of exposure to laboratory light sources, Part 3: Fluorescent UV lamps. 1st ed 1994 01. Genève, Switzerland: International Organization for Standardization, ISO 4892-3.
11. CIE Color Standard Illuminants for Colorimetry. Genève, Switzerland: International Organization for Standardization, ISO/CIE 10526:1996/CIE S005/E-1998.
12. Gigahertz Catalog E05/00, Gigahertz-Optik Inc. Newburyport, MA, U.S.A.
13. International Light. 17 Graf Road, Newburyport, MA 01950, U.S.A.
14. Newport Corporation. "The Book of Photon Tools," Oriel Corporation.
15. Webster's New Collegiate Dictionary, 1977 ed., published by G. and C. Merriam Co., Springfield, MA, U.S.A.
16. Piechocki JT. Light stability testing: a misnomer. ICH Second International Conference on Harmonisation, Orlando, FL, U.S.A., 1993.
17. Beaumont T. Photostability testing. In: David J Mazzo, ed. International Photostability Testing. Buffalo Grove, IL, U.S.A. Interpharm Press, 1999.
18. ASTM Method G 155–98. Standard Practice for Operating Xenon-Arc Type Light-Exposure Apparatus on Nonmetallic Materials. Philadelphia, PA 19103, U.S.A. American Society for Testing and Materials.
19. Wyszecki G. Development of new CIE standard sources for colorimetry. Die Farbe 1970; 19(1/6):43–76.
20. Hunt RWG. Measuring Colour. 2nd ed. West Sussex, UK: Ellis Horwood Ltd., 1991.
21. Fundamentals of Color and Appearance. New Windsor, New York: Macbeth, Division of Kollmorgen Instruments Corporation.
22. Energy Policy Act of 1992, U.S. Congress, PL 102-486. Washington, DC: U.S. Government Printing Office.
23. Riehl JP, Maupin CL. On the choice of a light source for the photostability testing of pharmaceuticals. PF 1995; 21(6):1654–1663.
24. Rindone GE. The Formation of Color Centers in Glass by Solar Radiation. Paris: Travaux du IV Congres International du Verre, 1956:373–389.
25. Exelby R, Grinter R. Phototropy (or photochromism). Chem Rev 1965; 65:247–260.
26. Dessauer R, Paris JP. Photochromism. In: Noyes WA Jr., Hammond G, Pitts JN Jr., eds. Advances in Photochemistry. Vol. 1. New York: Interscience, 1963:275–321.
27. Bouas-Laurent H, Dürer H. Organic Photochromism. Pure Appl Chem 2001; 73(4):639–665.
28. Sanvordeker DR. Photochromism of an anticonvulsant, 1-diphenylmethylmethyl-4-(6-methyl-2-pyridylmethyleneamino) piperazine, in the solid state. J Pharm Sci 1976; 65(10):1452–1456.

29. Houser KW. Thinking photometrically, Part I. Light Fair International 2001, Las Vegas, NY, USA. May 2001.
30. ICH 2 Quality Six Party Drafting Group on Light Stability Testing. Kyoto, Japan meeting, June 16–17, 1993 (notes).
31. International Organization for Standardization. Photography processed photographic colour films and transmittance, solar direct, total solar energy transmittance and ultraviolet transmittance, and related glazing factors. ISO 10977:1993(E). Genève, Switzerland: International Organization for Standardization.
32. Sequeira F, Vozone C. Photostability studies of drug substances and products. Pharm Tech 2000; 24(8):30–35.
33. Luzchem Research Inc., Ottawa, ON, Canada.
34. Cole C, Forbes PD, Davies RE. Different biological effectiveness of blacklight fluorescent lamps available for therapy with psoralens plus ultraviolet. A J Acad Dermatol 1984; 11(1):599–606.

4 Basic Principles of Drug Photostability Testing

John M. Allen and Sandra K. Allen
Department of Chemistry, Indiana State University, Terre Haute, Indiana, U.S.A.

INTRODUCTION

Many substances are known to undergo a variety of chemical reactions as the result of exposure to electromagnetic radiation (EMR); drugs are no exception. EMR having wavelengths from 290 to 400 nm is of particular interest in this regard, although radiation of longer visible wavelengths (i.e., 400–800 nm) may also be of some concern. Drug formulations are exposed to a variety of EMR sources during manufacture, packaging, distribution, on pharmacy shelves, and at the point of use. These exposures may include Hg vapor lamps, fluorescent and incandescent lighting, window-filtered sunlight and direct sunlight.

Drug photostability testing is conducted in order to quantify the extent to which photoreactions affect the composition of drug formulations and in some cases, to determine the mechanisms of drug photoreactions. In routine photostability testing, drug formulations are illuminated under carefully controlled conditions in order to determine if they are photostable and to quantify the extent of photo-degradation for those drugs that are shown to be photolabile. The basic principles of drug photostability testing have been described in some detail by Moore (1) and by Tonnesen (2).

Drug photoreactions can lead to a loss of the drug and thus a loss in efficacy, formation of toxic stable products, and formation of toxic short-lived intermediates (e.g., free radicals). There is also the potential for drug photoreactions in vivo since EMR reaches the interior of the eye and to a lesser extent into blood vessels and tissues. However, in vivo photoreactions are outside the scope of the present discussion.

Drug photostability testing provides a means of screening drug candidates early in the development process and allows identification of particularly photolabile (photochemically unstable) drugs prior to a large investment in development and testing. Once a photolabile drug has been identified, it may be possible to develop a strategy for modification of its molecular structure or some other means of photoprotection or further development of the drug may be abandoned if stabilization of a photolabile drug is unsuccessful.

BASIC CONCEPTS

The presence of any of several functional groups is likely to impart photolability to drug molecules. These include carbonyl (C=O), nitroaromatic, -N-oxide, -alkene (C=C), aryl chloride, weak C–H and O–H bonds, sulfides, and polyenes. Some of these functional groups impart photolability as a result of their chromophoric properties (e.g., carbonyl) and some of them impart photolability by virtue of their weak covalent bonds. (e.g., O–H bonds). A list of several common bonds and their respective bond energies (E_b) and the corresponding wavelengths (λ_{max}) are presented in Table 1.

As an illustration, upon absorption of a photon having a wavelength equal to or shorter than 332 nm, a drug molecule that incorporates a C–O bond absorbs

Table 1

Bond	E_b (kJ/mol)	λ_{max}
O–H	465	257
H–H	436	274
C–H	415	288
N–H	390	307
C–O	360	332
C–C	348	344
C–Cl	339	353
Cl–Cl	243	492
Br–Br	193	620
O–O	146	820

enough energy to rupture that bond and undergo photoreaction. Of course, there are a variety of competitive dissipative pathways for the absorbed energy (discussed below) that may prevent photoreaction from occurring.

Several classes of drugs contain members with significant photolability including: anti-inflammatory, analgesic, immunosuppressant, central nervous system, cardiovascular, diruretic, hemotherapeutic, gonadotropic steroid, synthetic estrogen, and dermatological drugs, chemotherapeutic agents, and vitamins. The mechanisms by which photochemical reactions involving many of these drugs proceed has been described in detail previously (3).

PHOTOREACTIVITY

The first law of photochemistry, which was articulated by Grotthus in 1817 and Draper in 1843, states that only radiation that is absorbed by a molecule can be effective in producing photochemical changes in that molecule. An appropriate expanded statement of the first law in the context of drug photostability testing is that only radiation that is absorbed by a system (e.g., drug formulation) can be effective in producing photochemical changes in that system. Therefore, a drug formulation that absorbs no incident EMR will not undergo photodegradation.

Consider the following description of absorption of ultraviolet (UV) and visible photons (hv) by drug molecules:

$$D + hv \rightarrow D^*$$

where D is the drug molecule, hv is EMR of a given energy, frequency, and wavelength, D* is the drug molecule in an electronically excited state. The drug molecule can dissipate the absorbed energy in several ways.

Direct Drug Photoreactivity

Direct photoreactivity refers to the situation in which drug molecules absorb UV or visible EMR resulting in loss of the drug and the formation of products:

$$D + hv \rightarrow D^*$$

$$D^* \rightarrow products$$

$D^* \to D + heat$ (internal conversion)

$D^* \to D + hv'$ (fluorescence)

$D^* \to D + hv''$ (phosphorescence)

$D^* + M \to D + M^*$ (energy transfer)

where D is the drug molecule, hv is EMR of a given energy, frequency, and wavelength, D* is the drug molecule in an electronically excited state, hv' and hv" are photons emitted via fluorescence and phosphorescence, respectively, M is an acceptor molecule, and M* is the acceptor molecule in an electronically excited state subsequent to accepting energy from a drug molecule.

Dissipative pathways for the absorbed energy (i.e., internal conversion, fluorescence, phosphorescence, and energy transfer) are competitive with photoreaction. For example, a drug may absorb light strongly yet not undergo photoreaction because it fluoresces or transfers energy to another molecule present in the formulation very efficiently. Thus, a drug formulation that absorbs incident EMR may or may not undergo photodegradation; this is dependent upon the mechanism by which the absorbed energy is dissipated by the drug molecules and the presence of other ingredients in the formulation.

Indirect Photoreactivity

In the case of indirect or sensitized photoreactions, a component other than drug molecules (e.g., additive or impurity in formulation) absorbs UV or visible EMR resulting in the formation of one or more reactive species which then react with the drug molecules as indicated below:

$RC\text{-}CR + hv \to RC\text{-}CR^*$

$RC\text{-}CR^* \to RC. + .CR$

$RC\text{-}CR^* + O_2 \to RC\text{-}CR + {}^1O_2$

$RC. + O_2 \to RCOO.$

${}^1O_2 + D \to products$

$RCOO. + D \to products$

Here, RC-CR is an organic molecule present as an additive or impurity in the drug formulation that absorbs UV or visible EMR. In some cases, the RC-CR can undergo photolysis (split apart) producing RC., a reactive free-radical species that can combine with molecular oxygen (O_2) producing peroxyl radicals (RCOO) that can react and thus consume some drugs. In other cases, energy is transferred from the electronically excited (RC-CR*) to molecular oxygen producing singlet molecular oxygen (${}^1O_2$), a selectively reactive species that can also react and consume some drugs. The degree to which a given drug molecule is susceptible to undergoing reactions with peroxyl radicals and singlet molecular oxygen is dependent upon the structure of the drug molecule, the composition of the drug formulation, and the conditions under which the drug is exposed to EMR.

Self-Sensitizing Photoreactions

Alternatively, the drug may, in some cases, act as a self-sensitizer:

$$D + hv \rightarrow D^*$$
$$D^* + O_2 \rightarrow D + {}^1O_2$$
$${}^1O_2 + D \rightarrow products$$

Here, the electronically excited drug molecule (D*) transfers energy to molecular oxygen producing ${}^1O_2$ which then can, in some cases, react with and consume the drug.

KINETICS OF DRUG PHOTOREACTIONS

For dilute solutions exhibiting low absorbance (A < 0.12), the rate expression for consumption of the drug is given as follows:

$$-d[D]/dt = -k[D] = -2.303b[D] \int_{\lambda_1}^{\lambda_2} \Phi_\lambda I_\lambda \varepsilon_\lambda d\lambda$$

where [D] is the molar concentration of the drug, k is a first-order kinetic rate constant (s^{-1}), b is the path length (cm), Φ_λ is the quantum yield for photoreaction, I_λ is the photon flux (einstein L^{-1}s^{-1}), and ε_λ is the molar absorptivity for the drug (M^{-1}cm^{-1}). The limits on the integral (λ_1 and λ_2) are chosen so as to encompass the wavelength range of interest (e.g., 290–400 nm). The quantum yield for photoreaction can be defined as the number of drug molecules than undergo photoreaction divided by the number of photons absorbed.

In cases where the drug solution exhibits high absorbance (A > 1.3):

$$-d[D]/dt = -k = -\int_{\lambda_1}^{\lambda_2} \Phi_\lambda I_\lambda d\lambda$$

In this case, k is a zero-order kinetic rate constant (s^{-1}). An examination of the equations presented here reveals that the quantum yield (Φ_λ), photon flux (I_λ), and molar absorptivity (ε_λ) are all λ dependent. The rate of drug photoreaction is thus directly proportional to the quantum yield and the photon flux reaching the samples from the illumination source over all absorbed λ for which the quantum yield for photoreaction is nonzero. The value for I_λ is obtained from actinometry.

ACTINOMETRY

Carefully calibrated instrumental (e.g., spectral radiometry) and chemical methods may be used to perform actinometric measurements. Actinometry is ideally carried out under the same illumination conditions as the drug undergoing photostability testing. Actinometry allows determination of I_λ, the photon flux or "light dose" to which drug samples are exposed during photostability testing, whether by an instrumental approach such as radiometry or by chemical methods.

Physical Actinometers

Instrumental approaches employ two main types of detectors: thermal detectors which convert radiant energy into heat (bolometers, thermistors, and thermopiles), and photoelectric devices which convert radiant energy into an electric current (phototubes, photovoltaic cells, and spectral radiometers). A wide variety of these instruments are commercially available. Spectral radiometers allow energy reaching the detector from the illumination source to be measured as a function of wavelength over fairly narrow (e.g, 2nm) intervals. This constitutes a significant advantage relative to other actinometers because it allows corrections for any changes in lamp output due to ageing.

Chemical Actinometers

A chemical actinometer is typically a solution that is illuminated under the same conditions as the samples. There are two general approaches to chemical actinometry: the use of high optical density and low optical density solutions. Low optical density actinometers will not be discussed here due to their fairly limited applicability for drug photostability studies. A high optical density actinometer solution employs a chemical reaction whose quantum yield for photoreaction is well characterized and constant across a wide range of wavelengths. The solution should absorb all incident photons across the wavelength range of interest. The use of potassium ferrioxalate solutions for chemical actinometry is well established (4–6). The rate expression consumption of a high optical density actinometer is given by:

$$-d[A]/dt = -k = -\Phi_\lambda I_\lambda d\lambda$$

where $[A]$ is the molar concentration of the actinometer, k is a zero-order kinetic rate constant, Φ is the (known) quantum yield for photoreaction of the actinometer:

$$I_\lambda = k/\Phi_\lambda$$

Thus, the photon flux (I_λ), to which the actinometer (and samples) is exposed, can be determined by measuring the rate constant for photoreaction of the actinometer (k) and dividing by the quantum yield for photoreaction of the actinometer (Φ).

PHYSICAL COMPOSITION OF SAMPLES

Samples that are subjected to drug photostability testing may be illuminated as a dilute solution in which drug molecules are uniformly exposed to radiation, in concentrated solutions in which EMR is absorbed by a thin layer of drug molecules, or in a suspension in which a fraction of the incident EMR is scattered and reflected. The choice of solvent for preparing solutions can have a significant effect upon the absorption spectrum of the drug in solution (7) resulting in shifts in λ_{max} to shorter or longer values, shifts in absorbance at λ_{max} changes in the quantum yield for photodegradation, changes in the rate of photodegradation, and alterations in the photoproducts obtained.

Solid Tablets

EMR from the source is completely absorbed by a thin layer of molecules at the surface of tablets. Absorption of EMR by surface drug molecules can result in formation of photoproducts that absorb EMR of the same wavelengths as the drug slows down further reaction (filter effect). Alternatively, the photoproducts may not absorb EMR of the same wavelength as the drug allows successive layers of drug molecules in the solid to undergo photoreaction. The mechanism of photoreactions in solids may differ from that in solutions yielding different products. This can occur because free radicals can move about rapidly in solution but their motions are constrained in solids. Photostability testing of tablets has been described in some detail by Matsuda (8).

Powders

Many of the principles associated with photostability testing of solid tablets also apply to powders. Solids, whether in tablets or powders, can scatter, reflect, absorb, or transmit EMR. The degree to which any of these particular phenomena take place depends upon particle size, geometry, and angle of incidence and wavelength of the EMR. In general, shorter wavelength EMR tends to be absorbed and scattered while longer wavelength EMR tends to be transmitted to a greater depth.

Liquids

Of all sample forms, liquid solutions allow drug molecules to the greatest exposure to EMR. The consequence of this is that drugs tend to be more photolabile in solutions than in other sample forms such as solid tablets. This is particularly true for dilute solutions where the absorption and scattering by components in the solution other than drug molecules is minimized. A drug that is stable in powders or tablets may not be stable in liquid solutions. Conversely, a drug that is stable in liquid solution is likely to also be stable in powders and tablets.

PROTECTION OF PHOTOLABILE DRUGS

Protection of photolabile drugs can be accomplished by several means. The simplest approach is probably by external protection. The use of opaque containers constructed of brown or amber glass and colored plastic is common. In addition, tablets and capsules are frequently protected by aluminum foil blister packaging. Tablets may also be coated using titanium dioxide (TiO_2) which effectively scatters incident EMR and slows photodegradation. However, TiO_2 coatings on tablets can cause increased photodegradation of drugs in some cases; particularly when humidity is poorly controlled.

Internal protection of drugs is accomplished by the addition of excipients to the formulation, which are designed to protect the drug from photodegradation by competitive absorption. Food dyes are frequently used for this purpose. They are chosen so as to absorb EMR of the same wavelengths as the drug absorbs. Physical quenchers such as cyclodextrin are used as internal protective agents that physically quench the electronic excited states of drug molecules (D^*). This essentially drains off the energy from absorption of EMR before the drug molecule has the opportunity to undergo photodegradation.

INTERNATIONAL COMMITTEE FOR HARMONIZATION GUIDELINE

In order for drug photostability testing to have some predictive validity for the actual conditions under which drugs are produced and handled, the experimental conditions under which such testing is carried out must be well characterized. Of particular importance in this regard is the illumination source chosen for photostability testing. The International Committee for Harmonization (ICH) guidelines (9) specify acceptable illumination sources for drug photostability testing. These guidelines also specify the total energy that should be delivered to samples during drug photostability testing.

The ICH guidelines specify that drugs should be exposed to the following radiation sources during photostability testing:

Option 1. Artificial daylight fluorescent lamp, or Xenon arc lamp, or metal halide lamp.

Option 2. Cool white fluorescent lamp and near-UV fluorescent lamp.

Illumination Source

The illumination source chosen for photostability testing can have a significant effect upon the results. This is true because different illumination sources provide outputs having different spectral energy distributions. Thus, two illumination sources may deliver the same total energy to samples, but the energy may be distributed differently across the emission spectra. Selection of illumination sources for drug photostability testing has been discussed in detail previously (10–11).

Sample Placement

The placement of drug samples in illumination chambers is very important (12–13). The irradiance within a given chamber used for photostability studies can vary substantially as a function of position in the chamber. It is thus important to arrange samples so that they experience uniform irradiation (e.g., placement at equal distances from the source). The spectroradiometer sensor or the actinometer solution should be placed in the chamber so that it is exposed to the same irradiation as samples. The rates of photochemical reactions are generally independent of temperature. However, subsequent thermal reactions are temperature dependent. These thermal reactions may thus complicate photostability testing. Samples and the actinometer should therefore be irradiated under temperature-controlled conditions. This is sometimes quite difficult because many commercially available illumination chambers are not constructed in a manner that allows effective temperature control. Control of humidity is also important in the case of solid tablets or other samples that are exposed to EMR in open containers. Changes in humidity can significantly alter the results of photostability testing of tablets particularly when the tablets are coated with TiO_2.

CONCLUSIONS

Drug formulations are invariably exposed to EMR that is energetic enough to initiate a variety of photoreactions.

Some drug formulations that are exposed to EMR that they absorb undergo direct, indirect, or self-sensitized photoreactions. Photoreactions involving several classes of drug formulations have been reported. Photoreactions result in loss of the drug, and the potential for formation of toxic products. Photostability testing allows identification of drug substances that are photolabile in the early stages of development and marketed products that will require special protection from light. Results obtained from photostability testing are dependent upon several factors including: light absorption by the drug substance or formulation, the quantum yield for photoreaction, photon flux or "light dose", the spectral power distribution of the illumination system, physical composition of samples, placement of samples in illumination chambers, and humidity. Actinometric measurements provide a means by which intra- and interlaboratory photostability measurements can be quantitatively compared. Illumination systems whose output adequately simulates the EMR to which drug formulations are exposed in actual use are essential for meaningful photostability testing results. Several approaches to stabllization of drug formulations (e.g., opaque packaging, and additives that absorb EMR) can reduce or eliminate many drug photoreactions.

REFERENCES

1. Moore DE. Principles and practice of drug photodegradation studies. J Pharm Biomed Anal 1987; 5(5):441–453.
2. Tønnesen HH. Photostability of Drugs and Drug Formulations. London, U.K.: Taylor and Francis Ltd. ISBN 0-7484-04449-X.
3. Albini A, Fasani E. Photochemistry of drugs: an overview and practical problems. In: Albini A, Fasani E, eds. Photochemistry and Photostability. Cambridge, U.K.: Royal Society of Chemistry, 1998:1–65.
4. Hatchard CG, Parker CA. A new sensitive chemical actinometer. II. Potassium ferrioxalate as a standard actinometer. Proc Royal Soc Ser 1956; A235:518–536.
5. Calvert JG, Pitts JN. Photochemistry. New York: John Wiley, 1968.
6. Gordon AJ, Ford RA. The Chemists Companion. New York, 1972:362–364.
7. Matsuda Y. Solvent properties. In: Murow SL, Carmichael I, Hug GL, eds. Handbook of Photochemistry. 2nd ed. New York: Marcel Dekker, 1993:283–297.
8. Some aspects on the evaluation of photostability of solid-state drugs and pharmaceutical preparations. Pharm Tech Japan 1994; 10(7):7–17.
9. International Conference on Harmonization (ICH) Guideline for the photostability testing of new drug substance and products. Fed Regist 1997; 62:27115–27122.
10. Piechocki JT. Selecting the right source for pharmaceutical photostability testing. In: Photostability of Drugs and Drug Formulations. London, U.K.: Taylor and Francis Ltd., 1998:247–271.
11. Riehl JP, Maupin CL. On the choice of a light source for the photostability testing of pharmaceuticals. Pharm Forum 1995; 21(6):1654–1663.
12. Clapham D, Sanderson D. The importance of sample positioning and sample presentation within a light cabinet. Photostability of Drugs and Drug Products. Presented at Photostability '99, 3rd International Meeting, Washington D.C., Jul 10–14, 1999.
13. Baertschi SW. Practical aspects of pharmaceutical photostability testing. Photostability of Drugs and Drug Products. Presented at Photostability '99, 3rd International Meeting, Washington D.C., July 10–14, 1999.

5 Spectra

Joseph T. Piechocki
Piechocki Associates, Westminster, Maryland, U.S.A.

INTRODUCTION

Several types of spectra are referred to in the pharmaceutical photostability literature, including emission, absorption, activation, action, and spectral power distributions (SPDs). Each of these terms has a specific meaning and use. For example, the absorption spectrum of a typical compound is often assumed to represent its degradation sensitivity, which is not always the case. This assumption arises because of the fact that in order for a compound to react with radiation, it must first absorb it and a good first indicator of these active wavelengths are its absorption spectra.

Another false assumption is that the light emitted from a source is equivalent to that which the sample will be exposed to during the test. This assumption neglects the effect of materials of construction, lenses, mirrors, filters, diffusers, sample containers, etc., which are often not considered.

The objective of this chapter is to increase the user's understanding of the various types of spectra, their uses, and abuses. With this information, one can better understand the problems involved, and design one's tests, and interpret the results of others.

EMISSION SPECTRA

An emission spectrum is defined by the International Union of Pure and Applied Chemistry (IUPAC) as a "[p]lot of the emitted spectral radiant power (spectral radiant exitance) or of the emitted spectral photon irradiance (spectral photon exitance) against a quantity related to the photon energy, such as frequency ν, wave number σ, or wavelength λ. When corrected for wavelength-dependent *variations* in the equipment response it is called the corrected emission spectrum (1)."

Emission spectra are dependent on many factors, including the elements used in their construction, the impurities contained therein, and the conditions under which they are used. In some instances, e.g., fluorescent and metal halide lamps, pharmaceutical photochemists may be dealing with a combination of both primary and secondary spectra. In the case of fluorescent lamps, the radiation of the primary emitter, mercury, is absorbed by the fluorescent powder(s) which is (are) activated to a higher energy state. The excited state then decays, emitting radiation at a longer wavelength than was used for activation. This radiation then combines with the original wavelength(s), which are not completely blocked by the phosphor(s) and is emitted by the lamp.

For metal halide lamps, mercury is used to initiate and maintain conduction through the lamp. Its arc provides the heat for vaporizing the metal halide, ionizing it and raising the bulb temperature to aid in the revaporization of any deposited metal on its surface.

The warming up process of a metal halide lamp is easily visible in the mercury and sodium vapor lamps used for street lighting. Their shift in coloration during warm-up is easily seen.

Likewise, all lamps have containers of glass. This material has various impurities depending on its geographical origin and vessels used in its production and manufacturing process. Its composition has also been shown to differ from batch to batch for most common glasses.

Hirt et al. have shown that ultraviolet (UV) and natural sources can alter the properties of glass filters (2). Searle et al. also reported similar results for UV and xenon sources (3).

Plasma Lamps

All materials used in the construction of a source can contribute to the resultant spectrum emitted by that lamp. Moreover, these materials, some present at very low parts per million, can have either a transitory or permanent effect on the resultant spectrum. Some of these, e.g., the presence of iron or addition of cerium or titanium to glasses, have produced safer and/or more useful (self-cleaning) lamps.

The more elements used in the construction of a source, the greater will be the number of possible variables. The simplest lamp design possible would contain two elements: a bulb to contain an inert gas, the gas(es), and its glass container. Such lamps are called electrodeless discharge lamps. When irradiated with radio-frequency (microwave) energy, these lamps are heated to high ionizing levels and emit radiation. The primary problem with these lamps, although they are very bright, is containment of their very hot plasma. Cooling is a major problem.

Arc Lamps

The addition of electrodes and possibly an easily volatilized and ionized metal to the most basic lamp design creates an arc pathway through an ionizable gas. This produces a single element arc lamp like those of mercury, sodium, and xenon. Their emission spectra primarily is that of the pure element. Because of the intense heat generated by the arc, electrodes and impurities vaporized and ionized, all of which contribute to the overall spectral power distribution (SPD) of the lamp. This effect is more probable for short-arc (less than 5 mm distance between electrode tips) than the long-arc types.

Any emitting surface, cold-cathode, hot-cathode, filament, or arc lamp will develop vapor pressures from the elements/compounds of their construction when heated. These vapors will also manifest themselves particularly during the initial start-up (breaking in) time of the lamp. Early atomic absorption, multi-element lamps were shown to have this problem. Some of these additives specifically react with residual gases and elements that might have a deleterious effect on lamp life. The electronics industry has used what are called "getter" substances in their tubes to remove some of these substances and improve emission.

An article by Riehl and Maupin (4) illustrates one of these problems (Fig. 1). In this figure, the lamp was operated for a period of time (burn-in time) to stabilize its output, after which the loss of a small portion of the emission spectrum was noted. This loss can most likely be attributed to the normal aging of the lamp and establishment of equilibrium of all components within the lamp envelope. All

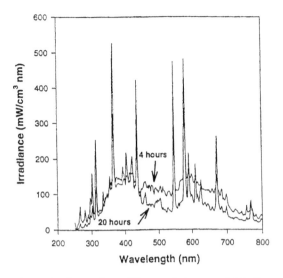

Figure 1 Comparison of the spectral power distributions of a metal halide lamp during its "aging" process. *Source*: Courtesy of Atlas Electric Devices, Inc.

lamps have a recommended "break-in time" to allow for these occurrences and reach a state of near constant emission intensity. The lengths of some equilibrium periods are longer for certain lamps, e.g., metal halides, whose intensities and spectra are related to the temperature of the bulb. The phenomenon of an element disappearing from a lamp's SPD was frequently observed for multi-element, hollow cathode lamps used in atomic absorption spectroscopy.

INCIDENT RADIATION

It is generally assumed that the SPD of a lamp represents the spectrum of the radiation seen by the sample(s). This may be in many cases a false assumption because it neglects the effect of filters, diffusers, protective glass panels, reflections, refractions, etc.

As shown in Chapter 13 of this book on chambers and mapping, each type of source (spot, bar, multiple spot, or multiple bar) has its own intensity map. Little attention has been paid to the effect of sources interacting with one another and their chambers. It is very probable that there would be both additive and subtractive phenomena observed in chambers with multiple lamps.

The intensity profiles found in Chapter 13 of this book all come from individual laboratories, not equipment suppliers. To date, no group or supplier has studied the effect of the various chamber designs on the results obtained.

Turner, in his spectroradiometric studies of one old self-developed photo exposure chamber, noted energy losses due to wall absorptions (5) and referred to them as "lost energy."

The International Conference on Harmonization (ICH) document advises users to make all dosage measurements in close proximity to the test sample to reduce possible intensity variations. This statement acknowledges the existence of the lack of uniformity of intensities on the test surface.

Chamber Effects

It is to be expected that all surfaces encountered by the emitted radiation, reflected or filtered by any device encountered between the lamp surface and the sample, will have an effect on its SPD reaching the sample. These effects may manifest themselves as reflections, refractions, focusings, losses of certain wavelengths, diffractions, etc. Some may be specific to a particular design and have specific functions, e.g., "hot mirrors," which are used to reduce the infrared radiation load on samples and lamps.

The use of noncommercial chambers made of materials other than aluminum or stainless steel is not recommended. As previously noted, Turner reported on absorptions caused by the paint of just such a chamber (5). He noted a definite loss in UV intensities.

Special Chamber Effects

Some manufacturers use turntables or carousels to rotate samples, and achieve what they believe are a uniform exposure of samples. In Japan (6), the use of turntables to rotate samples during exposure is common. Other manufacturers such as Luzchem (6) and UV Light (7) manufacture carousels to rotate samples and provide exposure to all intensities emitted by the lamp. One manufacturer, Weiss Gallenkamp (8), has designed a special diffuser to attenuate the regions of higher intensity emitted by the lamp to achieve a more uniform intensity profile over the test sample surface. Another manufacturer, Luwa, Environmental Specialties (ES) (9), advises users not to place their samples too close to chamber walls and may even mark lines on the chamber floor to make this task easier.

All of these effects are designed to expose the sample to all possible wavelengths and intensities of the source used, but do little to help define the source's properties and thereby assure its reproducibility.

The SPD of the source used for pharmaceutical photostability testing is the active agent, so it is important that it be well defined, reproducible, and controlled if reproducible results are our objective. As will be discussed in Chapter 6 of this book, the need for establishment of a single standard source, a problem recognized by many international groups such as the American Society for Testing Materials, International Organization for Standardization, Japanese Standards Association, and the American Association of Textile Chemists and Colorists, to mention just a few is essential if good science is to prevail.

ABSORPTION SPECTRA

As previously noted, there are a number of factors that can affect the emitted spectra before it reaches the sample. One of the most important of these items is the container, specifically its color, thickness, and geometry.

Colors and impurities obviously filter out certain radiation bands. Increasing the thickness of a container will obviously increase the absorption of any given absorbed band. One of the most difficult items to measure is the effect of the container geometry on the exposure. Depending on the physical shape of the container and the angle(s) of incidence of the radiation (cosine law effect), radiation can be reflected or refracted by flat surfaces and possibly focused internally by rounded corners.

All of these items make it difficult to determine the amount of radiation actually absorbed by a sample. For liquids, the best technique is to fill a duplicate container with a solution of a known actinometric solution, such as one of those developed by Gauglitz and Hubig, described in Chapter 8 of this book.

It is frequently assumed that the absorption spectrum of a test substance will accurately represent its activation or action spectrum because only absorbed radiation can bring about a photochemical change. Frequently neglected are solvent (bathochromic or hypsochromic) or matrix effects such as polarity, pH, complexation, dimerization, binding, self-filtering, and quantum efficiency differences between various absorption bands, etc., which may occur.

Often, one encounters comparisons of spectra taken in different media in the pharmaceutical photostability literature. An often encountered example is the failure to recognize the effect of oxygen capacity of the various mediums used on the results obtained, e.g., comparing the results obtained in aqueous to those obtained in methanolic media.

It is frequently assumed that all radiation passes uniformly through all samples, which even for liquids may not be the case. Self-filtering, the absorption of incident radiation by the outermost portions of the sample and/or its degradation products, and reducing the intensity of that seen by the subsequent inner layers can markedly affect the amount of radiation seen by deeper layers.

Activation Spectra

The term "activation spectra" is a common term used by polymer and photobiology chemists, but it is foreign to most pharmaceutical photochemists. Excellent reviews of this technique and "action spectra" and their applications have been published by Coohill (10,11), Caldwell et al. (12), Kleczkowski (13), Peak and Peak (14), Rundel (15), Sutherland et al. (16), Searle (17,18), Andrady (19), Andrady et al. (20), and Andrady et al. (21).

The power of these two terms comes from their ability to represent the effectiveness of incident radiation to change the test sample. They effectively provide the user with a map identifying which wavelengths produce the greatest photochemical change. Once this information has been obtained, the user can then determine which wavelengths might be most effective for a photoactivation or in need of suppression for a photosensitive sample.

An activation spectrum is "[a] representation of the relative effectiveness of various regions of a polychromatic source spectrum in causing a given type of photoreaction or photodamage" and is "… dependent upon the SPD the irradiating radiation (18)." The term "of various regions …" derives from the fact that the first applications used filter combinations, not spectrometers, to help isolate specific bands of radiation for these studies.

Determining factors for activation spectra according to Searle are (18):

1. spectral emission properties of the light source;
2. absorption properties of the material;
3. stability of the material to absorbed radiation; and
4. degradation parameter measured.

Activation spectra have found much use in the plastics and printing ink industries, but the activation spectra of these materials are highly source dependent and the sources used are designed for maximum effectiveness for polymerization. Their general application to other areas has been limited because most other areas are too heterogeneous, SPD wise. Surveys by Anderson et al. (22), Thoma and

Kerker (23), and McGreer (24) of sources found in the normal manufacturing, pharmaceutical, hospital, and home environment have revealed the lack of uniformity in the SPDs of these sources.

Activation spectra are special representations of the result of the absorption of radiation by a sample. These spectra are obtained by exposing a sample to a series of, if possible, equal bandwidths of the incident radiation for an equal amount of time, and plotting the noted percent decomposition against the bandwidths used. As previously mentioned, the plots are specific to a particular lamp. One advantage of this technique is that the operator need not determine the absolute value of the amount of radiation used. However, failure to titrate the dosage used could lead to over exposure and negatively affect the results obtained.

While it is true that for a sample to be photoactive it must absorb photonic energy, it is not true that all absorbed energy results in changes in the sample. The incident photons must be of a high enough energy to cause a change, and the reaction must have a large enough quantum efficiency for the reaction to be noted during the length of exposure used.

Many reactions have quantum efficiencies much less than unity, while some such as the ferrioxalate have efficiencies greater than unity. Likewise, some materials like ferrioxalate may have two concentration dependent photochemical reactions with differing quantum efficiencies.

Assumed in the definition of an activation spectrum is the fact that all of the lamps of a particular type, e.g., "cool-white," are very similar. This is not necessarily a valid assumption. Batch to batch variations, differences between geographical locations of the same company, manufacturers, or countries can affect the SPD of a lamp. Differences in "daylight" lamps and the UV content of "cool-white" SPDs have been noted.

Just as the media a sample is dissolved in can affect the spectrum of a sample, e.g., benzene in alcohol versus hexane, so too can the media affect action and activation spectra. Matsuda and Teraoka (25) reported that "Neither the action spectrum of ubidecarenone in Figure 2 (25) nor that in Figure 2 (25) resemble that of the intact ubidecarenone crystals observed in the previous investigation (26) by Matsuda and Mashara. These finding(s) suggest that the degradation of ubidecarenone is strongly controlled by the ingredient(s) added."

Akimoto et al. (27,28) showed clearly how pH can affect the photostability of CPT-11, a derivative of camptothecin, in aqueous solutions. In this case, the maximum wavelength of decomposition did correspond with the absorption maximum of the compound.

Action Spectra

Action spectra are plots of the amount of change noted in the test sample on a per photon basis. In practice, the sample is irradiated with a given number of photons, the amount of change in the test sample noted, the percent change per photon used calculated (quantum efficiency), and these values plotted versus the wavelength used.

There are two types of action spectra found in the literature, which this writer would classified as "true-action spectra" and "pseudo-action spectra." The basic difference between the two spectral types has to do more with scientific exactness. The pseudo-action spectra are easier to obtain and plot but they underestimate the true effect of lower wavelength photons, because fewer than the assumed numbers of photons actually reach the sample.

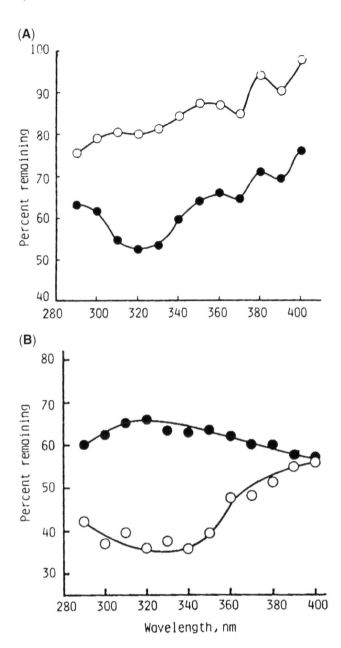

Figure 2 Effect of other ingredients (**A**) Tocopherol o, (**B**) Phytomadione o, on the pseudo-action spectrum of ubidecarenone (●). *Source*: From Ref. 29.

Pseudo-action spectra will probably suffice for most applications. To date, no reports of problems with this approach, which does save users a good amount of calibration time, have been reported.

True-Action Spectra
The term "action spectra" has been used since early in the 19th century, but was not widely used until the 1940s. A good review of the history of this term, from the photobiological aspect, can be found in the Presidential Address of Coohill, given at the 18th Annual Meeting of the Society for Photobiology (10).

An action spectrum is a plot of an effect, biological or chemical, as a function of wavelength, using a constant number of photons at each wavelength or reduced to the per incident photon standard. This definition assumes that a photon reacts on a molecular, one to one basis for most known pharmaceutical photostability examples at each wavelength. This phenomenon is sometimes referred to as the "wavelength sensitivity" or the "photon efficiency" of a test substance at each wavelength measured. By definition, action spectra are source independent.

In practice, a sample is exposed at a specific wavelength until a previously chosen number of photons are incident upon the sample. Then the effect of this exposure on a prechosen end point is measured. The procedure is then repeated at each wavelength, using a constant bandwidth, over the wavelength range of interest. Because each exposure is done for a constant number of photons, any contributions from the lamp SPD are essentially negated.

The benefits of making this measurement are:

1. it identifies the most effective wavelengths for photodecomposition;
2. the results obtained are not source dependent;
3. it may or may not correlate with the absorption spectra; and
4. the results obtained can aid in the development of more effective protective measures.

The disadvantages of this technique are:

1. it assumes that all individual reactions are independent and/or additive;
2. phenomena such as photochromism, isomerization, etc., will give false results; and
3. it is concentration and matrix dependent.

Action spectra have also been reported to vary with the concentration of the chromophore (10,11).

Pseudo-Action Spectrum
A more practical but less accurate plot of the action spectrum is what I would term the pseudo-action spectrum. The primary difference between normal and pseudo-action spectra derives from how they are obtained and plotted. Normal-action spectra are plots of the effect of a constant number of photons per wavelength versus wavelength, whereas pseudo-action spectra are obtained using a constant energy input, watts/meter^2, per nanometer.

Figure 3 is a plot of the energy per photon versus wavelength. As one can see, a photon at 300 nm is about 2.8 times more energetic as one at 800 nm. Thus, if one uses a constant energy at each wavelength to obtain an action spectrum, one will underestimate the per photon effect at 300 nm by about 66%.

The usefulness of action spectra can readily be seen in Figure 4. In these plots, it is readily demonstrated that the absorption spectrum is not a good indicator of the photosensitive wavelengths of nifedipine. These plots also demonstrate the

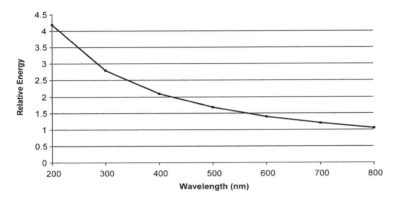

Figure 3 Plot of energy of photons at various frequencies.

effect different wavelengths of radiation can have on the result obtained. Other application of this technique can be found in the papers of Matsuda et al. (30,31), Teraoka and Matsuda (32) and Hanson and Simon (33).

SPECTRAL POWER DISTRIBUTIONS

The SPD, called the spectral radiant exitance (M_λ) by IUPAC (1), is a plot of "[t]he radiant exitance, M, at wavelength λ per unit wavelength interval in a commonly used unit, Wm^2nm^{-1}." It is the gold standard for the measurement of the output of any source.

Spectroradiometers, often dual-monochromator spectrophotometers, allow for the rapid measurement of all wavelengths that are used for obtaining SPDs. Several single monochromator spectroradiometers are also available on the market. These units are less expensive, smaller, of lesser resolution, and have higher amounts of stray light (Table 1) but are sufficient for most pharmaceutical photostability purposes.

Reference to almost any lamp catalog will show a number of lamp SPDs. SPDs allow users to compare one lamp to another for its degree of similarity. Since,

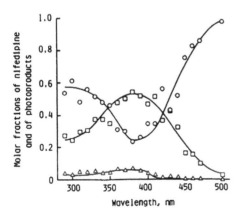

Figure 4 Effect of wavelength on the photodegradation of (O), nifedipine; (□), nitroso-derivative; (△) nitro-derivative.

Table 1 Spectroradiometers

Company	Model no.	Wavelength range (nm)	E-mail address
Control Development	CDIPDA/200–800/1.5	200–800	http://www.controldev.com
Custom Sensors	Zeiss MCS 52 × UV	215–800	http://www.customsensors.com
Ocean Optics		200–1110	http://www.Oceanoptics.com
International Light	1L700/760D/780	300–800	http://www. intl-light.com
Luzchem Research, Inc.	SPR-01	230–860	http://www.luzchem.com
Macam	SR9020	300–800	http://www.macam.com
Spectral Instruments, Inc.	430/440	350–980/ 190–980	http://specinst.com
World Precision Instruments	Spectromate	230–850	http://www.wpiinc.com
World Precision Instruments	S2000	250–800	http://www.wpiinc.com

Source: From Ref. 34.

as history has shown (see Chapter 3 of this book), one cannot expect suppliers to supply the SPDs of the specific lamps one may purchase, the only solution is to do it by oneself or have it done by a reliable third party.

The ICH document errs in advising users to depend on their lamp suppliers for this information. At best, users will receive a copy of a typical lamp's SPD, not even a copy of one from the batch their lamp came from. It is ultimately the user's responsibility, in the eyes of regulators, to assure the accuracy of data they accept.

Additionally, the only valid method of mapping a test chamber is with a spectroradiometer. This is the only instrument capable of showing the effect of reflections, refractions, spectral overlaps (for multiple source units), and absorptions by the materials used to construct the chamber. These effects were noted by Turner (5,6) and are the reason why many chamber manufacturers suggest that samples should not be placed too close to chamber walls.

A brief survey by this writer, indicates that many chamber manufacturers do not do their own chamber mapping but rather subcontract the work out and could/ would not supply any details of the work. Chapter 13 of this book on chambers and mapping discusses the problems with this approach in more detail.

SUMMARY

The knowledge of the different types of spectra found and used in pharmaceutical photostability testing allows users to better understand the parameters governing the photoactivation/photodegradation process. With this knowledge, one can better design dosage forms for photoactivation applications, and prevent adverse effects that may be a result of photodegradation, whether they are due to manufacturing or in-use causes.

REFERENCES

1. Braslavsky SE, Houk KN. Glossary of terms used in actinometry. Pure Appl Chem 1988; 60(7):1055–1106.
2. Hirt RC, Schmitt RG, Searle ND, Sullivan AP. Ultraviolet spectral energy distribution of natural sunlight and accelerated test light sources. J Opt Soc Am 1960; 50(7):706–713.
3. Searle NZ, Giesecke P, Kinmonth R, Hirt RC. Ultraviolet spectral distributions and aging characteristics of xenon arcs and filters. Appl Opt 1964; 3(8):923–927.
4. Riehl JP, Maupin CL. On the choice of a light source for the photostability testing of pharmaceuticals. PF 1995; 21(6):1654–1663.
5. Turner N. Photostability Testing—Let there be Light. IBC Conference on Stability Testing. The Dorchester Hotel, London, UK, Feb 23–24, 1998.
6. Suga Test Instruments Co., Ltd., TAKA AI Co., Ltd., 3 12, Nihonbashi, Chuo-ku, Tokyo 103, Japan. 160–0022 the Tokyo Shinjuku Ku Shinjuku 5-4-14.
7. Luzchem Research Inc., 5509 Canotek Rd, Unit 12, Ottawa, Ontario, K1J 9J9 Canada.
8. Weiss Gallenkamp, 900 Arlington Heights Road. Suite 320, Itasca, IL 60143.
9. Luwa, Environmental Specialties (ES), 4412 Tryon Rd. Raleigh, NC 27606.
10. Coohill TP. Photobiology school—action spectra again? Photochem Photobiol 1991; 54(5):859–870.
11. Coohill TP. Action spectra for mammilian cells in vitro. In: Smith KC, ed. Topics in Photomedicine. New York: Plenum, 1984:1–37.
12. Caldwell MM, Camo LB, Warner CW, Flint SD. Action spectra and their key role in assessing biological consequences of solar UV-B radiation. In: Worrest RC, Caldwell MM, eds. Stratospheric Ozone Reduction. Solar Ultraviolet Radiation and Plant Life, 1986:87–111 Springer-verlag, NY.
13. Kleczkowski A. Action spectra and their interpretation. In: Galo U, Santamaria L, eds. Research in Organic Biological and Medical Chemistry. New York: North Holland Publ. Co. Vol. 3 (Part II), 1972:48–70.
14. Peak JJ, Peak JG. Use of action spectra for identifying molecular targets and mechanisms of action of solar UV light. Physiol Plant 1983; 58:360–366.
15. Rundel RD. Action spectra and estimation of biologically effective UV radiation. Physiol Plant 1983; 58:360–366.
16. Sutherland BM, Dekihas NC, Sutherland JC. Action spectra from ultraviolet light-induced transformation of human cells to anchorage-independent growth. Cancer Res 1981; 41:2211–2214.
17. Searle NZ. Wavelength sensitivity of polymers. In: International Conference on Advances in the Stabilization and Controlled Degradation of Polymers, at Lucerne, Switzerland, May 1985. Vol. 1. Lancaster, PA: Technomic Publ. Co., 1985.
18. Searle ND. Spectral Sensitivity of Materials—Action and Activation Spectra versus Spectral Power Distributions. Presented at Photostability '99, 3rd International Meeting, Photostability of Drugs and Drug Products, Washington, DC, July 10–14, 1999.
19. Andrady AL. Wavelength sensitivity in polymer photodegradation. In: Andrady AL, Narasimhan B, Pascault J-P, et al., eds. Advances in Polymer Sciences. Vol. 128. New York: Springer-Verlag, 1997:49–94.
20. Andrady AL, Fueki K, Torikai A. Photodegradation of rigid PVC formulations. Wavelength sensitivity of light induced yellowing by monochromatic light. J Polym Sci 1989; 37:935–946.
21. Andrady AL, Torikai A, Kobatake T. Spectral sensitivity of chitosan photodegradation. J App Polym Sci 1996; 62:1465–1471.
22. Anderson NH, Johnston D, McLelland MA, Munden P. Photostability testing of drug substances and drug products in UK pharmaceutical laboratories. J Pharm Biomed Anal 1991; 9(6):443–449.
23. Thoma K, Kerker R. Photostabiltät von Arzneimitteln ein vernachlässigtes Qualitätskriterium? Poster, Arbeitsgermtinschaft für Pharmazeutische Verfahrenstechnik (APV) Symposium Regensburg, 14 April 1992.

24. McGreer M. Getting the light right: a comparison of the spectral power distribution between artificial light sources and real world environments. 4th International Meeting on the Photostability of Drugs and Drug Products, Research Triangle Park, NC (USA), July 16–17, 2001.

25. Matsuda Y, Teraoka R. Improvement of the photostability of ubidecarone microcapsules by incorporating fat soluble vitamins. Intern J Pharmaceut 1985; 26:289–301.

26. Matsuda Y, Mashara R. Photostability of solid state ubidecarenone at ordinary and elevated temperatures under exaggerated UV irradiation. J Pharm Sci 1983; 72(10):1198–1203.

27. Akimoto K, Akiko K, Ohya K, Sawada S, Aiyama R. Photodegradation reactions of CPT-11, a derivative of camptothecin 11: chemical structure of main degradation products in aqueous solution. Drug Stab 1997; 1:118–122.

28. Akimoto K, Kawai A, Ohya K. Photodegradation reactions of CPT-11, a derivative of camptothecin. II. Photodegradation behaviour of CPT-11 in aqueous solution. Drug Stab 1996; 1:141–146.

29. Matsuda Y, Teraoka R. Improvement of the Photostability of ubidecarone microcapsules by incorporating fat soluble vitamins. Intern J Pharmaceut 1985; 26:298–301.

30. Matsuda Y, Itooka T, Mitsuhashi Y. Photostability of indomethacin in model gelatin capsules: effects of film thickness and concentration of titanium dioxide on the coloration and photolytic degradation. Chem Pharm Bull 1980; 28(9):2665–2671.

31. Matsuda Y, Ito M. Photostability of sulphisomidine tablets: action spectra for colouration and photolytic degradation. Asian J Pharm Sci 1979; 1:107–118.

32. Teraoka R, Matsuda Y. Stabilization-oriented preformulation study of photolabile menatetrenone (vitamin K2). Int J Pharm 1993; 93:85–90.

33. Hanson KM, Simon JD. The photochemical isomerization kinetics of urocanic acid and their effects upon the in vitro and in vivo photoisomerization spectra. Photochem Photobiol 1997; 66(6):817–820.

34. Matsuda Y, Teraoka R, Sugimoto I. Comparative evaluation of photostability of solid state nifedipine under ordinary and intensive light irradiation conditions. Intern J Pharmaceut 1989; 54:211–221.

6 Sources

Joseph T. Piechocki

Piechocki Associates, Westminster, Maryland, U.S.A.

INTRODUCTION

Of all of the topics in the International Conference on Harmonisation (ICH) Q1B document, none is more important than that related to the source. The source is the active agent in photostability studies. Without control of the source, results are questionable.

The purpose of this chapter is to present the various options presented by the ICH document, their positive and negative aspects, and outline reasons for the adoption of single standard as has been done in other allied fields.

INTERNATIONAL CONFERENCE ON HARMONIZATION OPTIONS

The ICH document (1) offers users a choice of any of five different options. Table 1 is a listing of these options. These options arose from the results of a Japanese Pharmaceutical industry survey (2). No effort was made to determine which procedure was best. These results were presented to the Q1B Expert Working Group (EWG). The ICH EWG accepted what the Japanese had done and, after debate, adopted it as their guideline.

Figure 1 presents the spectral power distributions (SPDs) of the various lamps cited in the ICH document and used in the Japanese Pharmaceutical industry. One should note the difference between these lamps and the D_{65} standard also presented there.

The exact goal of the Q1B EWG is unclear based on the comments reported by the Rapporteur for the group Beaumont (3). They decided earlier on not to seek the advice of outside experts in this field to aid in the development of this guideline and this shortcoming is evident in their decision-making.

Further evidence of this lack of expert input is found in the opening line of Option 1, "Any light source that is designed to produce an output similar to the D_{65}/ID_{65} emission standard." Such a standard does not exist. D_{65} is a Commission Internationale de L'Eclairage (CIE) (4) standard and ID_{65} (5) an International Organization for Standardization (ISO) standard. The SPDs of these two standards are different; one is a filtered version of the other.

Figure 2 clearly illustrates the difference between D_{65} and ID_{65}. One may argue that the small shift in the ultraviolet (UV) is a very small and unimportant part of the spectrum, but it is also the most energetic region and where most substances absorb and sunburn does occur. The importance of this region is indicated by the number of companies adding sunscreens to their topical products. This region, UV-B is very important, and has generated a new industry—sun protection products. It is also important in immunosuppression as indicated in articles by Roberts et al. (6) and Gil and Kim (7).

The ICH needs to decide just which of these standards it is going to recommend, D_{65} or ID_{65}. Logic would dictate the former, D_{65}. Eventually, all substances

Table 1 Summary of ICH Source Options

Option	Description	Lamp type	Additional requirements
1	D_{65}/ID_{65}	Not identified	NA
1	Artificial daylight	Fluorescent	NA
1	D_{65} metal halide	Short-arc	Filter to remove >320 nm radiation
1	D_{65} xenon	Short-arc Long-arc	Filter to remove >320 nm radiation
1	ID_{65}	Fluorescent	ISO 10977:1993(E)
	Cool-white	Fluorescent	ISO 10977:1993(E)
2	Near ultraviolet	Fluorescent	Bandwidth: 320–400 nm; λ_{max} = 350–370 nm; 2 significant proportion of UV between 320–360 nm and 360–400 nm

Source: From Ref. 1.

will be exposed to it via daylight or sunlight after dosing. Many drug substances already bear warnings against such exposure.

Additionally, the ICH document specifies that "The intrinsic photostability characteristics of new drug substances and products should be evaluated." This statement means that all wave lengths should be tested. Only sources, such as D_{65}, which meet these criteria, should be used for testing. Papers advocating other choices and options such as Riehl and Maupin (8) and Nema (9) are not in conformance with the guidelines focus of developing the "intrinsic photostability characteristics."

INDIVIDUAL INTERNATIONAL CONFERENCE ON HARMONIZATION LAMP CLASSES

Fluorescent Lamps

Of the five options offered in the ICH document, fluorescent lamps account for three of them. These lamps have been very popular since their initial introduction

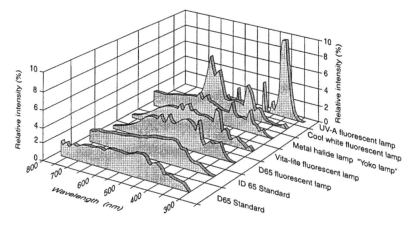

Figure 1 The relative spectral power distribution of D_{65} standard, ID_{65} standard and various lamps. *Abbreviation*: UV, ultraviolet. *Source*: From Ref. 10.

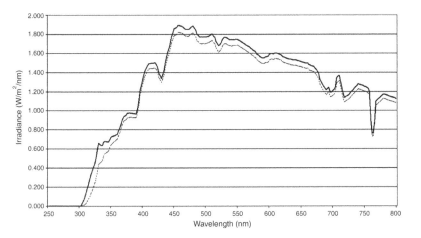

Figure 2   Spectral power distributions of D_{65} and ID_{65} standards. *Source*: Courtesy of M. McGreer, Atlas Electric Devices, Co.

into commerce. Lachman et al. (10) advocated the use of cool-white fluorescent lamps for pharmaceutical photostability testing because they were the then, best available source for reproducing the solar SPD. Based on their work, many cabinets incorporating these lamps were constructed and some are still in use today.

Because of their wide use, low cost and ready availability, some have advocated that fluorescent lamps be used for determining the "intrinsic photostability" of pharmaceutical substances and products even though there are problems with their SPDs and control. Tønnesen and Moore were quick to point out one such problem (11), the gap in the SPD between "... 380 and 430nm" when a UV-A and cool-white are used either separately or together. This writer prefers to refer to it as the "Tønnesen Gap" for ease of citation.

Fluorescent lamps are complex. They are composed of a filament, cathode emitter, inert gas and drop of mercury enclosed in a phosphor(s) coated glass tube with metallic end caps for electrical connections. When energized, the filament heats the cathode causing it to begin emitting electrons that volatilize and ionize the mercury vapor in the tube. This leads to the establishment of an arc between the cathode and anode of the tube increasing current flow though the tube which further raises the tube temperature and increases mercury ionization and UV emissions, while raising the tube temperature. The mercury UV emissions react with the phosphor(s), which in turn reemit some of the radiation at a longer wavelength(s).

The emission characteristics of a particular lamp are dependent on tube current, mercury pressure, phosphor composition, phosphor coating procedure, tube composition, and the frequency of the ballast used. Since these lamps are produced for a very competitive commercial market, much of the production information is considered proprietary and not published. Changes can be made, sometimes without the knowledge of customers.

Some lamps, like the new triphosphor cool-white lamps, Vita-Lite and daylight lamps have more than one phosphor. The use of more than one phosphor presents additional problems of homogeneity of the coating mixture, uniformity of coating if a multiple coating method is used, variations of the quantum efficiency between batches of the same phosphor, emissions blocking and the independent aging of each phosphor. Any one of these items can change the SPD of the lamp. The lamp glass bulb also plays a role in determining the initial and long-term lamp SPD since it can solarize, as will be discussed in Chapter 7 of this book.

Fluorescent lamps are inexpensive, readily available and in use world-wide. They also vary from batch to batch, company to company and between different geographical locations. They are designed to adequately render colors—not necessarily reproduce very accurately particular SPD. To what extent the SPD of a lamp can differ and still achieve its color reproducibly objective can be seen by comparing the old and new versions of the very popular cool-white lamp.

Figure 3 shows the SPD of the old (A) and new (B) cool-white lamps. The new triphosphor lamp, based on phosphors developed for television picture tubes is much better at rendering color than the old monofluorophosphor (12). From their SPDs, one could hardly call them equivalent. They do however meet the criteria of color rendition, to "fool the eye", very well.

Fluorescent lamps are rated in catalogs by two indices, their correlated color temperature (CCT) and color-rendering index (CRI). The CCT is the temperature of a black body whose chromaticity most nearly matches that of the subject light source. Because a fluorescent lamp only approximates a "black body," it is called correlated to distinguish it from actual. The CRI on the other hand is a subjective method developed by the CIE in which eight test colors are viewed under the test and a reference lamp(s) and its ability to reproduce the test colors numerically rated. Both numbers are not absolute but useful in selecting lamps for color rendering applications.

Wyszecki (13) proposed a "goodness-of-fit" criteria for the CIE in an effort to establish a better method for testing light sources with respect to their suitability to serve as a standard daylight source for colorimetry. While his method

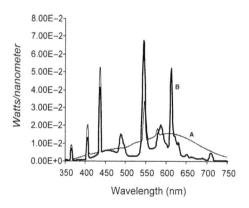

Figure 3   Spectral power distributions of old (**A**) monofluoro and new (**B**) trifluorophosphor cool-white lamps (12). Courtesy R. Levin, Osram Sylvania Inc.

failed in its colorimetric objective, it is a good method for comparing the SPDs of different lamps.

The danger of using these indices or a lamp's common name like cool-white or "daylight" for a scientific standard lies in the fact that they are not specific but general criteria for use in the color and lighting industries. A good example of the problem this approach can cause can be found in the work of Sequeira and Vozone (14). They and associates who were comparing two different exposure cabinets, one a bench top and the other a walk-in. They ordered visible (VIS) and UV lamps from the same distributor. They assumed that the ICH language was clear and explicit and the distributor assumed that since they were using familiar terms they knew exactly what they wanted. They received two different cool-white lamps and UV-A lamps. Needless to say, the results obtained were not exactly duplicates of one another.

Figure 4 shows the problem that Sequeira and Volzone, and associates encountered when trying to determine what UV exposure time to use. The spectral overlap is not the same, so they got values for the lamps that differed by fivefold. The radiometer was no help because it too had its own response curve and gave a third value. Even when the values obtained were used for the studies, the results were not exactly duplicates of one another.

This paper points out a serious problem with the ICH document, the use of quinine as an actinometer. It does not absorb radiation over the wavelength specified in the ICH document. Moore (15) has discussed this problem in detail.

Two other instances regarding lamp selection are even more important because they involved major instrument manufacturers. In one instance, the wrong UV-A lamp was distributed to customers, and in the other, the wrong VIS lamp (the trifluorophosphate) was substituted for the monoflurophosphor lamp and distributed with equipment.

The reason for mentioning these incidents is that the ICH document (1) states that "For options 1 and 2, a pharmaceutical manufacturer/applicant may rely on the SPD specification of the light source manufacturer." Unfortunately, users do not

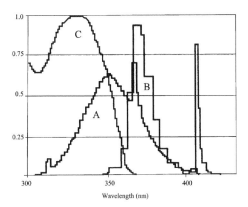

Figure 4 Spectral power distribution of 340 nm (A) and 360 nm (B) UV-A fluorescent lamps overlaid on the absorption curve of the quinine actinometer (C). *Source*: From Ref. 14.

have any ICH published reference spectra against which they can compare their lamps SPDs. Additionally, many users are not aware that the SPD they receive will most likely be that of a typical lamp, not the one they receive—nor even of the batch their lamp(s) came from. Only one firm is willing to supply the SPD for a particular lamp (16).

At the 1999 Pittsburgh Conference on Analytical Chemistry, Piechocki presented a paper titled "Pharmaceutical Photostability Testing—Black Box Chemistry" (17) in which these and other problems were highlighted. For the most part, the challenge for instrument suppliers posed in this presentation was ignored. Stand alone, relatively inexpensive, self-contained, broadband, spectroradiometers, such a Luzchem's SPR-01 (16) are rare.

Ultraviolet Lamps

The UV portion of the electromagnetic spectrum is very important because that is where most substances absorb. UV rays are used to sterilize substances and products as well as to give us the tans we are so proud of attaining during summer months. Their aid in the development of skin cancer is well proven as is that of immunosuppression.

The ICH document calls for a "... lamp combining VIS and UV outputs..." and a "... near UV fluorescent lamp having a spectral distribution from 320 to 400 nm with a maximum energy emission between 350 and 370 nm; a significant portion of the UV should be in both bands of 320 to 360 nm and 360 to 400 nm."

Figure 5 presents the lamp SPD for the lamp meeting ICH specifications, commonly called a UV-A blue light (BL), which is similar in UV output to the UV-A BL blue (BLB) lamp. The BLB lamp is specially coated to filter out VIS emissions.

Figure 6 displays the SPDs of two UV lamps commonly used for photostability testing and recommended in American Society for Testing and Materials (ASTM) G "Standard Practice for Operating Fluorescent Light Apparatus for UV Exposure of Nonmetallic Materials" (18) for mimicking either the D_{65} or ID_{65} UV portion of these SPDs. Their goodness-of-fit at the lower wavelengths is evident. Visual comparison with Figure 5 shows that neither the BL or BLB lamp is as good a fit. Therefore, they should not be used if results similar to D_{65} or ID_{65} are desired, as prescribed by the guideline.

Option 2 of the ICH document allows for the use of a "... cool white fluorescent lamp as defined in ISO 10977: 1993 (E), (5)" and "A near UV fluorescent light". This combination of lamps presents certain problems. We have previously noted the mix up of lamp types (old vs. new phosphor) and the "Tonnesen gap." In addition to these problems, there is the question of whether the exposure should be with both lamps illuminating at the same time, and if not, then which lamp should be used for the first exposure.

The ASTM cautions against simultaneous exposures with different types of lamps. In G 154-04 The Standard Practice for Operating Fluorescent Light Apparatus for UV Exposure of Nonmetallic Materials (18) they caution, "Do not mix different types of lamps. Mixing different types of lamps in a fluorescent UV light apparatus may produce major inconsistencies in the light falling on the samples, unless the apparatus is specifically designed to ensure a uniform spectral distribution." No chamber manufacturer, to date, has presented such information. Most do claim equal energy distributions but as was already

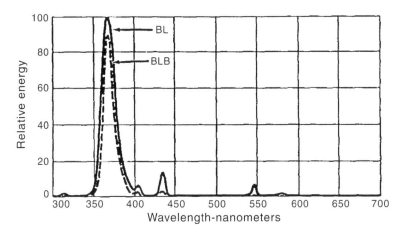

Figure 5 Spectral power distributions of the blue light (BL) and blue light blue (BLB) UV-A fluorescent lamps. *Source*: Courtesy of Philips Lighting Company.

presented in Chapters 2, 10 and 13 of this book, such information should be cross-checked.

From a chemical standpoint, simultaneous exposure is the only procedure to use. Any test substance can undergo *cis*-trans isomerism such as demonstrated by cinnamic acid and its derivatives or photochromism. Both of these phenomena can be reversible, presenting an equilibrium situation. If one uses sequential exposures—or even worse, separate samples—for each source, the results obtained may be biased.

Visible Lamps
VIS fluorescent lamps used for lighting are selected for their initial cost, intensity and color reproducibility. Their ability to meet ICH standards (5), not having scientifically acceptable reproducible SPD, are not the lighting engineers goals. They may also consider certain physiological effects associated with certain lamps, such as warm-white or cool-white when making their choices. On one hand, warm-white lamps give users the feeling of warmth and have SPDs with wavelengths in the red, orange, or yellow. Cool-white lamps, on the other hand, produce the feeling of coolness and have wavelengths in the blue and green.

A number of different lamp common names have been developed through the years. Figure 7 displays a number of the common lamp names and their associated SPDs from one manufacturers catalog. Not all catalogs will be identical.

All of the lamps shown in Figure 7 are composed of the same phosphors in different ratios. The antimony containing phosphor has a peak maximum at about 480nm whereas the manganese phosphor fluoresces at about 580nm. As a general rule, the "Deluxe" or higher-power lamps will have a higher wattage and color rendition number and be closer to a CIE standard, but not be in compliance with the ISO 10977 SPD. The reason for selection of the ISO 10977 lamp was that it was the most common lamp, used worldwide until the development of the tri-phosphor lamp. It was used in many places including laboratories, manufacturing areas, pharmacies, and homes.

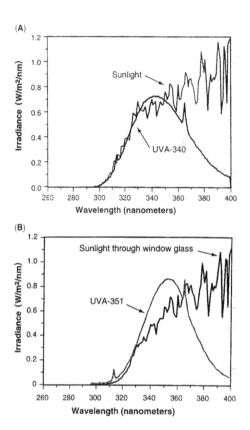

Figure 6 Spectral power distributions of the (**A**) 340 nm Ultraviolet (UV)-A lamp overlaid on that of sunlight and the (**B**) 351 nm UV-A lamp overlaid on that of sunlight through window glass. *Source*: Reproduced with permission from Q-Lab corporation.

This information, gathered from several surveys, led the EWG to focus on what was most used rather than which source was most scientifically valid. This information and the time constraints on the ICH process, which did not allow for studies of any problem, led to the conclusions drawn. An example of the debate that did take place can be found in the words of Sager "… if the ICH fails to harmonize on the protocol, it will still become an FDA document" (19).

One problem with all fluorescent lamps is their UV emissions. Because the mercury lines are frequently omitted from the SPDs shown in catalogs, people are falsely led into believing that they are not there. This is not exactly the case. Kobza et al. (20) and Cole et al. (21) have shown that UV wavelengths below 300 nm are emitted by bare bulbs and Kobza et al. showed they could be harmful. Eventually these findings and those of others led in the United States to the new ANSI/IESNA (American National Standard Institute/Illuminating Engineering Society of North America) standards (22–24) regulating UV lamp emissions.

Xenon Lamps

Option 1 allows for the use of Xenon lamps as a radiation sources. These lamps are simple in construction, available in a range of intensities and the basis of

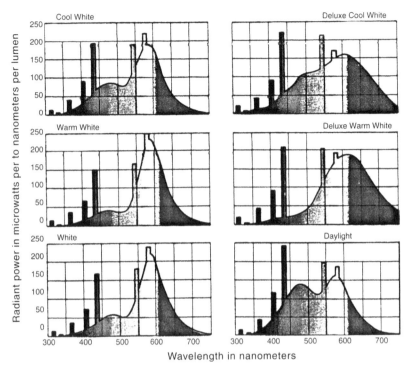

Figure 7 Spectral power distributions of some visible lamps. *Source*: Courtesy of General Electric.

international standards. From their initial discovery, it was recognized that their SPD resembled that of the sun (25). Initially they were troubled by the high-cost of Xenon, ozone generation and tendency to explode due to their high internal pressure and cracking glass seals. All of these negatives have since been overcome.

Xenon lamps come in a variety of different shapes and sizes and as pure Xenon and Mercury doped. The latter lamps, although still called Xenon lamps are not true solar replicators. Figure 8 shows the difference between the Xenon Mercury (A) and the Xenon lamp. While the Xenon Mercury lamps are brighter, they do not have the same SPD.

Xenon lamps are the basis of many international standards for replicating ID_{65} and ID_{65} testing conditions. They are valued not only for their high correlation with actual solar radiation but also for their high intensity which translates into reduced testing times and more efficient use of both human, equipment and physical resources.

The most negative comment regarding these lamps is the heat they generate. The amount of heat generated is a function of several factors including the wattage, distance from the source and time of exposure. There is a paradox regarding the heat problem and use of Options 1 and 2. Baertschi (26,27) first presented this paradox and Baertschi and Thatcher discuss this matter in Chapter 10 of this book. Long term, lower temperature, fluorescent lamp exposure may actually be more complicated than short-term high intensity Xenon or metal halide lamp exposure.

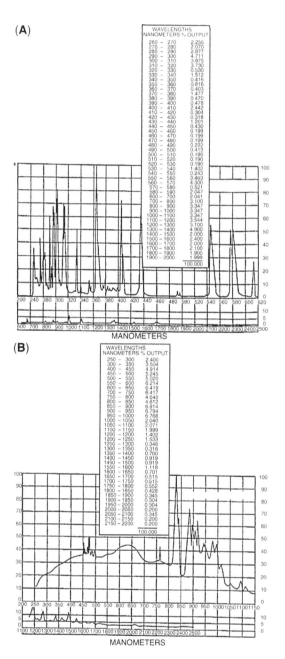

Figure 8 Spectral power distributions of (**A**) short-arc Hg-xenon and (**B**) a normal xenon lamps. *Source*: Reproduced with permission from the Xenon Corp.

Figure 9 is a representation of the unfiltered SPDs of a Xenon flash tube run at high (1) and low (2) power. The large decrease in the infrared component when this lamp is used in the high power mode is readily evident in this figure. The increased

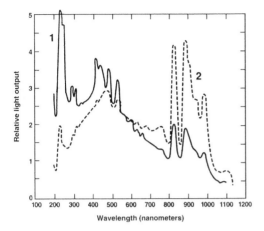

Figure 9 Spectral power distribution of a Xenon flash operated in high-power (**1**) and low-power (**2**) modes. *Source*: Reproduced with permission from the Xenon Corp.

UV output has proven to be useful for pulse sterilization of pharmaceutical products (28,29). The usefulness of this technique lies in the fact that the pulse height, length of exposure, time between pulses and total number of pulses per test can be regulated.

Figure 10 is a pictorial diagram of the relationship between these various parameters. The ability to control the heat exposure and cooling period greatly reduces the heat burden on samples. The time scale is most impressive because it means that sample exposures can be attained in seconds not hours or days. The ability to have all samples ready for analysis at the same time eliminates any question

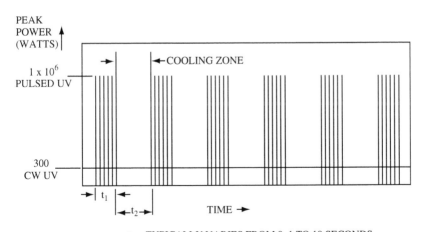

t_1 = TYPICALLY VARIES FROM 0–1 TO 10 SECONDS
t_2 = DEPENDS ON APPLICATION AND SYSTEM COOLING

Figure 10 Illustration of the relationship between pulse height, pulse width, number of pulses and heat generated in a UV exposure. *Abbreviation*: UV, ultraviolet. *Source*: Reproduced with permission from the Xenon Corporation.

with regard to storage and is an assurance that they are all analyzed under the same conditions.

The reproducibility of operating conditions for the exposure unit is also an added bonus. There is little likelihood that the equipment will drift over the short periods of time it is used for a test. This means that the same area for sample placement can be used for the series of tests performed, an advantage offered by no other approach. To date no one has used this approach.

The most commonly used Xenon lamp forms used for photostability testing are presently the short- and long-arc varieties. Short-arcs are those 5 mm in gap length or less and long-arcs are greater than 5 mm. They also come in air and water-cooled varieties, depending on their wattage and application. The short-arc lamps project an illumination pattern in the shape of a bulls eye, mimicking a point source, whereas the long-arc mimic bar radiators.

These lamps, when used with the appropriate filters, mimic the solar spectrum. Being high intensity sources, they do "solarize" faster than less intense sources and must therefore be more closely monitored. These lamps have been reported to produce results about 10 times faster than fluorescent lamps.

The cost of these units is higher than the fluorescent lamp varieties but this one negative factor is easily overcome by their speed and "full continuous spectrum" which provides the maximum amount of information per test.

Metal Halide Lamps

A metal halide lamp consists of a quartz bulb containing an inert gas, metal iodide, mercury and electrodes. In operation, a voltage is applied to the electrodes ionizing the mercury vapor and causing a current to flow through the tube. The current flow raises the internal temperature of the bulb and vaporizes the metal halide salt. The salt, in turn, is ionized by the arc into its components, while it emits its characteristic radiation (5). In normal operation, the metal would deposit on the colder, inner surface of the bulb. This deposition is prevented and/or reversed by the highly reactive ionized iodine, which reacts with the metal, revaporizing it. The revaporized salt is then free to begin the cycle anew. This operation helps the lamp retain its brightness for long periods, and increasing its longevity.

There are many varieties of metal halide lamps, the most familiar being the tungsten-halide lamps found in many homes, or the sodium and mercury vapor lamps found on streets and in auditoriums, convention centers and large space commercial buildings. Many have seen the change in color given off as these lamps warm up to their operating temperature.

Only a few of these lamps have been used for pharmaceutical photostability testing and they are the mercury, sodium, iron and thallium doped dysprosium varieties.

Figure 11 shows the SPDs of these lamps. It is evident that even the Iron, represented by the Dr. Hönle lamp, and thallium doped dysprosium lamps contain more line spectra than the xenon lamp. Once again, they are good replicators of color but not as good approximators of the solar SPD.

The sodium lamp is frequently used to illuminate areas where photosensitive substances are used.

SOURCE SELECTION CRITERIA

One interested in the ICH EWG objectives regarding photostability testing should read Beaumont's chapter on "Photostability Testing" in the book "International

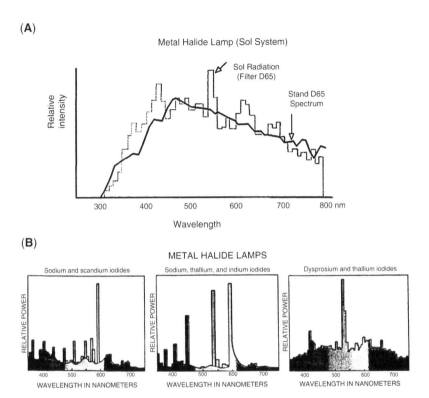

Figure 11 Spectral power distributions of typical metal halide lamps used in photostability. (**A**) Dr. Hönle Iron-based Sol System and (**B**) sodium and scandium iodides, sodium, thallium and indium iodides and dysprosium and thallium iodides lamps. *Source*: Reproduced with permission from Honle UV America and the Atlas Electric Devices, Co.

Stability Testing" edited by Mazzo (3). Other references are the articles by Sager et al. (30), which gives Food and Drug Administration's (FDA's) and the U.S. industry perspective and Tønnesen and Karlsen, which give the European perspective on this topic (31).

A review of these publications reveals a host of misstatements such as "After evaluation of the various lamps, none could be declared superior to others in terms of variability or operating conditions" (30) or "Although acceptable lamps may have different SPDs, these differences should not effect the yes/no outcome of the testing procedure…" (30), the previously cited quotation from Beaumont (3) or the statement of Wooton that "If a material is going to be exposed to different light sources it is unlikely that a single test will provide sufficient data to establish the stability of a material to all sources (32)."

The argument of Sager et al. (30) that pharmaceutical photostability testing is a yes/no procedure is false. The ICH document, of which Q1B is a sub-part, has established limits for degradation products, which do include numerical limits. Without specifying a definite SPD and allowing for other admitted nonequivalent SPDs the guideline is meaningless and nonscientific.

For decades researchers have recognized that the solar spectrum is the standard to use. This is evident in Chapter 1 of this book. Most research and development laboratories use Xenon systems for their evaluation of potential formulations.

For nearly a century, as outlined in Chapter 1 of this book, many people have sought to standardize pharmaceutical photostability testing procedures. Allied fields have worked for decades to establish the use of a single standard and have been successful. A good example of this effort is the recent ISO 10526/CIE S005/E-1998 CIE Standard Illuminants for Colorimetry (33).

Initially, surveys by Kerker (34), Thoma and Kerker (35), Anderson et al. (36) and the Japanese (37) were performed to determine the (current) industry practices. Subsequent to these, other surveys were performed in the United States, Europe, and Japan under the aegis of the ICH EWG and reported at the 1993 in Kyoto, Japan (38).

More recently, comprehensive studies of typical areas of exposure of samples from synthesis, through manufacturing, packaging and in-use have been performed and reported by McGreer (39), Reed et al. (40) and Baertschi et al. (41). The results of these studies show the variety of sources samples are exposed to during their lifetime. A source that is broadband and continuous from about 300 to 700 nm will cover the spectrum of all natural light except for the infrared, and meet the guidelines objectives.

Based on their study, Reed et al. (40) suggested the development of a "light-burden" survey for photosensitive products. The purpose of this survey would be to allow developers to identify the source of any problem and take the appropriate action.

It is interesting to note that in the ICH surveys, no attention was paid to surveying all areas of test sample exposure, including the outdoors, only the home or pharmacy were studied.

In addition, the surveys erred in being directed primarily to the Quality Control not R&D groups. Even under these limited conditions, not all sources were considered, for example, the increased presence of quartz halogen and new trifluorphosphor sources in home lighting.

The survey by McGreer (39) included a limited visit to a pharmaceutical manufacturer and showed some of the diversity of sources found in an industrial situation. Sources found in this survey included window-glass filtered daylight, monofluorophosphor cool-white, triphosphor cool-white fluorescents and mercury metal halide lamps. His visit to pharmacies revealed that both varieties of cool-white lamps were used, the old monofluorophosphor and the newer triphosphor.

Essentially, all surveys reported that a number of different sources were being used in most facilities and that R&D groups were most likely to use the more intense sources, as opposed to Quality Control. The R&D selection is based more on the time saving gained when using these sources rather than their scientific correctness, although in this case both parameters may have been taken in to account.

The only solution to this current morass is for all parties to agree (as they have in allied fields such as the automobile, paint, dye, colors and textile industries) upon a universal standard, such as the CIE D_{65} and select the best lamp available for its simulation, the Xenon arc. As with any standard, there must be restrictions, on this one, too. Restrictions are what define a standard, its purity, and strength.

Cost of equipment is not an issue since many contract laboratories do exist and can afford to supply this service to many customers. With the current trend toward out-sourcing testing, such an approach is economical.

Adopting the Xenon-arc-based D_{65} standard will mean removing most of the current options from the current ICH document, since they are non-equivalent and do not meet the current agreed to CIE/ISO/ASTM standards and compliance criteria. There is no need to add anything to the current standard but rather, there does exist a necessity to clarify and make it more specific to its assigned task. This will ease the regulatory review burden and speed up the process.

Table 2 lists the criteria that should be used when selecting a source for photostability testing. It does not contain the often-cited criteria "of cost." "Equivalence" should be the first criteria for source selection. "Cost" should only enter into the equation when the scientific criteria have been met. After all, the ICH process was started to reduce the need for duplicating studies and speed up the drug development and review process, based on sound scientific principles. Only the application of good science assures compliance with these goals.

When one factors in the extra time spent monitoring storage conditions waiting for results and the need to store and track samples and the loss of space, the cost of equipment becomes a minor factor.

The argument that Xenon-arc testing is equivalent to information "overkill" is not valid. It is the only source capable of covering the entire range of interest with no gaps. It has no "overlap" problems. Its reproducibility has been proven by decades of use in allied fields.

Table 3 lists the reasons for adopting a xenon lamp-based D_{65} standard. It is easy to see that there are more advantages than disadvantages for its adoption. The fact that this lamp is already a universal standard makes it amply clear that little needs to be done to validate it for use in pharmaceutical photostability testing. It being a specialty lamp may not seem to be an advantage but it does assure that it will be made to more exacting standards than the commodity lamps. The manufacturer's primary goal is to make a lamp of the desired SPD, in compliance with the CIE standards. This is similar to all other standards in use in the pharmaceutical industry.

Having the least number of components makes manufacturing a lot simpler and reduces the number of possible variables. This makes for cheaper manufacturing costs, ease of manufacture and reproduction.

Availability in a large number of wattage sizes makes these lamps useful for illuminating tabletop units or use in animal studies in whole rooms. It gives the most information per test because it covers all wavelengths, in the bandwidth specified, continuously, without gaps. No wavelength is missed and any other lamp can be simulated from its results, through computer manipulation, using already available software.

Table 2 Lamp Selection Criteria

1. D_{65} equivalance
2. Simplicity of design
3. Ready availability
4. Reproducibility
5. Stability
6. Validation

Table 3 Pros and Cons for Adopting a Xenon Lamp–Based D_{65} Standard

Advantages	Disadvantages
It is already a universal standard	Initial investment high
It is a specialty lamp	Generates heat
It has the least number of components	Most expensive lamps
It is available in a large number of wattages from 100 to 6500 W	
It gives the most information per study	
Fastest results	
Least labor intensive	
Gives the most data per test	

Because of its intensity, results have been reported that these are more than 10 times faster than those produced by fluorescent lamps. This result is not only time saving but there are also other resource savings. This technique also holds the promise of even greater savings when employed in the "flash mode" as is already used in some sterilization situations.

In the flash mode, the sample is exposed to more intense radiation than conventional Xenon-arc lamp exposure for only a few milli-second full spectrum energy at a time and with a cooling period between pulses. This results in complete UV-VIS exposure in a matter of fractions of a second without any large heating of the sample, common in other techniques.

From a practical standpoint, this means that a complete photostability study could be performed in a very short time. By controlling the intensity and number of pulses, a dosage study could easily be performed. Since there is no need for calibrating a number of different exposure spots for samples, all can use just one spot. In addition, with all samples being available for testing within a short period, there will be no need for storing samples or having to validate their storage conditions. The savings in labor, record keeping, sample tracking and analytical costs should be great.

Table 4 lists the often-cited reasons for not adopting a true D_{65} standard. Cost is always cited as the first reason something which is not "the done thing." It is self-evident that all things have a cost but the alternative is to do nothing and risk possible exploitation by a competitor. As outlined previously, costs can be minimized by standardization and applying good science. Continuing with the current guideline is not cost-effective for the industry, the public or the regulators.

Years of doing things the same way does not validate correctness. Many chemical tests have been proven incorrect over a period of time. Newer techniques like thin layer chromatography (TLC), high performance liquid chromatography (HPLC), capillary electrophoresis (CE) and mass spectrometry (MS) have revolutionized our standards of purity. So, too, should this testing procedure prove to be.

It is well to respect what others have done but the fact that someone for whatever reason chooses not do something in a correct scientific way should not validate it as an option in an international guideline. Failure to lead by standardization via *the international guideline* essentially greatly diminishes its value. In its present state, it is merely a collection of worldwide practices with many conflicts.

Table 4 Nonscientific Reasons for a D_{65} Standard

Cost of equipment

This is the way we have done it for many years and we haven't had any problems so there is no reason to change

It is our way and we choose to accept all approaches (options)

We have already made our choice and to change now would mean to admit we made the wrong choice and possibly wasted the company's money

Let sleeping dogs lie. We may discover something detrimental about our current or future product(s)

The hardest thing to correct is a previous decision or commitment. As with every new guideline, manufacturers rush in and try to apply it to their product(s). An eager industry also rushes to comply and makes decisions based on the supplier assurances, even though both may be unsure of what the true science behind the guideline might be. The incident mentioned earlier in this chapter of a supplier advocating the use of the wrong cool-white lamp is a good example of this problem. Several other similar incidents have occured but go unreported.

It is hard to admit to a mistake and such admissions can, under adverse situations, be career killing. But to take the other alternative of deciding not to make the correct change, increases the number of potential problems and potential liability for the individuals involved and their companies. Like cancer, the sooner it is discovered and treated, the better it will be.

Discovery of new things about old products is not a bad thing. By knowing the origin of degradation of products, one may discover the origin of an adverse effect. This could lead to a safer, more potent drug, essentially putting new life into an old product. Another possibility is that one might discover that an old drug has a previously unrecognized use as a photoactivatable drug. Searching drug archives for new uses is not foreign to the industry. Previous statements regarding "flash lamps" are applicable to automation for mass screening of such libraries.

All of these items reflect more the political reality of life rather than the science. To accept current approaches is to subject products to less than full development and, products and patients to possible future harm. Ultimately the science will prevail. Eventually the choice to be pro- or reactive will have to be made. Doing it right the first time saves money and time while identifying possible problems and hidden benefits of products.

LAMP STUDIES

Most assuredly, a number of sources may prove that a drug substance/product is photolabile as studies have shown (42,43). However, will they generate the same degradation products and in the same amounts? Studies by Thoma and Kerker (44,45), Sequeria and Volzone (14), Merrifield et al. (46) and Baertschi et al. (41) have shown this not to be the case. One cannot ignore the SPD of the source or whether the dosage measurement device, be it a radiometer or chemical actinometer, measures all of the incident radiation or only part of it as pointed out in Chapter 3, Figures 3 through 7 of this book.

Most previous studies have been done as limit studies i.e., after exposure for a predetermined time. No attention was paid to the fact that photolysis is a self-limiting phenomena, generally limited to the first several hundred micrometers as shown in Chapter 16, Figure 4 of this book. The degradation products absorb at longer wavelengths and serve to block further penetration of incident rays rendering further exposure meaningless. The only thing that the current ICH protocol proves is that the sample can be photodegraded. Essentially, tests run using the ICH protocol over-expose samples. Intermediate sampling times should be taken.

The intermediate testing points will not only speed up the procedure but also help to identify the quantum efficiency of the reaction. This information will allow one to better select the appropriate preventive measures and provide better directions for use, if needed; for instance in an operating room where vials are separated from their outer cardboard cartons.

ALTERNATIVE LAMPS

There are a number of different lamps available on the commercial side that claim to be "natural" or "daylight" or "sun lamps" for reproducing some or all of the benefits of natural sunlight, even D_{65}. Most are poor substitutes.

Three lamps that have been used for photostability testing are shown in Figure 12, which clearly illustrates the difference between fluorescent artificial daylight lamp manufacturers and the D_{65} standard. Two of these lamps (A and B) are manufactured, like all fluorescents for general lighting not to scientific standards. The third lamp is specifically manufactured to give the correct ICH radiometric and photometric readings, even though these readings do not actually measure all of the incident radiation.

As was previously mentioned, specialty lamps available from only one source can be a problem. Lamp A and C manufacturer's original lamps disappeared from the marketplace for a time and changes were made in their replacements. Sole source lamps are not very reliable international standards.

A number of basic questions regarding sole source lamps arise. Who will validate the lamp(s)? Will these lamps be evaluated as a "yes/no" problem as advocated by Sager et al. (30) or on a qualitative and quantitative scientific basis as any other analytical method for impurities?

CONCLUSION

The ICH guideline (1) states that "The intrinsic photostability characteristics of a new drug substance and products should be evaluated." Having little or no knowledge of which wavelengths are most important, it is prudent to evaluate all. The best lamp selection is a well-characterized, continuous, broadband source to assure that all wavelengths are tested, the CIE D_{65} in its latest revision CIE 85-1989 which replaced CIE 20. This choice will assure that the sample has been tested with all possible wavelengths, thus assuring that all possible sources wavelengths are tested.

If the ICH guideline is to reach its objective of reducing the unnecessary duplication of efforts and speed up the review process, revision of the guideline is of course necessary and the use of broad qualifiers such as "similar" must be replaced by definite standards as is common in all compendia.

(A)

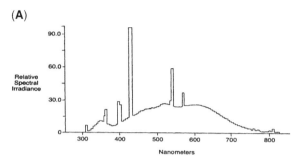

(B)

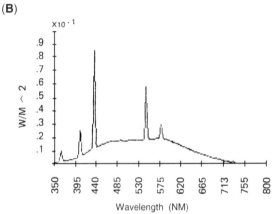

(C)

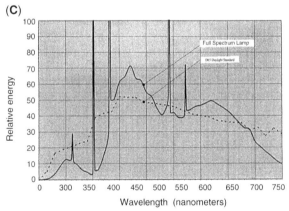

Figure 12 Spectral power distributions of specialty lamps from (**A**, **B**, **C**) three different manufacturers used for photostability testing.

REFERENCES

1. International Conference on Harmonisation (ICH). Guideline for the Photostability Testing of New Drug Substance and Products. Federal Register 1997; 62:27115–27122.
2. Matsuo M, Machida Y, Furuichi H, Nakamura K. Study on establishing light stability testing. Iyakuhin Kenkyu 1988; 19(6):1028–1052.

3. Beaumont T. Photostability testing, Chapter 8. In: Mazzo DJ, ed. International Photostability Testing. Colorado: Interpharm Press 1999:59–71.
4. Commission Internationale de L'Eclairage (CIE). International Commission on Illumination publication, Colorimetry. Official Recommendations of the International Commission on Illumination. Paris 16, France: Bureau Central de la CIE.
5. ISO 10977:1993(E), Photography processed colour films & transmittance, solar direct, total solar energy transmittance and ultraviolet transmittance and glazing factors. International Organization for Standardization (ISO). Case Postale 56, CH-1211 Genève, Switzerland.
6. Roberts LK, Beasley DG, Learn DB, Giddens LD, Beard J, Stanfield JW. Ultraviolet spectral energy differences affect the ability of sunscreen lotions to prevent ultraviolet-radiation-induced immunosuppression. Photochem Photobiol 1996; 63(6):874–884.
7. Gil EM, Kim TH. UV-induced immune suppression and sunscreen. Photodermatol Photoimmunol Photomed 2000; 16:101–110.
8. Riehl JP, Maupin CL. On the choice of a light source for the photostability testing of pharmaceuticals. PF 1995; 21(6):1654–1663.
9. Nema S, Washkuhn RJ, Beussink DR. Photostability testing: an overview. Pharm Tech USA 1995:170–185.
10. Lachman L, Swartz CJ, Cooper J. A comprehensive pharmaceutical stability testing laboratory III: a light stability cabinet for evaluating the photosensitivity of pharmaceuticals. J Am Pharm Assoc 1960; 49(4):213–218.
11. Tønnesen HH, Moore DE. Photochemical degradation of components in drug formulations. Pharm Tech Intl 1993; 5:27–34.
12. Srivastava AM, Sommerer TJ. Fluorescent lamp phosphors. Electrochem Soc Interface 1998:28–31.
13. Wyszecki G. Development of new CIE standard sources for colorimetry. Die Farbe 1970; 19(1/6):43–76.
14. Sequeira F, Vozone C. Photostability studies of drug substances and products. Pharm Tech USA 2000; 24(8):30–35.
15. Moore DE. Photosensitization by drugs: quinine as a photosensitizer. J Pharm Pharmacol Commun 1980; 32:216–218.
16. Luzchem Research Inc. Unit 12, Ottawa Ontario, Canada K1J9J9.
17. Piechocki J. Pharmaceutical photostability testing—black box chemistry. Pittcon 99 the 1999 Pittsburgh Conference on Analytical Chemistry in Orlando, FL, U.S.A.
18. American Society for Testing and Materials (ASTM). G 154–04 Standard Practice for Operating Fluorescent Light Apparatus for UV Exposure of Nonmetallic Materials. Philadelphia, PA: American Society for Testing and Materials, USA.
19. Sager N. The Pink Sheet 1995; 28; T&G-5.
20. Kobza A, Ramsay CA, Magnus LA. Photosensitivity due to 'sunburn' ultraviolet content of white fluorescent lamps. Brit J Derm 1973; 89:351–359.
21. Cole C, Forbes PD, Davies RD, Urbach F. Effect of indoor lighting on normal skin. Ann New York Acad Sci 1985; 453:305.
22. ANSI/IESNA RP-27.1–00. Photobiological Safety for Lamps and Lamp Systems—General Requirements. American National Standards Institute, NW, Washington, D.C., 20036.
23. ANSI/IESNA RP-27.2–00. Photobiological Safety for Lamps and Lamp Systems—Measurement Systems. American National Standards Institute, NW, Washington, D.C., 20036.
24. ANSI/IESNA RP-27.3–96. Photobiological Safety for Lamps—Risk Group Classification and Labeling. American National Standards Institute, NW, Washington, D.C., 20036.
25. Schulz P. Elektrische entladungen in edelgasen bei hoken drucken. Reischsberichte Phys 1944; 1:147–153.
26. Baertschi SW. ICH Option 1 and Option 2. Sources: potential for different drug photodegradation pathways. Fourth International Conference on the Photostability of Drug Substance and Drug Products, July 16–19, 2001, Research Triangle Park, NC (U.S.A.), July 16–19, 2001.
27. Baertschi SW. Practical aspects of pharmaceutical photostability testing. Photostability '99. Third International Conference on the Photostability of Drug Substance and Drug Products. Washington, DC. July 10–14, 1999.

28. Bushnell A, Cooper JR, Dunn J, Leo F, May R. Pulsed light sterilization tunnels and sterile-pass throughs. Pharm Eng 1998:49–58.
29. Panico LR. Instantaneous Surface Sanitization With Pulsed UV. Presented at the Global Conference on Hygienic Coatings. Brussels, Belgium: July 8–9, 2002.
30. Sager N, Baum GD, Wolters RJ, Layloff T. Photostability testing of pharmaceutical products. PF 1998; 24(3):6331–6333.
31. Tonnesen HH, Karlsen J. Photochemical degradation of components in drug formulations. Pharmeuropa 1995; 7:137–141.
32. Wooton AB. Design, installation and commissioning of photostability test cabinets, IBC Technical Services Ltd. Conference on Stability Testing Design and Interpretation of Data for International Registration of Pharmaceuticals, London, April 27–28, 1994.
33. ISO 10526/CIE S005/E-1998. CIE Standard Illuminants for Colorimetry available from International Organization for Standardization (ISO), Case Postale 56, CH—1211 Genève, Switzerland.
34. Kerker R. Untershungen zur Photostabilität von Nifedipin, Glucocorticoiden, Molsidomin und ihren Zubreirungen. Ph.D. dissertation, Ludwig Maximillians Universität, München, Germany, 1991.
35. Thoma K, Kerker R. Photostabiltät von arzneimitteln ein vernachlässigtes qualitätskriterium? Poster presented at APV Symposium Regensburg, April 14, 1992.
36. Anderson NH, Johnston D, McLelland MA, Munden P. Photostability testing of drug substances and drug products in UK pharmaceutical laboratories. J Pharm Biomed Anal 1991; 9(6):443–449.
37. Yatani K, Shimizu R, Ueno M, Matsuo M, Tsunakawa N, Murayama S, Takeda K. Investigation to establish testing procedures for light stability test. Iyakuhin Kenkyu 1988; 19:1028–1053.
38. ICH 2 Quality Six Party Drafting Group on Light Stability Testing—Notes. Kyoto, Japan Meeting, June 1993.
39. McGreer M. Getting the light right: a comparison of the spectral power distribution between artificial light sources and real world environments. PPS '04, Fourth International Meeting on the Photostability of Drugs and Drug Products, Research Triangle Park, NC (U.S.A.), July 16–17, 2001.
40. Reed RA, Harmon P, Manas D, Wasylaschu W, Bergquist P, Hunke W, Ip D. The role of excipients and package components in the photostability of liquid formulations. PPS '03, Third International Meeting on the Photostability of Drugs and Drug Products, Washington, D.C., July 10–14, 1999.
41. Baertschi SW, Kinney H. Snider B. Issues in evaluating the "in-use" photostability of transdermal patches. PPS '99, Third International Meeting, Photostability of Drugs and Drug Products, Washington, D.C., July 10–14, 1999.
42. Drew HD, Thornton LK, Juhl WE, Brower JF. An FDA/PhRMA Interlaboratory Study of the International Conference on Harmonisation's Proposed Photostability Testing and Guidelines. PF 1998; 24(3):6334–6346.
43. Matsuda Y. Some aspects on the evaluation of photostability of solid-state drugs and pharmaceutical preparations. Pharm Tech Japan 1994; 10(7):7–17.
44. Thoma K, Kerker R. Photoinstability of drugs. Part 1. The behaviour of UV-absorbing drugs in simulated daylight. Pharm Ind 1992; 54(2):169–177.
45. Thoma K, Kerker R. Photoinstability of drugs. Part 2. The behaviour of visible light-absorbing drugs in simulated daylight. Pharm Ind 1992; 54(3):287–293.
46. Merrifield D, Carter P, Clapham D, Sanderson F. Addressing the problem of photolysis instability during formulation development. In: Tonnesen H, ed. Photostability of Drugs and Drug Formulations. London, U.K: Taylor & Francis, 1996:141–154.

7 Glasses, Filters, and Containers

Joseph T. Piechocki
Piechocki Associates, Westminster, Maryland, U.S.A.

INTRODUCTION

Glass plays an important role in pharmaceutics as well as everyday life. It functions as a barrier against the environment, protects us from harmful rays and preserves our foods and medicines. It is useful but can be a problem for some products. Unfortunately, a clear definition of the glass required for each of these purposes is still lacking. To date, no compendium has included a definition even though they are replete with warnings to "protect from light."

As documented in Chapter 12 of this book, glass impurities have been shown to contribute to the photolability of otherwise photostable materials/products. These same impurities can affect the results obtained during photostability testing via their alteration of the absorption curves of the glasses used.

Filters have generally been used with much regard as to their absorption curves, stability, or reproducibility. Ill-defined terms such a "window-glass filters" have arisen with little attention being given to their true scientific meaning by the pharmaceutical community. As we shall see not all window-glasses are equivalent.

Containers are made of varieties of different formulations, inorganic and organic. Singly or combined, they provide protection of the product. One unique container, often forgotten is the human body. Once ingested/dosed, all of the cleverest packaging and formulation protective efforts are lost and the drug substance is exposed in its generally most vulnerable form, unprotected by any formulation and dilute, in the capillaries in the skin.

This chapter will try to make users aware of the various glasses, filters, and containers (glass and plastic) encountered in normal processing, packaging, and dosing and day-to-day life of pharmaceuticals. Hopefully, this information will allow one to better design and protect materials, products and users from substances that may be photodegradable.

GLASSES

There are many different types of glasses made that can have an impact of the photodegradation of pharmaceuticals and their testing results. Among these types are; fused quartz, soda-lime, borosilicate (e.g., Pyrex) and a variety of colored glasses. These glasses come in a variety of different shapes such as float, sheet, plate, and cylinder blown glass. They may be chemically treated, coated or laminated with plastic, clear or tinted.

For pharmaceutical purposes the United States Pharmacopoeia, (USP), European Pharmacopoeia (EP) and Japanese Pharmacopoeia (JP) grade glasses are used for pharmaceutical packaging. They do, as noted before, fail to define the spectral characteristics of the glasses to be used in a particular monograph, even though they caution to "protect from light." Their current warning implies that only visible (VIS) light need be protected against, even though a majority of pharmaceutical substances are photosensitive to the ultraviolet (UV) region of the spectrum, based on the correct definition of "light."

In the International Conference on Harmonization (ICH) Q1B document (1) little attention is paid to the problem of the different glasses used for photostability studies or the differences in their physical shapes. In Option 1 the use of an "appropriate filter" to remove "significant radiation below 320 nm" is specified. No details are given as to the characteristics of said filter, nor concern that to use such a filter would mean that the specifications for the D_{65} standard could not be met.

Under the Annex section of the document, two different options for the use of the quinine actinometer are allowed. A cylindrical Japanese Industrial Standard (JIS) R3512 (1974), 20 mL, Japanese ampoule (Option 1) or a 1 cm quartz cell (Option 2) are both allowed, even though they both have distinctly different UV absorption characteristics and geometries. In addition, the thickness of the vial is not specified in the diagram in the guideline.

Glass has played an important role in the development of spectroscopy and its sources. It is a container for some of its harshest environments and its sources. It is also important to remember that glass is not inert, photochemically. Its alkalinity and other impurities can contribute to the photolability of a product as pointed out in Chapter 12 of this book.

Glass can solarize, that is as defined by Rindone "... the net effect of the UV, visible and infrared components of solar radiation in producing changes in the light transmission of glass" (2). This phenomenon, reported by many researchers including White and Silverman glass (3), Swarts and Cook (4), Starkle and Turner (5) and Boettner and Miedler (6) causes the blue color of glass to change from its naturally blue color to its more familiar green color when exposed to UV radiation.

In Figure 1 we see the effect of UV spectrum of an untreated and UV treated glass panel. The effect of this treatment on the UV cut-off and transmittance of both the UV and VIS regions of the spectrum is readily evident.

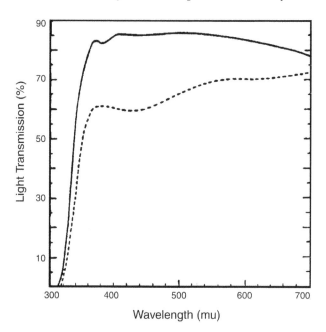

Figure 1 Spectral absorption of Fe^{3+} in soda-silica glass.
Note: —— before irradiation, – – – after irradiation. *Source*: From Ref. 3.

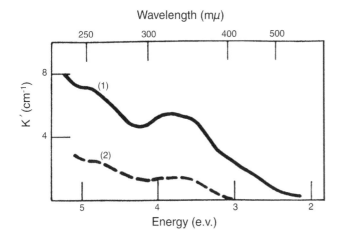

Figure 2 Effects of bleaching by tungsten light upon absorption induced by two hours x-irradiation in graphite-reduced glass containing 0.1% Fe. (*Curve 1*) Original absorption. (*Curve 2*) Absorption remaining after eight minutes exposure to intense tungsten source. *Source*: From Ref. 4.

Figure 2 demonstrates how this solarization phenomenon can be reversed by exposing the glass to high-intensity visible light and heat (4). Regional and batch-to-batch variations in the spectral characteristics of glass can be attributed to differences in impurities and their amounts, particularly iron.

Figure 3 shows the spectra of many common glasses. Curve C is of particular interest because it represents a major step forward in the development of

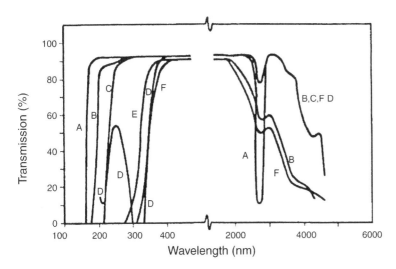

Figure 3 Optical transmission of envelope materials, 1 mm thick from ILC technology. (**A**) Synthetic fused silica. (**B**) Natural fused silica. (**C**) Titanium doped natural fused silica. (**D**) Cerium doped natural fused silica. (**E**) "Pyrex" glass. (**F**) "Nonex" glass. *Source*: From Ref. 7.

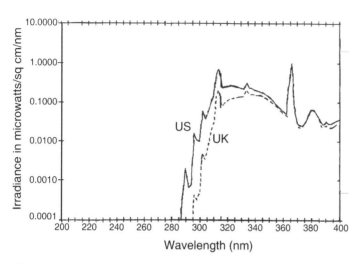

Figure 4   Comparison of 32 watt, T8 4100 K, SP 41, fluorescent lamps of the same type produced by GE, U.S. and GE U.K. *Source*: From Ref. 8.

safer high-intensity lamps. The addition of titanium to glass produces "ozone-free glass" by cutting off the ozone producing radiation, which hindered early lamp development (7).

Figure 4 shows the effect of using glasses from different geographical regions on the spectral power distribution (SPD) of a cool-white type lamp. European glass has long been noted for its naturally higher iron content (8,9). What makes this particular figure distressing is that it represents a difference within a brand name, internationally. This difference only became important in the commercial market because of the adoption of regulations, by some countries, restricting the amount of UV radiation that can be emitted by commercial lamps. In the United States, the American National Standards Institute (ANSI)/Illuminating Engineering Society of North America (IESNA) 29.1-29.3 (10–12) and the new energy restrictions (13) have changed the wattages and/or SPDs of commercial lamps.

Figure 5 shows the effect of the dopant cerium on the spectral power distribution (SPD) of a typical cool-white fluorescent lamp. This dopant has been useful in reducing UV emissions from the emissions of household quartz, halogen, and other lamps, particularly the desktop variety.

Quartz

Synthetic, pure, quartz, SiO_2 glass (also known as fused quartz) does not absorb UV radiation above approximately 150 nm. Because of its high transparency, melting temperature and resistance to corrosion, it is the material of choice for UV optical components and as an envelope material.

It was recognized early on, in the development of high-intensity lamps, that pure quartz enveloped UV lamps produced ozone in unacceptable quantities. Most facilities using them, particularly for fluorescence spectroscopy, were required to

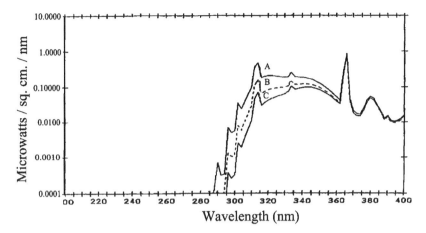

Figure 5 Comparison of 32 watt, T8 4100 K, SP 41, fluorescent lamps of the same type doped with various amounts or cerium. A = 0 ppb, B = 250 ppb, and C = 500 ppb
Source: From Ref. 8.

provide special venting. The development of titanium doped "ozone free" quartz glass solved this problem. The effect of the addition of titanium to quartz is illustrated in Figure 2.

Soda-Lime Glass
The most common form of glass is soda-lime glass. It is produced by fusing pure silica with sodium carbonate or potassium carbonate and calcium oxide. These materials reduce the working temperature of the glass, making it more amenable to further processing.

This type of glass has found use in pharmaceutical photostability testing as a filter for high-intensity sources.

Borosilicate Glass
Borosilicate glass is soda-lime glass to which boric oxide has been added. It is known under the brand names of Pyrex™ and Kimax™. The addition of boron to the glass produces a product with superior durability, chemical, and heat resistance.

This type of glass because of these superior qualities is more preferred as a modern lamp filter and other applications.

Aging
All glasses age because of chemical attack, thermal shock, and radiation exposure (solarization). This aging process also causes optical changes in the glass (14–16). As previously noted in Figure 1, these changes not only affect the UV cut-off wavelength but also the transmittance of the glass in both the UV and VIS ranges.

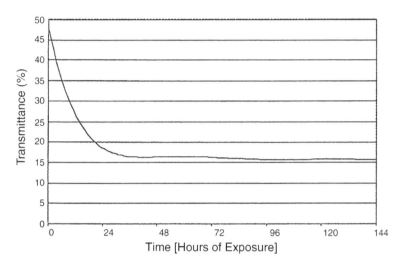

Figure 6   Transmittance of borosilicate filter versus time of xenon exposure.

Also, because, as previously noted, the UV solarizing effect can be reversed by VIS radiation and heat, a broadband irradiation experiment will result in an equilibrium in the glass—a balance between the solarizing and desolarizing phenomenas.

Figure 6 shows how a borosilicate filter will change its transmittance dramatically during the first 24 hours of operation and remain fairly stable thereafter. For this reason many manufacturers such as Atlas (17) and Q-Panel (18), among others, offer "preaged" for use in their equipment. This problem is not as critical for quartz filters so they are not "preaged." It is reported that $LiO-BeO-SiO_2$ glasses do not solarize (3).

The SPD of a lamp changes with aging. Not all of the effects are due to changes in the arcs or phosphors some are container related. Lamp aging effects are due to a number of factors including dirt and solarization, both reversible items. Dirt, as simple as a fingerprint, can cause changes and failures in high-intensity lamps, as many consumers are aware of from package labeling. The off gassing of test samples can also contribute to the problem. This latter can be solved by adequate ventilation of the sample area.

Depending on the particular lamp system used—air- or water-cooled—the filter solarization problems will be different. Water-cooled systems used for the more intense system age differently from air-cooled and must deal with the problem of water purity and additional filtering by the water jacket.

Window (Plate) Glass

Window-glass has evolved through the centuries and is still changing today with the development of newer types of glasses, the latest of which is "self-cleaning" and "smart" glasses. All of these new developments are useful for their intended purposes, which may be room darkening, UV protection or thermal mediation or protection. However, from a scientific standpoint they are all different and will bias the obtained results.

It is well to note that, though many have reported measurements of the energy behind "window-glass filtered daylight" or "room light," few have recorded the type(s) of glass or diffuser in use at the measurement time. This failure makes such measurements questionable.

Practically all plate glass, today, is produced by the "float-glass process" developed by Sir Alastair Pilkington, in the 1950s. In this process, molten glass is floated on a bath of molten tin and allowed to cool as it is passed along as a continuous ribbon. The glass plate produced is approximately 6 mm thick due to the surface tension of the molten glass. Thicker or thinner plates can be produced by either compressing or pulling the mass while still in a semiliquid state.

The glass plate produced by the "float-glass" process has a thin layer of tin metal on one side of the plate. This layer is useful for applying other coatings such as silver to the glass.

Layered Window-Glasses

Laminated Glass
One of the first developments of a layered glass was the invention of laminated glass in 1903 by French chemist Edouard Benedictus. His inspiration came when a glass flask in his lab which had become coated with cellulose nitrate was accidentally dropped and did not break into pieces.

Today a film, typically polyvinyl butyral (PVB) or an equivalent, is incorporated between layers of glass plate. This same film, PVB, also removes the UV rays from daylight and protects the interiors from fading. These laminates are typically used in building glasses and where falling shattered glass might be a problem.

Most people are not aware of the optical properties of these films until they wear photochromic lenses and work in or travel in automobiles illuminated through this type of glass. This type of glass also is used to save many colored objects from fading.

Low-Emissivity Glass
Low-emissivity glasses were developed to reduce the ability of insulated glass to transfer thermal energy. These glass "sandwiches" typically have one panel coated with a thin metallic-based coating to reflect thermal energy. Frequently incorporated between these layers is an inert gas of a low heat conductivity to aid in its insulating ability.

Insulated Glass
Insulated glasses are layered glasses with a vacuum or inert gas sandwiched between them. The exact composition varies with the manufacturer. As with the low-emissivity windows they may also contain a gas of low heat conductivity.

Self-Cleaning Glass
One of the most recent developments in the glass industry has been the introduction of "self-cleaning glasses" (19). These glasses have a coating of titanium on their surface which, when activated under aerobic conditions by visible light, oxidizes any organic material in contact with it. The Japanese have reported a more efficient version of this material is possible by treating it with nitrogen (20).

What, Then, Is "Window-Glass?"

From the previous presentation, it is evident that the term "window-glass" is more of a general term. It represents not one particular glass but a family of glasses. Defining "window-glass" for scientific purposes is difficult. Evidence of this problem can be found in American Society for Testing and Materials (ASTM) publications such as G 24–97. This standard, for example, states "5.1.3." Unless otherwise specified, the glass cover shall be good grade, clear, flat-drawn sheet glass, free of bubbles or other imperfections. Typically, "single strength" glass that is 2 to 2.5 mm thick is used. (21). The International Organization for Standardization (ISO) 10977:1993(E) (22) cited JIS Standard JIS Z8902-1984 which requires the use of 6 mm thick glass (23).

ASTM Standard G 24–97 Standard Practice for Conducting Exposure to Daylight Filtered through Glass also reports, "at 320 nm the percent transmission of seven different lots of single thickness plate glass tested ranged from 8.4% to 26.8%" (21). This variation can make a significant difference in the results obtained using this type of glass.

Conclusion

The terms "solar-glass," "window-glass," and "an appropriate filter" are terms, used in photostability testing and appropriate to pharmaceutical photostability testing. As detailed previously, there are many different types of glasses, which may be used separately or in combination to produce what is collectively termed "window-glass." For this reason, the use of a windowsill for a sample exposure surface should be avoided unless one has mapped that surface spectroradiometrically before, during, and after testing.

While this last statement may seem excessive, one must remember that the SPD of real window-glass filtered daylight can change with the time of day, weather, season, albedo, etc. Also, studies under "in-use hospital conditions or other sun lighted situation should take into consideration the types of glasses used for glazing."

"Solar-glass" is that glass which when used in combination with a xenon-lamp source will meet the requirements of the latest solar irradiance spectra, such as ASTM G 155–05a (24). The SPD produced, when used with an appropriate lamp (generally a xenon arc), should produce a SPD equivalent to that of the internationally agreed upon, and published in Commission Internationale de L'Eclairage (CIE) Publication Number 85 (25) or its associated ISO, ASTM, ANSI, and JIZ standards.

Because of the difficulty in controlling the impurities in glass, the development of a single glass filter with closely defined optical characteristics may be economically prohibitive. When one also considers the positive and negative effects of dopants on solarization and the desire by companies to differentiate products, progress toward this goal on an international scale may be very slow.

FILTERS

A filter in its simplest sense is a device to remove or attenuate a specific item. The item may be physical, chemical, or optical. In pharmaceutical photochemistry, it includes

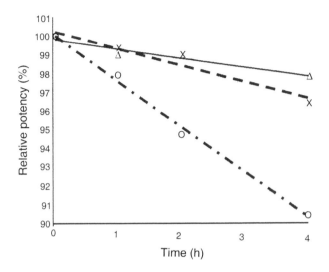

Figure 7 Loss of potency of xanomeline prototype transdermal patch after exposure to simulated sunlight. *Source*: From Ref. 27.

all of these categories. Radiation attenuators may be used to reduce certain areas of high-radiation intensity, chemical solutions to filter out unwanted bands or heat from sources, optical to remove absorb bands of UV-C or dichroic to transmit unwanted infrared while reflecting desired UV and VIS radiation. For a better understanding of the fundamental principles of filters one should refer to such catalogs as that produced by the Newport Corporation (26).

Many items around us act as filters but go undetected as such. For instance, the atmosphere around us, the bulbs surrounding the radiation sources we use, the plate glass in our windows, our eyeglasses and our clothing and skin act as filters and attenuators of radiation. It is important to be aware of all of these factors when evaluating photostability test results. Baertschi et al. addressed many of these issues in their paper "Issues in Evaluating the In-Use Photostability of Transdermal Patches" (27).

In Figure 7 we see the attenuating effect of clothing, both black and white, on the photostability of a photolabile product. This is important to note because clothing is not a 100% protector against solar radiation.

When using filters one should keep in mind that a filter's actual optical characteristic is dependent on a number of items such as, for example, the angle of incidence of the incoming beam (cosine effect), albedo, and temperature of the filter, to name just three. This is why many international standards are based on the performance of the filter in a system, not its absorption characteristics.

Absorption

Absorption filters function to remove radiation by absorbing the radiation internally. This produces heat, which can change the filters dimensions and optical properties.

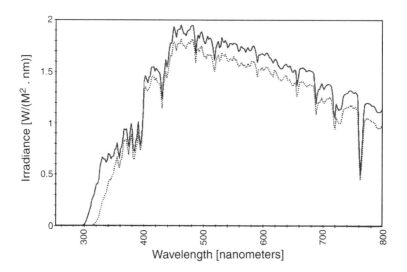

Figure 8 Plot of solar irradiance (——) in Miami, Fl versus the same irradiation measured through an Atlas window-glass filter (- - -). *Source*: From Ref. 17.

These filters are frequently used in applications requiring lower light intensities such as spectrometers, spectrophotometers, etc. Because they may age on storage (depending on their elements of composition), they should be checked on a regular basis, especially before use. The most frequent use of these filters in pharmaceutical photochemistry are as solar and window-glass filters for xenon and metal-halide lamp illuminators.

Figure 8 is displayed the SPD of solar radiation in Miami both and filtered through an Atlas window-glass filter. It is evident that because one represents D_{65} (Normal Solar Radiation) and the other the ID_{65}, standard, there is no equivalence between the two. The Expert Working Group for ICH Q1B erred in specifying the "D_{65}/ID_{65} emission standard." No such emission standard exists.

Interference Filters

Interference filters are produced when a thin transparent spacer is placed between two semireflective coatings, producing multiple reflections and interferences, which can be used to narrow the transmitted frequency band. When the spacer is a half wavelength of the desired wavelength, other wavelengths will be attenuated by destructive interference.

Dichroic

Dichroic filters, a special category of interference filter, have the unique ability to both reflect and transmit incident radiation. In this particular instance separating heat and light. These lamps generally reflect about 90% of the energy they do not transmit. Their first major application was in reducing the heat burdens produced by projector (PAR) lamps.

At least one instrument manufacturer has incorporated such a filter into an instrument to reduce the heat burden on the sample surface area.

Plastic

Plastics are used for many purposes including the exposures and protection of pharmaceutical products. Good examples of this book of these usages can be found in Chapters 10, 15, and 16 of this book.

Figure 9 presents the UV absorption spectra of many of the plastics used for packaging. Some of these plastics are used as UV filters. The ICH document allows for the use of a plastic film to cover samples during testing but gives no recommendations as to which are acceptable. Baertschi and Thatcher as reported in Chapter 10 of this book and Joshi et al. (28) studied the problem of some plastic films usable for photostability testing. The conclusion of both studies was that the 0.02 mm polyethylene films were the least UV absorptive.

One frequently neglected filter is the source filter (diffuser or cover). The effect of these devices are frequently neglected when making lamp measurements. Ina publication by Cole (29), the effect of a typical K-12 acrylic filter on the SPD of a warm-white fluorescent lamp is clearly illustrated. All wavelengths below 360 and above 650 nm are essentially filtered out. While some labs and offices may still use diffusers, many have been eliminated in favor of "egg-crate" types to compensate for use of the new lower wattage, Energy Act compliant lamps.

Some of the new lower wattage lamps may also be coated for safety purposes with clear plastic which may be absorptive and effect results. One of the problems with using bare fluorescent lamps is that they do not filter out all incident UV radiation used to make the phosphor fluoresce. These lamps are made with the

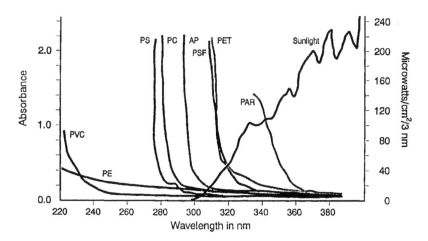

Figure 9 Ultraviolet spectral absorption of 2 mil polymer films and spectral irradiance of sunlight. *Abbreviations*: PVC, Polyvinyl chloride; PE, polyethylene; PS, polystyrene; PC, polycarbonate; AP, aromatic polyester; PSF, polysulfone; PET, poly(ethylene terephthalate); PAR, polyacrylate. *Source*: From Ref. 16.

Figure 10 Bank of 4 ft cool-white fluorescent lamps in banks of six lights, all placed with the label on one side of the bank and with adjacent banks alternated top to bottom. Figure illustrates the variation of brightness along the length of long fluorescent tubes. *Source*: Joinsill and Sanofi-Winthrop.

thinnest possible phosphor coating because the phosphor is the most expensive item in bulb construction. There is generally a 20% variation in the phosphor layer thickness from one end to the other for large fluorescent tubes, which also contributes to the possibility of UV leakage.

Figure 10 is a photograph of a large portion of a wall of 4 ft fluorescent cool-white lamps, two high, set up in banks of six bulbs each, with adjacent banks rotated 180° to one another, used for photostability studies. These lamps were set up in banks of six and their label side was always on the same side of the fixture, to compensate for coating uniformity differences. The differences in intensity along the length of these lamps are readily evident in this figure.

The problem with coating uniformity, especially for the long lamps, is well known in the industry. The variation in coating thickness along the length of the lamp is reported to be about 20% (obtained through private communication). This proves to be no problem for the commercial market but for scientific purposes should be of concern.

Figure 11, presents some transmission curves of the typical filters offered by one manufacturer for their equipment. Most manufacturers offer similar varieties of filters.

Figure 12 shows the effect of filter thickness on the resultant SPD water-cooled, quartz jacketed xenon arc lamp. As expected the UV cut-off point is shifted to longer wavelengths with increasing filter thickness.

CONTAINERS

Containers play an important role in all of science. Sample containers have been discussed elsewhere in this book, namely, Chapters 10, 15, and 16. This section will

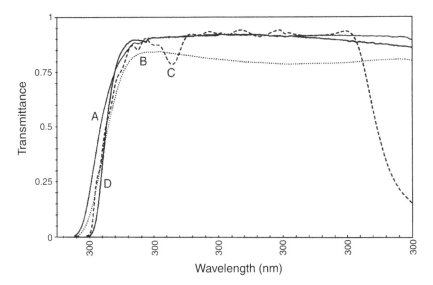

Figure 11 Typical filters used in an Atlas Sunchex air-cooled exposure chamber. (A) 280 nm, (B) 300 nm, (C) infrared, (D) window-glass equivalent. *Source*: From Ref. 17.

deal mainly with all types of containers including ampoules, vials, bottle, and lamp bulbs (containers) and their effects on the spectral power distributions of lamps.

As previously stated, lamp impurities can affect the absorption spectra of glass. Ubiquitous of all impurities is iron. The major difference between European and U.S. glass is due to native iron contents.

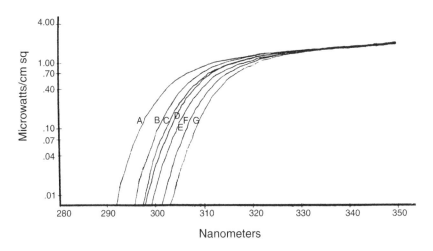

Figure 12 Atlas xenon with quartz water jackets filtered through varying thicknesses of Schott WG 320 filters. *Source*: From Ref. 28. (A) 0.64 mm, (B) 1.00 mm, (C) 1.30 mm, (D) 1.70 mm, (E) 2.00 mm, (F) 3.00 mm, (G) 4.00 mm. *Source*: From Ref. 17.

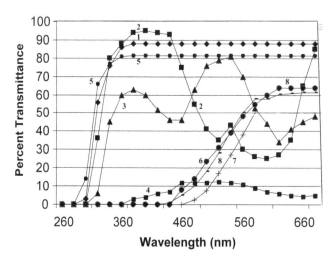

Figure 13 Spectrophotometric transmission curves of colored glass containers studied by Taub and Steinberg. (1) Flint A: 2.8 mm; (2) Blue A: 2.5 mm; (3) Green A: 2.8 mm; (4) Green B: 2.5 mm; (5) Green C: 2.5 mm; (6) Amber A: 2.4 mm; (7) Amber B: 2.5 mm; (8) Amber C: 2.4 mm. Source: From Ref. 29.

We saw in Figure 4 the differences caused by the two different sources of glass on the SPD of lamps of otherwise identical composition. The UV region is most affected. Because of this lower UV content, until most recently, European lamps have been preferred for lighting photolabile objects in museums, libraries, etc. (Private Communication).

Figure 13 shows some of the first reported spectra of some of the glasses used for pharmaceutical photostability testing and packaging (30). Early on, the value of colored glasses in protecting photolabile drugs and drug products was recognized.

Currently, the two most popular forms of glass packages are clear and amber glass. The use of amber glass is generally avoided whenever possible because of the inspection problems it poses. The official compendia specifications pay little attention to specifying just what the real spectral characteristics of each type of glass should be. Their current specifications should be updated to include the absorption spectra of acceptable glasses.

Figures 12–15 shows some of the glasses that have been used and reported in the scientific literature (31–36) particularly the one vial PJA used for the ICH Q1B quinine actinometer. Large variations in glass wall thicknesses have been reported. This can affect the amount of protection provided by a container. The point of container thickness variation was emphasized by Thoma and Aman, in Chapter 15 of this book.

SUMMARY

Glass plays an important role in everyday life and pharmaceutical photostability testing. Appreciation of its characteristics allows pharmaceutical photostability

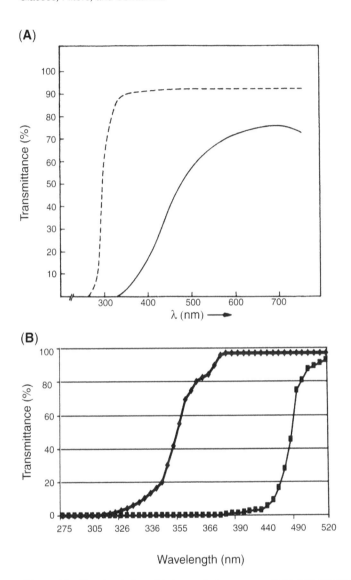

Figure 14 Spectrophotometric transmission curves of German pharmaceutical glass. (-- -- --) white glass, (——) brown glass. Spectrophotometric transmission curves of U.K. pharmaceutical glass (31). (**A**) Clear glass, (**B**) Amber glass. *Source*: From Ref. 30.

chemists to gain a true insight into the intrinsic properties of their drug substances and products. Such information allows for the identification of potential problems, development of safer products, choosing of the most effective and economical protective method, and recognition of other potential uses (photoactivatable drugs).

(A)

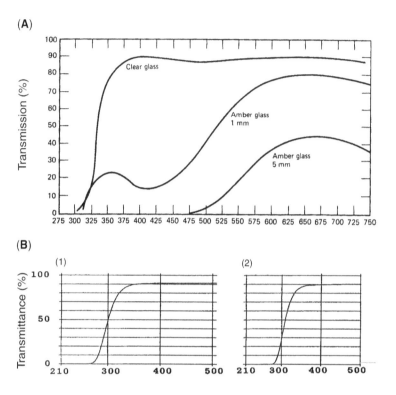

(B)

Figure 15 (**A**) Spectral transmission curves showing effect of glass type and thickness. (**B**) Transmittance curves of ampoules (*B1*) and vials (*B2*) used in original and subsequent Japanese photostability studies. (*B1*) 20 mL colorless ampoules (JIS R3512, No. 5), (*B2*) 15 mL colorless vials (N-16, Maruemu, Osaka). *Source*: From Ref. 5.

REFERENCES

1. International Conference on Harmonization (ICH) Guideline for the photostability testing of new drug substance and products, Federal Register 1997; 62:27115–27122.
2. Rindone GE. The formation of color centers in glass by solar radiation. In Travaux du IV Congress International du Verre, Paris; 1956, VII$_4$, 373–389.
3. White JF, Silverman WB. Some studies on the solarization of glass. J Amer Cer Soc 1950; 33(8):252–257.
4. Swarts EL, Cook IM. Photosensitivity of reduced silicate glasses containing iron. 7th International Conference on Glass, June 26–July 3. 1965.
5. Starkle D, Turner WES. The influence of ferric oxide content on the light transmission of soda-lime-silica glass with special reference to the ultra-violet. J Soc Glass Tech 1929; 324–332. Read at the Bournemouth Meeting Sept. 21, 1928.
6. Boettner EA, Miedler LJ. Transmittance changes in glasses induced by ultraviolet radiation. J Opt Soc Amer 1961; 51:1310–1311
7. An overview of flashlamps and cw arc technology, Technical Bulletin 3. ILC Technology, Sunnyvale, CA.
8. Bergman RS, Parham TG, McGowan TK. UV emissions of general lighting lamps. IES Conference, Miami August 7–11, 1994.

9. Audin, L. "Ultraviolet Light Output from High Efficiency Light Sources: Concerns, Field Measurements and Analysis," World Energy Engineering Conference in Atlanta, GA, October 26, 1993.

10. ANSI/IESNA RP-27.1-96 Photobiological Safety for Lamps and Lamp Systems—General Requirements. American National Standards Institute, 1819 L Street, NW, Suite 600, Washington, DC 20036.

11. ANSI/IESNA RP-27.2-00 Photobiological Safety for Lamps and Lamp Systems—Measurement Systems. American National Standards Institute, 1819 L Street, NW, Suite 600, Washington, DC 20036.

12. ANSI/IESNA RP-27.3-96 Photobiological Safety for Lamps - Risk Group Classification and Labeling. American National Standards Institute, 1819 L Street, NW, Suite 600, Washington, DC 20036.

13. Energy Policy Act of 1992, U.S. Congress, PL 102–486. Washington, DC: U.S. Government Printing Office.

14. Ketola WD, Skogland TS, Fischer RM. Effects of Filter and Burner Aging on the Spectral Power Distributions of Xenon Arc Lamps, In: Durability Testing on Nonmetallic Materials, ASTM STP 1294, ed. By Robert J Herling. American Society for Testing and Materials, Philadelphia, PA, 1995.

15. Ketola WD and Robbins JS. UV Transmission of Single Strength Window Glass. In: Accelerated and Outdoor Durability Testing of Organic Materials, ASTM STP 1202, Warren D Ketola. and Douglass Grossman, eds. American Society for Testing Materials, Philadelphia, PA, 1993.

16. Searle NZ, Giesecke P, Kinmonth R, Hirt, RC. Ultraviolet Spectral Distributions and Aging Characteristics of Xenon Arcs and Filters. Appl Opt, 1964; 3(8):923–927.

17. Atlas Material Testing Technology LLC, 4114 N, Ravenswood Ave., Chicago, IL 60613.

18. Q-Panel Lab Products, 800 Canterbury Road, Cleveland OH 44145.

19. Mortensen P. Light-Reactive Coating Cleans Surfaces. Photonics Spectra 2001; 38.

20. Asahi R, Morikawa T, Ohwaki T, Aoki k, Taga Y. Visible-Light photocatalysis in nitrogen-doped titanium oxides. Science 2001; 283:269–271.

21. ASTM G 24-97 Standard Practice for Conducting Exposures to Daylight Filtered through Glass. 1–4, published by American Society for Testing and Materials, 100 Barr Harbor Dr., West Conshohocken, PA 19428.

22. ISO 10977:1993(E), Photography Processed Photographic Colour Films & transmittance, solar direct, total solar energy transmittance and ultraviolet transmittance, and related glazing factors. International Organization for Standardization, Case Postale 56, CH 1211 Genève, Switzerland.

23. JIS Z8902-1984, Xenon Standard White Light Source. Japanese Industrial Standards Committee (JISC), 1-3-1 Kasumigaseki, Chiyoda-ku, Tokyo 100-8901, Japan.

24. ASTM G 155-05a Standard Practice for Operating Arc Light Apparatus for Exposure of Non-Metallic Materials. Published by American Society for Testing and Materials, 100 Barr Harbor Dr., West Conshohocken, PA 19428.

25. CIE, Publication Number 85. Commission Internationale de L Eclairage (CIE), Bureau Central de la CIE, 4 Av. du Recteur Poincaré, 75 Paris 16, France.

26. Newport Corporation. Optics & Filters. Stratford, CT 06615.

27. Baertschi SW, Kinney H, Snider B. Issues on evaluating the in-use photostability of transdermal patches. Pharm Tech 2000; 24:70–80.

28. Joshi HM, Formicola S, Pazdan JL, Carpenter P. Photostability and transmittance of four commercially available plastic wraps. Pharm Tech 2000; 24(4):74–129.

29. Cole C, Forbes PD, Davies R, Urbach F. Effect of indoor lighting on normal skin. Ann NY Acad Sci 1985; 453:305–316.

30. Arny HV, Taub A, Steinberg A. Deterioration of certain medicaments under the influence of light. J Am Pharm Assoc 1931; 20(10):1014–1023.

31. Thoma K. Arzneimittelstabilität. Eigenverlag, Frankfurt, Germany, 1980:114.

32. Bhadresa B, Sudgen JK. Light Transmission through Amber Glass Medicine Bottles.

33. Mendenhall DW. Stability of parenterals. Drug Dev Ind Pharm 1984; 10(8 and 9): 1297–1342.

34. Oxidation and Photolysis. In: Chemical Stability of Pharmaceuticals: A Handbook for Pharmacists. Kenneth A, Connors, GL Amidon, VJ Stella, eds. NY: Wiley, 1986:113.
35. Yoshioka S, Ishihara Y, Terazono T, et al. Quinine actinometry as a method for calibrating ultraviolet radiation intensity in light stability testing of pharmaceuticals. Drug Dev Ind Pharm 1994; 20(13):2049–2062.
36. Porter WR, Bauwens SF. Variation in light transmission properties of amber oral syringes. Amer J Hosp Pharm 1986; 43:913–916.

8   Chemical Actinometry

Günter Gauglitz
*Institut für Physikalische und Theoretische Chemie, Universität Tübingen,
Tübingen, Germany*

Stephan M. Hubig
*Department of Chemistry, University of Houston,
Houston, Texas, U.S.A.*

INTRODUCTION

The photodegradation of drugs is one of the main problems in connection with storage of chemicals and especially the availability of drugs. Exposure to radiation, as well as elevated temperatures, causes photodegradation reactions to occur which reduce the quality of drugs over time. For this reason, not only a qualitative information on the photostability is desirable, but also detailed knowledge on the quantitative aspects of the photoreactions is essential to determine time scales at which drugs will decompose. In principle, any chemical substance can be exposed to radiation and its overall change in concentration during the photochemical reaction can readily be determined. However, detailed quantitative examinations require more than measurements of the photodegradation turnover. Energy of radiation or intensity of a radiation source is correlated to the number of photons emitted at each particular wavelength of the radiation. For this reason, the simple statement that light will degrade chemicals at a certain rate does not lead to reproducible results.

As will be discussed later, photochemical reactions depend on the wavelength of the radiation, the number of photons absorbed (not incident photons), and some other parameters, which may affect the amount of absorption of the sample. Thus, to be able to give normalized information on the photostability of any drug, certain prerequisites for the determination of photodecomposition rates are essential. For this reason, in the following section, various parameters such as the photochemical quantum yield, the amount of light absorbed, primary photochemical processes, and some fundamental photokinetic equations will be discussed. First, basic parameters are defined with respect to their units. Then, the determination of radiation by physical measurements is described, and the advantages and disadvantages of chemical actinometry as compared to the photophysical methods are reviewed. Thus, we will demonstrate the necessity of chemical actinometers for easy and accurate light-intensity measurements.

The determination of photochemical quantum yields is not a simple task, and, in some cases, approximations are required. Nevertheless, according to the parameters chosen, various well-known photochemical reactions can be used to measure irradiance, which is an essential quantity in the field of photokinetics. Finally, some selected chemical actinometers will be discussed with respect to their pros and cons and their best areas of application. At the end, special applications of actinometry such as measurements of polychromatic light and high-intensity light sources (lasers) will be described. The overall aim of this chapter is to help the reader to choose the best actinometers out of the numerous examples in the literature and avoid technical mistakes.

MEASUREMENT OF LIGHT FLUX (INTENSITY)

Physical Methods

A common practice for the determination of the intensity of irradiation is the use of physical devices that are based on either the internal or the external photo effect. The internal photo effect is exploited in semiconducting materials. Thus, photons absorbed by these materials cause a charge transfer from the valence band to the conducting band of the semiconductor. The resultant increase in conductivity is measured and is related to the number of photons impinging on the surface of the photodiode. Many types of photodiodes are known, the specific properties of which are tailored to their particular applications (for details, see Chapter 9). Thus, different sizes, response times, and signal-to-noise levels have been achieved. Moreover, such devices are inexpensive; however, as will be discussed later, they

- need frequent recalibration,
- exhibit wavelength dependency,
- show inhomogeneous spatial distribution of the sensitivity on the surface, and
- can be damaged by high-intensity UV irradiation.

The same problems apply to photomultipliers, which are based on the external photoelectric effect. A somewhat different type of a doped semiconducting material is used. Photons hit the surface of a photocathode. If their energy is high enough, electrons of the material take up the energy, overcoming the Coulombic work term. Thus, the electrons are able to leave the material and accelerate toward a dynode, which has a slightly more positive potential. This type of scintillation procedure leads to a multiplication of the number of electrons, depending on the number of dynodes present. Normally, photomultipliers consist of more than 10 dynodes ("electrodes" between the photocathode and the final anode) supplying an internal multiplication effect similar to that of internal preamplifiers. Thus, at the final electrode (anode), a self-amplified signal is obtained. Besides the drawbacks mentioned above for photodiodes, additional problems result from thermal electrons being amplified as well, which reduces the signal-to-noise ratio.

Chemical Methods

Because the amount of decomposition of a well-characterized photochemical reaction depends on the photochemical quantum yield (a constant) and the amount of radiation absorbed by the sample, the intensity of unknown irradiation sources can be determined very accurately by measuring the amount of decomposition, if the quantum yield of the photoreaction is known. Among the large number of well-characterized photoreactions, only a few are suitable for actinometry. Thus, only photoreactions with very simple mechanisms are less sensitive to the experimental conditions of the irradiation. Well-defined experimental conditions and easy monitoring are important requirements for a suitable actinometric system if reproducible results are to be obtained. Examples of acceptable reaction types include photodegradations, photoisomerizations, photooxidation, etc., as discussed later in detail (vide infra).

An analogy to the rate constant in a thermal reaction is the photochemical quantum yield that defines the rate of a photoreaction. This quantity is not always constant, and is dependent on wavelength and—in some cases—on irradiation intensity, which may critically affect the convenient use of a photoreaction as a chemical actinometer. Thus, a well-defined correlation between the measured amount of decomposition and the number of photons absorbed is a prerequisite.

Comparison of Physical and Chemical Methods

Physical and chemical methods exhibit various advantages and disadvantages. Even though physical methods are generally rather easy to use for measurements of relative intensities, they may cause some problems to the photochemist:

- High-intensity UV irradiation damages the detector surface (vide supra), which needs to be taken into account when high-intensity sources such as lasers are measured.
- Nonhomogeneous spatial distribution of the sensitivity over the detector surface causes geometric problems when broad or diverging light beams are examined.
- The linear dynamic range is limited.
- Multiple reflections in solutions or thin layers influence the effective intensity of such samples.
- The sensitivity of physical devices depends on the wavelength.
- In general, the sensitivity is very low in the UV region.
- Correction factors as well as adjustment plots in operation manuals for physical devices provide an accuracy of less than ±0.5%.
- Calibration curves lose their validity during use due to photodegradation of the detector-head material as well as aging of the electronics.
- Absolute determinations of radiant flux densities or radiant fluxes need frequent and tedious recalibration.

Consequently, a number of photoreactions have been proposed in recent years as chemical actinometer systems, which exhibit the following advantages:

- Determination of absolute light flux (number of photons incident onto the sample).
- Inexpensive detection systems, which are easily replaced in case of damage by irradiation.
- Reusability, in the case of (reversible) photochromic systems.
- Suitability for photochemical investigations, because the actinometric system is easily replaced by the sample of interest without changing the irradiation geometry or other experimental conditions. (This is especially valid for liquid samples.)
- Immediate availability and needs no calibration.

On the other hand, the use of photochemical systems as actinometers requires greater expenditures of manpower and equipment than the simple application of a photodiode.

PHOTOKINETIC PRINCIPLES

Basic Photokinetic Equations

For the most simple photoreaction

$$A \xrightarrow{h\nu} B$$

the change in concentration of the educt is given by the following differential equation (concentration of reactants symbolized by small letters):

$$\frac{da}{dt} = \dot{a} = -\Phi \cdot I'_{absA} \tag{1}$$

In this equation, Φ represents the photochemical quantum yield and I'_{absA} the amount of light absorbed by the educt. Thus, the quantum yield is a measure for the ratio between the change of concentration of the reactant of interest and the number of incident photons that are absorbed by the reactant during the measuring time. This definition has caused some inconvenience in the past. Thus, to simplify the evaluation of this most simple-looking equation, integrations of the intensity of the radiation source over time have been undertaken and compared with the change in concentration of the analate during the same time period. However, such evaluation procedures merely provide a so-called "apparent quantum yield," which does not represent a constant quantity and thus cannot be used for the desired quantitative determinations.

First, we note that the number of photons absorbed rather than the number of incident photons has to be taken into account. Second, integrations over extended time periods most likely bear substantial errors because the intensity of the source may fluctuate or drift. As a consequence of this, the only exact measure for the efficiency of a photochemical reaction is the true differential quantum yield, which needs to be determined for each step of the reaction. Similar to thermal reactions, photochemical reactions may be complex. Accordingly, the only correct measure is the so-called partial (true differential photochemical) quantum yield, which is defined for each linearly independent step of the reaction.

In kinetic treatments and in particular in photokinetics, the degree of change of the reaction is preferably used instead of the change of concentration with time, and it is defined for each partial reaction as x_k. This degree of change corresponds to the stoichiometric change of concentration with respect to any reactant. If this quantity is included in the definition of the photochemical quantum yield, even reactions with a complex mechanism can be evaluated quantitatively and accurately. Using this approach, the quantum yield is independent of the moment at which it is measured, and—together with the absorption coefficients at the wavelength of irradiation—it provides a good measure of the change even in complex reactions. Thus, the most accurate definition of the quantum yield is depicted as follows:

$$\varphi_k^A = \dot{x}_k / I'_{absA} \tag{2}$$

where the index A symbolizes the reactant, which starts the photoreaction, and k refers to the partial steps of the reaction (an equilibrium reaction results in two partial reactions).

Amount of Radiation Absorbed

The amount of radiation absorbed is a rather complex quantity, which is not at all constant as frequently considered in photokinetic examinations. It is based on Lambert–Beer's law, which gives the exponential dependence of the incident to the

transmitted intensity to path length. This law requires various assumptions, which include that

- the incident light beam is parallel and perpendicular to the front surface of the sample,
- the incident radiation is homogeneously distributed over the entire front area of the sample,
- the sample does not scatter the light, and
- the back wall of the sample does not reflect.

Using an absorption coefficient ε_λ at the wavelength λ, the incident radiation $I_0(\lambda)$ is defined as

$$I_d(\lambda) = I_0(\lambda) \cdot 10^{-\varepsilon_\lambda \cdot a \cdot d} \tag{3}$$

The fact that the absorption coefficient depends on the wavelength critically affects the determination of the amount of intensity absorbed. In addition, Lambert–Beer's law is limited to dilute solutions. In Eq. (3) only the absorption of the reactant A is considered. However, the decrease in intensity is caused by all light-absorbing components in the solution. For this reason, the amount of light absorbed by all reactants in the solution is given by

$$I'_{abs} = I_0 - I_d = I_0(1 - 10^{-A'}) \tag{4}$$

and I'_{abs} is given in units of $mol\,photons\,cm^{-2}s^{-1}$. For abbreviation the symbol' is used to show that the intensity is measured at the wavelength of irradiation. Thus, the absorbance A' at the wavelength of irradiation λ' represents the overall absorbance at this wavelength according to

$$A' = d \cdot \sum \varepsilon'_i \cdot a_i \tag{5}$$

In general, only one component is photoexcited and initiates the photochemical process. Therefore, its absorbance $(d\varepsilon'_A a)$ needs to be compared with the total absorbance (A'), and—as shown elsewhere (1)—the amount of light absorbed is defined as

$$I'_{absA} = I_0(1 - 10^{-A'}) \cdot \varepsilon'_A \cdot a / A' \tag{6}$$

from which the factor

$$F(t) = \frac{1 - 10^{-A'(t)}}{A'(t)} \tag{7}$$

is extracted as a time-dependent reaction parameter. This factor takes into account that the concentrations of the reactants and thus the absorbance of the solution will vary during the photoreaction. Therefore, the rate of a photoreaction does not linearly depend on the concentration of the reactants as observed in thermal reactions, and thus there is no constant proportionality factor.

Finally, it has to be taken into account, that in all equations the area of irradiation is given in units of cm^2, whereas the concentrations are given in units of $mol\,L^{-1}$. Accordingly, a factor of 1000 is introduced for the conversion from liters to cm³. Thus, the final differential equation for a simple photochemical reaction is given by

$$\dot{a} = -\varphi_1^A \cdot I'_{absA} = -1000 \cdot I_0 \cdot \varphi_1^A \cdot \varepsilon'_A \cdot a(t) \cdot F(t) \tag{8}$$

This equation is the basis of all actinometric measurements and needs to be adapted according to the particular mechanisms and irradiation conditions of the photoreactions in the various actinometers. Such adjustments may be rather difficult. In most applications either photoisomerization or photooxidation reactions are used. The details of which are described in section "Special Applications" of this chapter and various approximations are applied as follows.

Approximations

A successful approximation for Eq. (8) is the introduction of total (complete) absorption. Under such conditions, the photokinetic factor $F(t)$ is replaced by A'^{-1}, and Eq. (8) reduces to

$$\dot{a} = -1000 \cdot I_0 \cdot \varphi_1^A \cdot 1/d \tag{9}$$

Because of this, the rate becomes time independent if only the educt absorbs. An additional approximation is based on the fact that at the beginning of the reaction the amount of products and thus their absorption is negligible. Therefore, for small conversions, only the educt will contribute to the overall absorption of the solution.

A different approximation is based on the use of very dilute solutions in which the absorbance is smaller than 0.02 units ($A' \leq 0.02$). In this case, the photokinetic factor is considered constant as an expansion into series. Accordingly, the factor exhibits a value of 2.303, and the rate depends on the concentration in a similar way as observed in first-order (thermal) reactions. Assuming a partial absorption during the photoreaction and taking into account that the photoproducts will also absorb at the actinic wavelength, the photokinetic equations become more complicated. There are a large number of such different equations, each tailored to a specific problem, as demonstrated in the various examples in the following sections.

Monitoring of Actinometer Response: Spectrophotometric Analysis

There are various approaches to determine the percent of decomposition of a photochemical reaction. Although most analytical methods measure the overall change in concentration, after a certain period of irradiation, by means of conventional instruments (GC, HPLC) (2), other methodologies, based on changes in the ultraviolet or visible wavelength region as evaluated by photometric methods (3–6) make it possible to continuously monitor the photoreaction.

Combined irradiation and measurement devices (7) have been constructed, which permit actinometric measurements and photokinetic experiments to be carried out at the same cuvette position. [One of these devices is available commercially (8).] Such experimental setups allow a time-resolved monitoring of the photoreaction (spectrally or at a single wavelength) and thus represent the most accurate way to perform actinometric measurements. However, only very few laboratories are equipped with such devices, and, generally, irradiation intensities are integrated over an extended time period after which the amount of photodegradation is determined in a single measurement. As mentioned above, such methods imply a variety of approximations and contain many possible sources of error.

Ideal Actinometers

Photochemical reactions are suitable for actinometry if the following conditions are:

- Simple mechanism of the photoreaction
- Quantum yield is independent of intensity
- Amount of light absorbed constant or measurable
- Mechanism and reaction independent of temperture

The progress of the reaction may be monitored by easy methods preferable utilizing UV/VIS absorption spectroscopy. The better the equipment and the more accurately defined the photoreaction are, and the more the actinometric measurement is carried out following photokinetic procedures, the more accurate is the determination of the radiation intensity using chemical actinometers. Moreover, the user should be well aware of all approximations applied, including their consequences.

Upon careful examination of an actinometric reaction, boundary conditions are frequently found at which the reaction can be evaluated using rather simple equations. For example, under certain conditions a linear relationship between the irradiation intensity and the change in photoproduct concentration over time is found as depicted in the general equation below:

$$I_0 = F \cdot (\Delta a / \Delta t) = W \cdot (\Delta A / \Delta t) \tag{10}$$

where F or W are factors based on quantum yields, absorption coefficients, and other parameters. These factors need to be calibrated for each actinometer and for specific well-defined conditions.

In the case where polychromatic radiation sources are used, the photochemical quantum yields should be independent of the wavelength. However, this is a condition that is very difficult to achieve, and requires specific conditions that are discussed in the following sections.

SELECTED CHEMICAL ACTINOMETERS

The requirements for chemical actinometers, listed in the previous sections [such as sensitivity, reproducibility, (thermal) stability, ease of analytical procedure, etc.], reduce the seemingly unlimited number of actinometers reported over the years (9) to a short list of well-established and highly recommended chemical actinometers. In this section, we first present a selected list of such reliable actinometers, which all operate in the liquid phase. After that, a few controversial systems are described and their potential error sources discussed. Finally, actinometric procedures suitable for solid-state applications, polychromatic sources, and laser irradiation will be introduced in section "Special Applications."

The Classic Ferrioxalate Actinometer

The potassium ferrioxalate actinometer developed by Hatchard et al. (10,11) in the 1950s is probably the most widely used and most thoroughly investigated solution-phase actinometer [Ref. (9) and references therein]. Irradiation of an aqueous solution (0.006–0.15 M) of $K_3Fe(C_2O_4)_3 \cdot 3H_2O$ with radiation between 250 and 470 nm (vide infra) results in a two-step photoreduction of iron (III) to iron (II) with quantum yields higher than unity, i.e.,

$$Fe^{III}(C_2O_4)_3^{3-} \xrightarrow{h\nu} Fe^{2+} + C_2O_4^- + 2C_2O_4^{2-} \tag{11}$$

$$Fe^{III}(C_2O_4)_3^{3-} + C_2O_4^- \longrightarrow Fe^{2+} + 2CO_2 + 3C_2O_4^{2-} \tag{12}$$

The photogenerated iron (II) is analyzed quantitatively as its red-colored ($\varepsilon_{510} = 1.11 \times 10^4 \, M^{-1} cm^{-1}$) 1,10-phenanthroline complex. A wavelength-independent photochemical quantum yield of $\Phi = 1.24 \pm 0.02$ is found for the irradiation range between 250 and 366 nm. In this wavelength region, the actinometer is highly sensitive. The wavelength-independent quantum yield allows measurements of broadband (polychromatic) UV/VIS light sources. Actinometric measurements are highly reproducible and reliable, even though being somewhat tedious and requiring time-consuming analytical procedure (complexation reaction under red light over an extended period of time) when carried out carefully following the recommended guidelines. [The reliability of various alternative actinometric procedures for ferrioxalate is discussed in Ref. (9) and references therein].

For wavelengths in the visible range, the photochemical quantum yield Φ for the photoreduction of iron (III) decreases somewhat, reaching a value of 0.93 at 468 nm. Beyond 470 nm, the reliability of the ferrioxalate actinometer seems questionable, owing to the fact that in this wavelength region even a 0.15 M actinometer solution (with a path length of $d = 1$ cm) does not completely absorb the incident radiation. Although a correction for the fraction of radiation absorbed has been suggested in such cases, the validity of quantum yields determined under these conditions is controversial because the lack of complete absorption of the actinic light strongly affects the linearity of the concentration/time diagrams utilized in the photokinetic evaluation of the actinometric measurement (12).

In summary, ferrioxalate is a highly reliable liquid-phase actinometer for the 250 to 470 nm wavelength region. It also can reliably measure the intensities of low-energy UV laser lines (see section "Special Applications"). However, its tedious, time-consuming analytical procedure renders it less attractive than other chemical actinometers, which have been developed more recently, for the same wavelength range (vide infra). Because of its well-established quantum yields, it is still highly recommended as a standard for the calibration of new (easier-to-use) chemical actinometers.

The Azobenzene Actinometer

The reversible E/Z photoisomerization of azobenzene upon UV/VIS irradiation, in methanolic solution, has been utilized to develop the first reusable chemical actinometer for the UV lines (between 250 and 450 nm) of mercury arc lamps (13–16), i.e.,

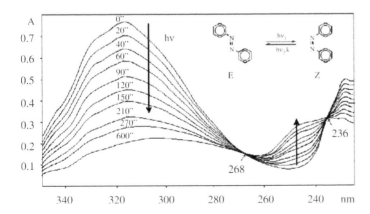

$$\text{(13)}$$

In practice, the actinometric measurements are carried out using a sealed, 1-cm quartz cuvette containing a 6×10^{-4} M solution of azobenzene (3 mL) in methanol under nitrogen atmosphere. The analytical procedure consists of making a series of spectroscopic measurements, using a UV/VIS spectrophotometer, to follow the irradiation dosage. The handling of solvents, pipetting, or other chemical procedures is not required. The reversibility of the isomerization, as indicated in Eq. (13), allows the original E-isomer actinometers to be regenerated by "back irradiation" (hv_2) or by thermal (k) $Z \rightarrow E$ isomerization. Thus, these actinometers are reusable.

Owing to the different absorption spectra of the E- and the Z-isomer of azobenzene (Fig. 1), the irradiation and evaluation procedures depend on the irradiation wavelength as follows:

1. In the 275–340 nm wavelength range, the concentrated solution of E-azobenzene absorbs the incident radiation completely (100% absorption). Thus the $E \rightarrow Z$ isomerization, with a quantum yield of $\Phi = 0.14$ (13,15), can be monitored spectroscopically by measuring the decrease in absorbance at 358 nm (ΔA_{358}) as a function of the irradiation time (t). Because under the conditions of complete absorption a linear correlation between ΔA_{358} and the irradiation time intervals (Δt) results, the intensity, I_0 (in Einstein cm^{-2} s^{-1}), of incident radiation is readily obtained using the following simple relationship (15):

$$I_0 = W \cdot (\Delta A_{358} / \Delta t) \tag{14}$$

where $W = 4.6 \times 10^{-6}$, 4.6×10^{-6}, 5.3×10^{-6}, and 3.6×10^{-6} Einstein cm^{-2} for the irradiation wavelengths (mercury lines) 280, 302, 313, and 334 nm, respectively (15).

Figure 1 Photoreaction of azobenzene. *Abbreviations*: hv, E, Z. E and Z isomers are shown in the figure (different structure).

2.   At 254 nm, the Z-isomer of azobenzene absorbs more strongly than the E-isomer (Fig. 1). Thus, for actinometric measurements of the 254 nm mercury line, the azobenzene actinometers are first preirradiated at 313 nm until a photostationary state (containing mostly Z-isomer) is obtained. The preirradiated actinometers are then exposed to the actinic radiation at 254 nm and the $Z \rightarrow E$ isomerization, with a quantum yield of $\Phi = 0.31$, is monitored at 358 nm. The light intensity I_0 (at 254 nm) is obtained using Eq. (14), where $W = 2.3 \times 10^{-6}$ Einstein cm^{-2} (15).

3.   The $Z \rightarrow E$ isomerization can also be utilized for actinometric measurements in the wavelength range between 370 and 500 nm owing to the relative absorptions of the Z- and E-isomer in this spectral region (Fig. 1). Preirradiation to a photostationary state (vide supra), however, is required. A different quantum yield, $\Phi \approx 0.6$, has been found (15) for this actinic wavelength range. The concentrated (6×10^{-4} M) azobenzene solution does not provide complete absorption of the incident radiation under these conditions, and thus the simple (linear) evaluation method using Eq. (14) is no longer applicable. Instead, a more sophisticated (graphical or arithmetical) kinetic evaluation suitable for conditions of partial absorption is recommended (13).

In summary, the azobenzene actinometer provides several features that are of great advantage for routine measurements of the UV lamps. They are

1.   The actinometer solution is stored in a sealed cuvette and no additional procedures, chemical or physical, are applied.
2.   The spectroscopic evaluation is simple, fast, and accurate.
3.   The actinometers are reusable. In addition, reliable results even with pulsed nitrogen, excimer, and Nd:YAG lasers have been reported (see section "Special Applications").

The Meso-diphenylhelianthrene Actinometer

This actinometer for visible radiation, between 475 and 610 nm, is based on the self-sensitized photooxidation of meso-diphenylhelianthrene (MDH) to its endoperoxide in aerated toluene solution (17).

The quantum yield of $\Phi = 0.22$ (18) for the formation of the endoperoxide in Eq. (15) is independent of the irradiation wavelength over the entire recommended wavelength range. The photoreaction is easily monitored spectroscopically by monitoring the growth of the endoperoxide at 429 nm (Fig. 2).

$$+ \; O_2 \quad \xrightarrow{h\nu} \qquad\qquad\qquad\qquad\qquad\qquad (15)$$

Under conditions of the complete absorption of the actinic radiation (MDH concentration $> 10^{-3}$ M), the increase in the absorbance at 429 nm (ΔA_{429}) is found to be

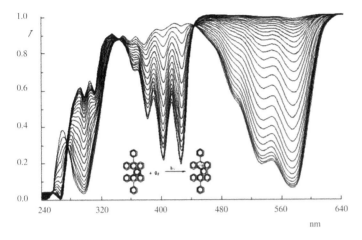

Figure 2 Photoreaction of meso-diphenylhelianthrene.

linearly related to the irradiation time [Δt, see Eq. (16)]. Thus, radiation intensities I_0, are readily obtained based on Eq. (6) with a calibration factor of $F(429)=4.1\times10^6\,\mathrm{M^{-1}cm^{-1}}$, which has been determined by three independent methods (17).

$$I_0[\text{Einstein cm}^{-2}\text{s}^{-1}] = F(429)\cdot(\Delta A_{429}/\Delta t) \tag{16}$$

The MDH actinometer is a highly reliable actinometer for visible radiation between 475 and 610 nm and may also be used as a quantum counter for polychromatic radiation sources in this wavelength region, owing to its wavelength-independent quantum yield. The photometric evaluation at 429 nm is fast, simple, highly accurate, and reproducible, if a freshly prepared actinometric solution (or the solution stored at $-15°C$, in the dark, for no longer than three months) is used. The MDH actinometers are not suitable for measurements of high-intensity pulsed lasers owing to the limiting amount ($\approx 2\,\text{mM}$) of oxygen that is available even in oxygen-saturated toluene solutions (Hubig SM, unpublished results). To extend the wavelength range of these actinometers further into the long-wavelength region, methylene blue (610–670 nm) and hexamethylindotricarbocyanine iodide (670–795 nm) have been recommended recently as sensitizers for the photooxygenation of MDH in chloroform (19,20).

The Aberchrome 540 Actinometer
The Aberchrome 540 actinometer (21) is based on the reversible photocyclization of the pale yellow furyl fulgide, **I** in Eq. (17), to its deep-red cyclized isomer (**II**), i.e.,

$$\tag{17}$$

I   **II**

For actinometric measurements, a 1cm cuvette containing a ≈5 mM solution of the furyl fulgide (Aberchrome 540) in toluene is used, which completely absorbs UV radiation between 313 and 366 nm. The color change from yellow to red due to the photoinduced (conrotatory) ring closure is monitored at 494 nm, and the resulting linear increase in absorbance (ΔA_{494}) with irradiation time (Δt) is evaluated using Eq. (18) to calculate the intensity (I_0) of the incident UV radiation, i.e.,

$$I_0[\text{Einstein s}^{-1}] = V \cdot N_A \cdot \Delta A_{494} / (\Phi \cdot \varepsilon_{494} \cdot \Delta t) \tag{18}$$

where V, N_A, and $\varepsilon_{494} = 8200$ M^{-1}cm^{-1} being the volume of the actinometer solution, Avogadro's number, and the molar absorption coefficient of the red photoproduct at 494 nm, respectively. According to Heller and Langan (21), the photoreaction in Eq. (17) occurs quantitatively with a quantum yield of $\Phi = 0.20$, as determined by ferrioxalate actinometer. Moreover, irradiation of the colored photoproduct with visible (436–546 nm) radiation restores the original (pale yellow) fulgide solution, allowing the actinometer to be used repeatedly (21).

The back irradiation of the colored product with visible radiation has also been explored as actinometer; however, its use is generally not recommended due to a strong dependence of the quantum yield on both the irradiation wavelength and the temperature (Heller HG, Langan JR, unpublished results). Recently, even the use of Aberchrome 540 as a UV actinometer has come under scrutiny. First, the repeated use of Aberchrome solutions after back irradiation (vide supra) is no longer recommended because photokinetic studies (22–24) revealed that there is an increase in the errors in the determination of I_0 with an increasing number of photocycles. These errors have been attributed to the $E \rightarrow Z$ isomerization of the fulgide, which competes with the ring closure reaction (22–24). Second, a recent reinvestigation (25) of the photokinetics of isolated Z- and E-isomers as well as the cyclized photoproduct leads to the conclusion that the photocyclization of Aberchrome 540 using 366-nm irradiation does not occur with quantitative conversion, as originally reported (21). As a consequence, the validity of the quantum yield for ring closure has been questioned, and thus a thorough recalibration of the Aberchrome 540 actinometer is advisable (25).

The Potassium Reineckate Actinometer

The potassium reineckate actinometer developed by Wegner and Adamson (26), in 1966, represent the first actinometer for long-wavelength visible radiation. It is based on a photoinduced ligand exchange in a chromium (III) complex, i.e.,

$$Cr(NCS)_4(NH_3)_2^- + n\,H_2O \longrightarrow Cr(NCS)_{4-n}(H_2O)_n(NH_3)_2 + n\,SCN^- \tag{19}$$

Thus, upon irradiation of an aqueous reineckate solution with visible radiation between 316 and 750 nm the release of one or more thiocyanate ions is observed with a quantum yield of $\Phi = 0.3$. The photogenerated thiocyanate can easily be quantified spectrophotometrically at 450 nm as its Fe^{III}(SCN)$_3$ complex. This actinometer is quite sensitive to visible light and covers a broad wavelength range. However, the quantum yield does show a wavelength dependence (27), which also varies with the temperature and the pH of the solution. Moreover, because the same thiocyanate release is also observed as a result of the thermal degradation processes (the rate constants depend on the pH of the solution), appropriate control experiments with foil wrapped samples similarly exposed need to be carried out in parallel to the actinometric measurements. Additionally, corrections may be necessary to account for inner-filter effects caused by the formation of $Fe^{III}(SCN)_3$, which also absorbs at 450 nm (vide supra).

In summary, actinometric measurements with reineckate require extreme care as these actinometers call for very tedious procedures (including controls in the dark), which add numerous potential error sources.

The Quinine Actinometer

The International Conference on Harmonization (ICH) guideline on the photostability testing of new drugs (28) introduces the photolysis of quinine hydrochloride (see formula below) as the primary actinometric procedure for monitoring exposure to near-UV light (320–400 nm) by cool-white fluorescent lamps, near-UV fluorescent lamps, and metal halide or xenon arc lamps. This recommendation is based on a study by Yoshioka et al. in 1994 (29) who studied the spectroscopic response of aqueous solutions of quinine hydrochloride at 400 nm after exposure to various photon sources typically used for the photostability testing of drugs. They reported a linear increase in absorbance (at 400 nm) of the irradiated quinine solution with increasing energy output (EO) of the lamps using a radiometer, as an actinometric reference. Accordingly, the ICH guideline (28) defines the absorbance change (ΔA_{400}) itself as a standardized measure of radiant energy (=intensity × exposure time) and accordingly recommends choosing exposure times for drug testing sufficiently long to ensure a change in absorbance of $\Delta A_{400} \geq 0.8$.

The choice of quinine hydrochloride as a standard for photostability testing in the ICH guidelines has stimulated a great deal of concern in the pharmaceutical community and has remained highly controversial. Most importantly, the photochemistry of quinine hydrochloride in water has not been investigated thoroughly. There are at least four reaction pathways known, including photoreduction resulting in deoxy-quinine in acidic media (30), photodecomposition to form methoxyquinoline and quinuclidine carboxaldehyde at neutral pH (31), photoinduced adduct formation in citrate buffer (32), and photosensitized oxidation reactions in acidic milieu, the latter being attributed to the generation of singlet oxygen via quinine sensitization (33).

Even though quantum yields are not available for any of the above photoreactions, quinine hydrochloride has nevertheless been promoted as a convenient and reliable actinometer that provides reproducible results within defined experimental conditions (34). However, the question as to whether such standard experimental conditions can be clearly defined and readily reproduced remains open at this stage, and the reliability and reproducibility of actinometric measurements with quinine hydrochloride have been questioned by several research groups (35–37) (Christensen KL, Christensen JØ, Frøkjær S, Langballe P, Hansen LL. The influence of temperature and light on the quinine chemical actinometric system. Unpublished results) for a variety of reasons as follows:

1. A major concern is the change in absorbance at 400 nm due to dark reactions. For example, it is known that quinine solutions are thermally unstable and that the rate of decomposition—as measured by an increase in absorbance over time—strongly depends on the temperature (35). Most importantly, the absorbance changes in the dark, or dark reaction, are much more pronounced after the quinine solution

has been exposed to light. Thus, the "photoreaction" seems to continue after termination of the irradiation procedure at a reduced but significant rate, which strongly depends not only on the temperature but also on the radiation exposure level prior to the dark reaction (36,37) (Christensen KL, Christensen JØ, Frøkjær S, Langballe P, Hansen LL. The influence of temperature and light on the quinine chemical actinometric system. Unpublished results). Consequently actinometric measurements utilizing quinine hydrochloride need to be carried out at a certain (standardized) temperature, and strict temperature control is required especially if strong (heat-producing) radiation sources are used. Additionally, infrared and water filters in front of the sources are recommended to reduce the heat exposure during the irradiation period. Changes in absorbance at 400 nm will be critically affected by the timing of the spectrophotometric measurement after termination of the radiation exposure; thus immediate spectroscopic analysis is highly recommended.

2.   Another matter of great concern is the pH dependence of the absorption spectrum of quinine. The free base predominant at pH > 8.5 (34) shows an absorption maximum at 330 nm and almost no absorption above 350 nm (Matsuo M, private communication). In contrast, the monoprotonated quinolinium species present in the pH range between 4.13 and 8.5 (34) shows a significantly broadened absorption band ($\lambda_{max} \approx 335$ nm; Matsuo M, private communication). At a pH < 4.13, the nitrogen atom in the quinoline moiety as well as the bridged aliphatic nitrogen atom are protonated (34), which results in an even broader absorption band with maximum near 350 nm, a shoulder near 320 nm, and a tail beyond 400 nm (Matsuo M, private communication). Thus, depending on the pH of the aqueous quinine hydrochloride solution, the UV output of different lamps, all with different spectral power distributions (SPDs), will be differently absorbed by the quinine actinometers. Moreover, it is known that the pH of the aqueous quinine solution is changing during photolysis (34), which consequently results in a change in the spectral response of the actinometer during the actinometric measurement. On the basis of these findings the pH conditions of the actinometer solution need to be clearly defined and stabilized during irradiation using a suitable buffer system. [Citrate buffer should be avoided (vide supra, (Ref. 32).] Unfortunately, as demonstrated by Moore (33), under conditions of controlled pH, simple buffered solutions of quinine do not exhibit actinometric properties.

3.   Because quinine also acts as a photosensitizer, producing singlet oxygen in aqueous solutions (33), actinometric measurements can be affected by the amount of oxygen in solution. However, observations on the effects of oxygen are controversial. Whereas one research group claims the linear photoresponse of the quinine to be unaffected by oxygen (34), others report differences by at least a factor of 2 in the absorbance changes at 400 nm when nitrogen and oxygen purged actinometer solutions are compared (36). In addition, the oxygen effect also strongly depends on the temperature, due to the temperature dependence of the solubility of oxygen in water. These differences are probably attributable to the quality of water used and the cleanliness of the vessels used. Moore (33) has demonstrated that it is possible to use a buffered solution of quinine as an actinometer if an appropriate, oxidizable substrate is present, such as 2,5-dimethylfuran, and the absorbance at 280 nm is monitored. This same effect and varying oxygen concentrations are most likely the reason for the various reports differing as to the oxygen effect. An instance of a quinine solution turning black

after irradiation, probably due to an insufficient cleaning, has been reported (personal communication).

4. Finally, owing to its rather narrow absorption profile between 300 and 400 nm the quinine actinometer shows very different responses depending on the UV source used. Comparative studies on two light sources specified by the ICH guideline (28), viz., a xenon lamp and a near-UV fluorescent lamp, will result in substantially different calculated exposure times necessary to reach the minimum prescribed absorbance change of $\Delta A_{400} = 0.8$ (37). These results clearly demonstrate that the increase in the absorbance at 400 nm of the quinine actinometer is not a rigorous measure of the total EO of a UV photon source, because ΔA_{400} is strongly affected by the SPD of the source intensity in the 320–400 nm wavelength range. In other words, sources of equal EO but different spectral profile will appear different in total EO if monitored by the quinine actinometers. In fact, the same finding applies to the efficiency of different UV sources in causing photodegradation of a certain drug under investigation, i.e., optimum overlap of the absorption spectrum of the drug with the wavelength profile (SPD) of the source will result in maximum photodamage. Needless to say, the photoresponse of colored drugs (with absorption spectra reaching further in the visible wavelength range) cannot be quantified at all with the quinine actinometers, which only "see" UV radiation below 400 nm or less.

A final comment needs to be made on the choice of quinine hydrochloride because of its low sensitivity, which—in general—allows its use as an integrating actinometer for long exposure times. Highly sensitive actinometers (such as ferrioxalate or the new generation of photometric actinometers, vide supra) are generally used to determine the light intensity (I_0, energy per time unit) rather than the integrated EO of a lamp. The total EO over a certain exposure time may then be calculated as the product of intensity and time, i.e., $EO = I_0 t$. To monitor the stability of a source over time, electronic devices such as photodiodes are ideal because photocurrent readings can be taken continuously or at certain time intervals and readily stored in a computer. Alternatively, sensitive chemical actinometers that only require very short exposure times for intensity measurements can be used repeatedly at several times during the photostability testing procedure to examine the long-term stability of the light source.

In summary, the recommendation of quinine hydrochloride as an integrating chemical actinometer in photostability testing procedures remains highly controversial. The ambiguities in the reaction mechanism, unknown quantum yields, and problems in standardizing and controlling the various reaction parameters, such as temperature, pH, and oxygen content of the solution make the use of these actinometers very unattractive and unreliable as compared to a variety of other chemical actinometers available for the same irradiation wavelength range (vide supra).

SPECIAL APPLICATIONS

The actinometers described in the section "Selected Chemical Actinometers" are mostly used to measure the intensity of monochromatic radiation emitted by conventional sources, such as mercury or xenon arc lamps, metal halide lamps, etc. In this section, special applications of chemical actinometer will be reviewed, including the measurement of intensities of lasers and polychromatic light sources. In addition, the application of chemical actinometers to the determination of quantum yields of solid-state photoreactions will be discussed.

The Measurement of Laser Intensities by Chemical Actinometry

With the recent advances in laser and pulsed-Xenon lamp technology and its numerous applications to various fields of science and medicine, it has become very desirable to design and calibrate chemical actinometers for laser radiation sources. In general, chemical actinometers that are well established for conventional sources are not necessarily suitable for measurements of laser sources, owing to the fact that lasers emit radiation of very high-intensity and/or very high pulse energy. As a result, chemical actinometers may need to be recalibrated for high-intensity lasers because their photochemical quantum yields may depend on the intensity of the light source. Moreover, great care must be taken to avoid temporary bleaching of the actinometer solution and thus to ensure complete absorption of the radiation during high-intensity irradiation or during the application of short and intense pulses (38). In the following sections, we will review the application of well-established acti-nometers (See section "Selected Chemical Actinometers") for high-intensity/pulse rate intensity measurements. In addition, chemical actinometers specifically designed for these sources will be described.

Laser Actinometry with Ferrioxalate

The classic ferrioxalate actinometers (as described in detail in section "Selected Chemical Actinometers") have been examined for the use with pulsed and CW sources of various wavelengths by several research groups (38–40). For example, actinometric measurements of nitrogen laser generating 5-ns laser pulses are reported in Ref. (39). Thus, comparative measurements with a 0.006 M ferrioxalate solution and commercial energy meters (Quantronix and Scientech) show a linear response of the ferrioxalate solution up to a total absorbed energy of 900 mJ (in 1000 laser shots). The photochemical quantum yield of ferrioxalate is reported to be independent of the laser intensity up to a pulse power of about 10^8 W cm^{-2}, which corresponds to energies up to 1 mJ/pulse. This result is in agreement with other studies (38), which confirm reliable results with the ferrioxalate actinometers for laser pulse energies up to 5 mJ. Deviations from the linear response for higher pulse energies have been attributed to inner-filter effects or secondary photolysis processes of the photoproducts as well as to redox disproportionation of the oxa-late anion radical [See Eqs. (11) and (12)] whenever it is photogenerated in high quantities (38).

The ferrioxalate actinometer has also been calibrated for the argon and kryp-ton ion laser lines using NBS-calibrated thermopile and pyroelectric detectors (40). Thus, irradiation of a 0.15 M ferrioxalate solution in a 5-cm cuvette, with a 20-W (all lines) argon ion laser at 457.9 nm results in a quantum yield of $\Phi = 0.84$, which is independent of the radiation flux density (intensity) between 1.7 and 280×10^{-6} Einstein cm^{-2} min^{-1}. Irradiation of a 0.006 M ferrioxalate solution with a krypton ion laser at 406.7 nm results in a quantum yield of $\Phi = 1.19$ for a laser intensity of 1.5×10^{-5} Einstein cm^{-2} min^{-1}. Actinometric measurements of the 363.8 nm argon ion laser line, using either a 0.006 M or a 0.15 M ferrioxalate solution, result in a quan-tum yield which varies between $\Phi = 1.14$ and $\Phi = 1.29$, depending on the actinometer concentration. The lower values obtained with the concentrated actinometer solution were found to be incorrect, and a quantum yield of $\Phi = 1.28$ is recommended for laser intensities up to 6×10^{-6} Einstein cm^{-2} min^{-1}. All quantum yields obtained with the ion lasers, are in good agreement with those obtained at similar wavelengths with other (conventional) radiation sources.

In summary, ferrioxalate can be used for actinometric measurements of high-intensity pulsed or CW sources, in the UV and visible wavelength region. However, reliable results are only obtained for pulse energies of less than 5 mJ, and intensities of less than 3×10^{-4} Einstein cm^{-2} min^{-1} for CW irradiations.

Laser Intensity Measurements with the Azobenzene Actinometer

The azobenzene actinometer has been calibrated for intensity measurements with pulsed nitrogen, excimer, and Nd:YAG lasers. Comparative actinometric measurements of the radiation intensity of a weak nitrogen laser (337.1 nm, 5 ns, and <5 mJ/pulse) utilizing ferrioxalate (vide supra) and a dilute (10^{-5} M) azobenzene solution resulted in an isomerization quantum yield identical to that obtained using the 334-nm mercury line (14). Thus, we conclude that a concentrated (6.4×10^{-4} M) azobenzene solution can be used in the same way as described for actinometric measurements of the mercury line at 334 nm (15) with an calibration factor of $W = 3.6 \times 10^{-6}$ Einstein cm^{-2} [see Eq. (14)].

The azobenzene actinometers have also been calibrated for measurement of pulse energies of the third harmonic output (at 355 nm) of Nd:YAG lasers (Doherty S and Hubig SM, unpublished results). To ensure complete absorption of the laser radiation at this wavelength, a 2.5 mM solution of azobenzene in methanol is needed ($A_{355} > 4$). The $E \rightarrow Z$ photoisomerization of azobenzene in such a concentrated solution is monitored at 436 nm. The overall photochemical conversion in all actinometric measurements is kept below 5% to avoid significant absorption by the photoproduct, which initiates the reverse photoisomerization. Under these experimental conditions, a linear increase in the absorbance at 436 nm, with increasing number of laser pulses, is obtained, and the energy E_0 of a single laser pulse in Einstein cm^{-2} pulse^{-1} can be calculated using the following equation:

$$E_0 = W \cdot \Delta A_{436} / \Delta p \tag{20}$$

where $\Delta A_{436}/\Delta p$ is the change in absorbance (monitored at 436 nm) with the number of laser pulses Δp, and $W = [1000 \, \Phi(\varepsilon_{Z,436} - \varepsilon_{E,436})]^{-1}$ is the calibration factor. A calibration factor of $W = 7.25 \times 10^{-6}$ Einstein cm^{-2} (which corresponds to a quantum yield of $\Phi = 0.2$) is obtained by comparison of actinometric measurements with azobenzene, ferrioxalate, and the transient benzophenone actinometer (vide infra). For the comparison with ferrioxalate, the laser energies should be kept below 5 mJ/pulse (vide supra). Comparison with the benzophenone actinometer confirms a reproducible calibration factor W independent of the laser pulse width (30, 200 ps, and 11 ns) and the pulse energy of up to 50 mJ/pulse.

Similarly, intensity measurements utilizing a pulsed XeCl excimer laser (308 nm, 15 ns, and 150 mJ/pulse) have been carried out using a dye laser unit as flow-through cell for a 10^{-3} M azobenzene solution (18). A linear correlation was found between the increase in absorbance at 358 nm and the number of laser pulses (Δp), and the energy (E_0) per laser pulse can be calculated using Eq. (21), i.e.,

$$E_0 \, [\text{Einstein cm}^{-2} \, \text{pulse}^{-1}] = W \cdot \Delta A_{358} / \Delta p \tag{21}$$

Using the same calibration factor determined for the 313-nm mercury line ($W = 5.3 \times 10^{-6}$ Einstein cm^{-2}), a pulse energy of $E_0 = 2.9 \times 10^{-8}$ Einstein pulse^{-1} is obtained, in good agreement with bolometer measurements, which yield a pulse energy of 150 mJ (18).

In summary, concentrated azobenzene solutions represent reliable chemical actinometers for pulsed UV sources, such as the nitrogen laser line at 337.1 nm, the third harmonic output of Nd:YAG lasers at 355 nm, or the XeCl excimer laser line at 308 nm. In contrast to the ferrioxalate actinometers, the azobenzene actinometers are not restricted to low-energy lasers but can measure laser pulse energies up to 150 mJ.

Laser Intensity Measurements with a Photooxygenation Actinometer

For high-power lasers in the 280–550 nm range, tris(2,2'-bipyridine)ruthenium(II) sensitized photooxygenation of tetramethylethylene (TME) to its peroxide has been recommended as a chemical actinometer (41). This actinometer is especially suitable for the argon ion laser lines, but it can also be used for high-power pulsed nitrogen lasers and most other pulsed or CW lasers with output wavelengths below 550 nm. The actinometer is based on a reaction of TME with singlet oxygen ($^1O_2$) generated by the Ru(bpy)$_3^{2+}$ photosensitizer, i.e.

$$Ru(bpy)_3^{2+} \xrightarrow{\ h\nu\ } Ru(bpy)_3^{2+*} \tag{22}$$

$$Ru(bpy)_3^{2+*} + O_2 \longrightarrow Ru(bpy)_3^{2+} + {}^1O_2 \tag{23}$$

$${}^1O_2 + TME \longrightarrow (CH_3)_2C(OOH)C(CH_3) = CH_2 \tag{24}$$

The complete actinometer system consists of a photolysis cell containing an oxygen-saturated methanol solution of TME and Ru(bpy)$_3^{2+}$, connected to a gas burette filled with oxygen. The conversion of the photooxygenation is monitored by the oxygen uptake, read directly from the burette. Using an oxygen pressure of about 1 atm and a TME concentration of ≈ 0.11 M, oxygen is consumed with an efficiency of ≈ 0.76 mol of O_2 per Einstein. This corresponds to an oxygen consumption of 5 mL/min for an applied laser radiation intensity of 1 W at 500 nm. Because the Ru(bpy)$_3^{2+}$ sensitizer shows strong absorption in the 280–560 nm wavelength range and the photooxygenation quantum yield is independent of the actinic wavelength, these actinometers can also be used for polychromatic light sources (viz. conventional lamps or multiple laser lines) in this wavelength region.

In summary, this photooxygenation actinometer is highly reproducible and easy to use. The chemical substrates are commercially available and no expensive analytical instrumentation, such as a spectrophotometer, is needed. Repetitive measurements can be made using the same actinometric solution. If a flow cell is used for the actinometer solution, laser intensities up to 1.5 W can be reliably measured. It is suggested that the wavelength range of this actinometer should be easily extended to longer wavelengths by replacing Ru(bpy)$_3^{2+}$ by other singlet-oxygen sensitizers with red-shifted absorption spectra (e.g., methylene blue).

Transient Chemical Actinometers
Transient chemical actinometers can be used to measure the intensity of single laser pulses if the laser source is connected to a time-resolved spectrophotometer, similar to those required for laser flash photolysis experiments. The most commonly used transient actinometers for UV laser pulses are the benzophenone actinometer (42). Thus, a benzophenone solution in benzene with an absorbance of $A_0=0.1$ at the laser wavelength is exposed to a single laser pulse in a laser flash photolysis experiment, and the amount of photogenerated benzophenone triplet is measured by monitoring the transient absorbance (ΔA_{530}) at 530 nm on the ns/μs time scale. The laser energy per pulse (E_0) in Einstein cm^{-2} pulse^{-1} can then be calculated using the following equation (42):

$$E_0 = \Delta A_{530} / [2303 \cdot A_0 \cdot \Phi_{ISC} \cdot \varepsilon_{530}] \qquad (25)$$

where $A_0=0.1$ is the absorbance at the laser wavelength, $\Phi_{ISC}=1.0$ is the triplet quantum yield (42), and $\varepsilon_{530}=7220$ M^{-1}cm^{-1} (42) or 7630 M^{-1}cm^{-1} (43) is the extinction coefficient of triplet benzophenone at 530 nm.

In a similar way, anthracene triplet ($\Phi_{ISC}=0.71$, $\varepsilon_{422}=64,700$ M^{-1}cm^{-1}) and the naphthalene triplet ($\Phi_{ISC}=0.75$, $\varepsilon_{414}=24,500$ M^{-1}cm^{-1}) in cyclohexane solution have been introduced as transient chemical actinometers for the third-harmonic (355 nm) and fourth-harmonic (266 nm) output of Nd:YAG lasers, respectively (44). In summary, transient chemical actinometers are ideal for accurately measuring the energy of single laser pulses, provided the quantum yields and extinction coefficients of the transients are well known (45–47). Thus, the well-established benzophenone actinometer (42–44) has been used as a reliable reference to calibrate the azobenzene actinometer (see section "Laser Intensity Measurements with the Azobenzene Actinometer"; Doherty S, Hubig SM, unpublished results) and the Aberchrome 540 actinometer (48,49) for intensity measurements with pulsed Nd:YAG and/or XeCl excimer lasers. However, such actinometer can only be used when a complete set of laser flash photolysis equipment including a kinetic spectrometer is available.

Chemical Actinometry of Polychromatic Light Sources
According to the ICH guideline (28), the photostability testing of new drugs is carried out with polychromatic photon sources, such as cool-white fluorescent lamps for visible radiation exposure or near-UV fluorescent lamps for exposure in the 320–400 nm wavelength range. Thus, to quantify the EO of such lamps, chemical actinometers for polychromatic radiation are highly desirable. Such actinometers would also be useful for measurements of sunlight intensities, which play an important role in the photolysis protocols of the U.S. Environmental Protection Agency, which are required by the Toxic Substances Control Act of 1976 (50,51).

Chemical actinometers are suitable for the determination of intensities of polychromatic radiation sources if the following two requirements are met:

1. The actinometer solution must absorb all the photons emitted by the source. Thus, the absorption spectrum of the actinometer must cover the entire wavelength range of the source.
2. The chemical actinometer must exhibit a wavelength-independent quantum yield over the entire spectral range of the source.

In other words, the actinometers respond in exactly the same way to the absorption of a photon from the blue or the red edge of the emission profile of the lamp. Because in this case the response of the actinometer is independent of the energy of the photons and only depends on the number of incident photons, such actinometers are also referred to as quantum counters.

The design of a quantum counter with a wide wavelength range that covers the entire spectrum of a particular polychromatic source represents a great technical challenge owing to the fact that the absorption bands of most chemical actinometers extend only over a limited wavelength region. Thus, only two of the actinometers introduced in the section "Selected Chemical Actinometers" are suitable for use as quantum counters, and then only for rather narrow wavelength ranges ($\Delta\lambda < 150$ nm). For example, ferrioxalate may be used for the actinometry of polychromatic (UV) radiation between 250 and 366 nm, and the MDH actinometer is a suitable quantum counter for visible radiation between 475 and 610 nm. In contrast, the potassium reineckate actinometer, which exhibits a rather wide wavelength response (316–750 nm), is not recommended for measurements of polychromatic radiation sources, owing to the significant variation its quantum yield exhibits within that irradiation wavelength range (27). Similarly, the use of quinine hydrochloride as a reliable quantum counter for the 320–360 nm wavelength range (28,29,34) is controversial and scientifically undocumented, as previously discussed (see the section "The Potassium Reineckate Actinometer").

To overcome the undesired wavelength dependency of chemical actinometers, luminescence quantum counters have been designed (52). For example, the laser dye rhodamine B exhibits a wavelength-independent fluorescence quantum yield of $\Phi = 1.0$ over a quite large range of excitation wavelengths from 250 to 600 nm. Thus, a rhodamine solution of high enough concentration to absorb all photons in this wavelength range will emit a reproducible spectrum the intensity independent of the wavelength distribution (SPD) of the excitation source, dependent only on the total number of photons absorbed. In other words, all incident photons, of varying energies, are converted into exactly the same number of emitted photons of one particular energy (or energy range), which can be quantified utilizing a calibrated detector.

While the general idea of such electronically integrating actinometers (53) is quite intriguing, numerous experimental problems need to be addressed. Most of them arise from a severe overlap of the absorption and emission spectra of the quantum counter. This leads to a significant reabsorption of the emitted radiation, yielding an underestimation of the number of photons emitted by the radiation source. Although front-face emission measurements can reduce the effects of reabsorption, reproducible corrections for reabsorption are difficult to establish because they depend on the penetration depth of the incident radiation, into the quantum counter solution (54). A better solution to eliminate the problems of reabsorption might be to choose whenever possible quantum counters that show negligible overlap between absorption and emission spectra. Such conditions are found, for example, in metal complexes, such as that of tris(2,2'-bipyridine) ruthenium(II).

Another approach for the design of chemical actinometers for polychromatic sources may lie in the combination of several substrates with different absorption spectra into one actinometric solution to provide a wide wavelength response. For example, a mixture of several dye sensitizers that generate singlet oxygen in solution may be combined with a singlet-oxygen probe (such as 1,3-diphenylisobenzofuran) (55,56) to produce a chemical actinometric system with a wavelength-independent response over a wide spectral range.

A final comment needs to be added on the use of polychromatic radiation sources for the photostability testing of new drugs. Even in the ideal case when the total EO of a polychromatic source can be reliably measured with a suitable chemical quantum counter, the photostability test carried out under these conditions will only yield an wavelength-averaged photoresponse of the drug, which may be caused by an infinite number of photoreactions. Crucial information on the efficiency of a particular photodegradation reaction of a drug can only be gained from monochromatic photolysis experiments, which reveal the wavelength dependence of the photochemical quantum yield and thus identify the critical wavelength range responsible for the most severe photodamage.

Chemical Actinometry for Solid-State Photoreactions
Measurements of photochemical quantum yields have proved to provide critical information for the determination of the mechanism of the photoreaction. In solution and in the gas phase, such quantitative measurements of photoreactions are readily carried out using the chemical actinometers described in the previous sections. The quantum yields of solid-state photoreactions are more difficult to obtain for two reasons:

1. Light scattering on the surface of the solid material causes only a fraction of the actinic light to be absorbed by the substrate.
2. If the photoproduct strongly absorbs the actinic light, the photoreaction will only occur on the surface of the solid material.

As the latter effect, which is comparable to the inner-filter effect during photolysis experiments in solution, does not always occur, radiation scattering problems are inherent to all solid-state photoreactions and are particularly relevant to the photostability testing of solid-state drugs, such as tablets, pills, or powders. Thus in this section, we will discuss three experimental approaches for the determination of solid-state photochemical quantum yields utilizing chemical actinometers.

The Matsuura Method for Approximate Estimates of Solid-State Quantum Yields
The Matsuura method (57) compares the photolysis result of a solution actinometer with that of a thin crystalline film of equal surface area after exposure. Thus, evaporation of a solution containing the photoreactive substrate results in a thin crystalline film on the glass wall of a test tube, which is subsequently exposed to actinic radiation in a merry-go-round type photolysis apparatus. To test for complete absorption of the incident photons within the crystalline film, the evaporation process is carried out at various concentrations of the substrate, which leads to films of different thickness. If the yield of photoproduct after a certain exposure time is independent of the concentration of the original solution before evaporation, complete absorption of all actinic photons is established. The quantity of the photons absorbed by the crystalline film is then estimated by parallel photolysis of a 0.1 M solution of 2,4,6-triisopropylbenzophenone in methanol solution, which has a well-established quantum yield of $\Phi = 0.52$ (58). The volume of this actinometer solution in the test tube is adjusted so that the crystalline film and the solution exhibit irradiated surfaces of identical size. In summary, this method provides approximate estimates of solid-state quantum yields; however, differences in the reflection of the

actinic light from the crystalline film and from the solution surfaces are completely neglected. In addition, the evaporation process may produce a crystal form and size not found in the substance/product, which may further complicate interpretation of the results obtained.

The Zimmerman Photolysis Cell

A specially designed photolysis cell is described by Zimmerman and Zuraw (59) for the accurate determination of the quantum yields of photoreactions in highly light-scattering solid materials. In this procedure, the solid substrate is deposited on the inner wall, deep inside a double-walled cylindrical photolysis cell containing ferrioxalate solution. Consequently, the substrate is largely surrounded by the actinometer solution that absorbs nearly all the scattered radiation from the sample. Actinometric measurements of the intensity (I_0) of the actinic radiation are carried out with and without a sample in the photolysis cell, and the difference in I_0 represents the absolute number of photons absorbed by the solid substrate. In other words, the intensity of the scattered light from the solid sample is measured with high precision by chemical actinometer and can thus be subtracted from the intensity of the incident beam to obtain the absolute amount of radiation absorbed by the solid sample. According to Zimmerman and Zuraw (59), quantum yields as low as 0.00003 can be determined within 10% error limits using this highly precise method.

The First Actinometric Reaction on a Solid Support

An experimental setup similar to the Zimmerman photolysis cell was utilized by Lazare et al. (60) to determine the quantum yield for the photoreaction of a substrate adsorbed on silica gel. The photolysis cell consists of an aluminum dish for the powdered silica gel sample, which is covered by a double-walled hemispherical Pyrex cap filled with ferrioxalate actinometers solution. The sample is irradiated through a quartz light pipe, which enters the photolysis cell through a hole at the top of the cap. Thus, nearly all scattered light from the silica gel sample is absorbed by the surrounding actinometer solution, and the amount of light absorbed by the substrate (which is adsorbed on the silica gel surface) is determined by a similar subtraction method as described in the Zimmerman experiment (vide supra).

This actinometric procedure was utilized to determine the photochemical quantum yield for the solid-state photoisomerization of the Diels–Alder adduct (I) (formed from 2,5-dimethylbenzoquinone and cyclopentadiene) to the pentacyclic diketone (II), i.e.,

$$(26)$$

(I)   (II)

Interestingly, identical photoisomerization quantum yields of $\Phi = 1.00$ are obtained for substrate I in solution and adsorbed on silica gel. Thus, the photoisomerization reaction on silica gel is proposed as a new solid-state actinometer with well-known quantum yield.

REFERENCES

1. Mauser H, Gauglitz G. Photokinetics—theoretical fundamentals and applications. In: Compton RG, Hancock G, eds. Comprehensive Chemical Kinetics. Vol. 36. Amsterdam: Elsevier, 1998:7–19.
2. Skoog DA, Leary JJ. Principles of Instrumental Analysis. Fort Worth, TX: Saunders College, 1992.
3. Steinfeld JJ. Molecules and Radiation: An Introduction to Modern Molecular Spectroscopy. Cambridge, MA: MIT Press, 1985.
4. Gauglitz G. Praktische Spektroskopie. Tubingen: Attempto Verlag, 1983.
5. Gauglitz G. UV/Vis-Spektrokopie. In: Gunzler G, ed. Ullmann's Enzyclopedia Technol. Chem. Weinheim: VCH, 1995.
6. Svehla G, ed. Comprehensive Analytical Chemistry. Vol. 14. Analytical Visible and Ultraviolet Spectroscopy. Amsterdam: Elsevier, 1986.
7. Gauglitz G, Luddecke E. Automatische digitale Meß- und Bestrahlungsanlage zur kinetischen Untersuchung von Photoreaktionen. Z Anal Chem 1976; 280:105.
8. AMKO, Tornesch, Germany.
9. For a comprehensive list of chemical actinometers, see: Kuhn HJ, Braslavsky SE, Schmidt R. Chemical actinometry (Report of the IUPAC Commission on Photochemistry). Pure Appl Chem 1987; 61:187–210.
10. Hatchard CG, Parker CA. A new sensitive chemical actinometer. II. Potassium ferrioxalate as a standard chemical actinometer. Proc R Soc London Ser A 1958; 235:518–536.
11. Calvert JG, Pitts JN Jr. Photochemistry. London: Wiley, 1967:783–786.
12. Gauglitz G, Hubig SM. Photokinetische Grundlagen moderner chemischer Aktinometer (Photokinetics of novel chemical actinometers). Z Phys Chem NF 1984; 139:237–246.
13. Gauglitz G. Azobenzene as a convenient actinometer for the determination of quantum yields of photoreactions. J Photochem 1976; 541–547.
14. Gauglitz G, Hubig SM. Azobenzene as a convenient actinometer: evaluation values for UV mercury lines and for the N_2 laser line. J Photochem 1981; 15:255–257.
15. Gauglitz G, Hubig SM. Chemical actinometry in the UV by azobenzene in concentrated solution: a convenient method. J Photochem 1985; 30:121–125.
16. Gauglitz G, Hubig SM. Reversibles chemisches Aktinometer (Reversible Chemical Actinometer). German Patent Pending DP 32 42 489.2, 1984.
17. Brauer HD, Schmidt R, Gauglitz G, Hubig SM. Chemical actinometry in the visible (475–610 nm) by meso-diphenylhelianthrene. Photochem Photobiol 1983; 37:595–598.
18. Hubig SM. Chemische Aktinometrie—Photokinetische Grundlagen, Entwicklung und Kalibrierung neuer chemischer Aktinometer (Chemical Actinometry—Photokinetics, Design and Calibration of Novel Chemical Actinometers). Ph.D. dissertation, University of Tubingen, Germany, 1984.
19. Adick HJ, Schmidt R, Brauer HD. A chemical actinometer for the wavelength range 610–670 nm. J Photochem Photobiol A Chem 1989; 49:311–316.
20. Adick HJ, Schmidt R, Brauer HD. Chemical actinometry between 670 and 795 nm. J Photochem Photobiol A Chem 1990; 54:27–30.
21. Heller HG, Langan JR. Photochromic heterocyclic fulgides. Part 3. The use of (E)-a-(2,5-dimethyl-3-furylethylidene)(isopropylidene)succinic anhydride as a simple convenient chemical actinometer. J Chem Soc, Perkin Trans II 1981:341–343.
22. Boule P, Pilichowski JF. Comments about the use of Aberchrome 540 in chemical actinometry. J Photochem Photobiol A Chem 1993; 71:51–53.
23. Guo Z, Wang G, Tang Y, Song X. Photokinetic study on the photochromic reaction of Aberchrome 540: a further comment about the use of Aberchrome 540 in chemical actinometry. J Photochem Photobiol A Chem 1995; 88:31–34.
24. Yokoyama Y, Hayata H, Ito H, Kurita Y. Photochromism of a furyl fulgide, 2-[1-(2,5-dimethyl-3-furyl)ethylidene]-3-isopropylidene succinic anhydride in solvents and polymer films. Bull Chem Soc Jpn 1990; 63:1607–1610.
25. Uhlmann E, Gauglitz G. New aspects in the photokinetics of Aberchrome 540. J Photochem Photobiol A Chem 1996; 98:45–49.

26. Wegner EE, Adamson AW. Photochemistry of complex ions. III. Absolute quantum yields for the photolysis of some aqueous chromium (III) complexes. Chemical actinometry in the long wavelength visible region. J Am Chem Soc 1966; 88:394–404.

27. Hubig SM. Zur Photokinetik chemischer Aktinometer (On the Photokinetics of Chemical Actinometers). Diploma thesis, University of Tubingen, Germany, 1980.

28. Guideline for the photostability testing of new drug substance and products. Federal Register 1996; 61:9309–9313.

29. Yoshioka S, Ishihara Y, Terazono T, et al. Quinine actinometry as a method for calibrating ultraviolet radiation intensity in light stability testing of pharmaceuticals. Drug Dev Ind Pharm 1994; 20:2049–2062.

30. Stenberg VI, Travecedo EF, Musa WE. A new photoreduction on quinoline alkaloids. Tetrahedron Lett 1969; 25:2031–2033.

31. Epling GA, Yoon UC. Photolysis of cinchona alkaloids. Photochemical degradation to 5-vinylquinuclidine-2-carboxaldehyde, a precursor to synthetic antimalarials. Tetrahedron Lett 1977; 29:2471–2477.

32. Laurie WA, McHale D, Saag K, Sheridan JB. Photoreactions of quinine in aqueous citric acid solution. Part 2. Some end-products. Tetrahedron 1988; 44:5905–5910.

33. Moore DE. Photosensitization by drugs: quinine as a photosensitizer. J Pharm Pharmacol 1980; 32:216–218.

34. Drew HD, Brower JF, Juhl WE, Thornton LK. Quinine photochemistry: a proposed chemical actinometer system to monitor UV-A exposure in photostability studies of pharmaceutical drug substances and drug products. Pharm Forum 1998; 24:6334–6346.

35. Sekine H, Ohta Y, Nakagawa T. Usefulness of pyridoxine hydrochloride for actinometry. Drug Stabil 1996; 1:135–140.

36. Baertschi SW. Commentary on the quinine actinometry system described in the ICH draft guideline on photostability testing of new drug substances and products. Drug Stabil 1996; 1:193–195.

37. Kester TC, Zhan Z, Bergstrom DH. Quinine chemical actinometry studies under two light sources specified by the ICH guideline on photostability testing. National Meeting of the American Association of Pharmaceutical Scientists, Seattle, Washington, 1996.

38. Demas JN. The measurement of laser intensities by chemical actinometry. In: Ware WR, ed. Creation and Detection of the Excited State. Vol. 4. New York: Marcel Dekker, 1976.

39. Gruter H. Measuring the pulse energy of a nitrogen laser with the potassium ferrioxalate actinometer. J Appl Phys 1980; 51:5204–5206.

40. Demas JN, Bowman WD, Zalewski EF, Velapoldi RA. Determination of the quantum yield of the ferrioxalate actinometer with electrically calibrated radiometers. J Phys Chem 1981; 85:2766–2771.

41. Demas JN, McBride RP, Harris EW. Laser intensity measurements by chemical actinometry. A photooxygenation actinometer. J Phys Chem 1976; 80:2248–2253.

42. Hurley JK, Sinai N, Linschitz H. Actinometry in monochromatic flash photolysis: the extinction coefficient of triplet benzophenone and quantum yield of triplet zinc tetraphenylporphyrin. Photochem Photobiol 1983; 38:9–14.

43. Bensasson R, Land EJ. Triplet-triplet extinction coefficients via energy transfer. Trans Faraday Soc 1971; 67:1904–1915.

44. Bensasson R, Goldschmidt CR, Land EJ, Truscott TG. Laser intensity and the comparative method for determination of triplet quantum yields. Photochem Photobiol 1978; 28:277–281.

45. Carmichael I, Hug GL. Triplet-triplet absorption spectra of organic molecules in condensed phases. J Phys Chem Ref Data 1986; 15:1–250.

46. Bonneau R, Carmichael I, Hug GL. Molar absorption coefficients of transient species in solution (Report of the IUPAC Commission on Photochemistry). Pure Appl Chem 1991; 63:289–299.

47. Murov SL, Carmichael I, Hug GL. Handbook of Photochemistry. 2nd ed. New York: Marcel Dekker, 1993.

48. Wintgens V, Johnston LJ, Scaiano JC. Use of a photoreversible fulgide as an actinometer in one- and two-laser experiments. J Am Chem Soc 1988; 110:511–517.

49. Becker RS, Freedman K. A comprehensive investigation of the mechanism and photophysics of isomerization of a protonated and unprotonated Schiff base of 11-*cis* retinal. J Am Chem Soc 1985; 107:1477–1485.

50. Dulin D, Mill T. Development and evaluation of sunlight actinometers. Environ Sci Technol 1982; 16:815–820.

51. Carroll FA, Strouse GF, Hain JM. A visual manifestation of the Norrish type II reaction. The cyclohexanone sunburn dosimeter. J Chem Educ 1987; 64:84–86.

52. Taylor DG, Demas JN. Light intensity measurements. I: large area bolometers with microwatt sensitivities and absolute calibration of the rhodamine B quantum counter. II: luminescent quantum counter comparator and evaluation of some luminescent quantum counters. Anal Chem 1979; 51:712–717, 717–722.

53. Amrein W, Gloor J, Schaffner K. An electronically integrating actinometer for quantum yield determinations of photochemical reactions. Chimia 1974; 28:185–188.

54. Ostrom GS, Demas JN, DeGraff BA. Luminescence quantum counters. Comparison of front and rear viewing configurations. Anal Chem 1986; 58:1721–1725.

55. Merkel PB, Kearns DR. Radiationless decay of singlet molecular oxygen in solution. An experimental and theoretical study of electronic-to-vibrational energy transfer. J Am Chem Soc 1972; 94:7244–7253.

56. Young RH, Brewer D, Keller RA. The determination of rate constants of reaction and lifetimes of singlet oxygen in solution by a flash photolysis technique. J Am Chem Soc 1973; 95:375–379.

57. Ito Y, Matsuura T. A simple method to estimate the approximate solid state quantum yield for photodimerization of trans-cinnamic acid. J Photochem Photobiol 1989; 50:141–145.

58. Ito Y, Matsuura T, Fukuyama K. Efficiency for solid-state photocyclization of 2,4,6-triisopropylbenzophenones. Tetrahedron Lett 1988; 29:3087–3090.

59. Zimmerman HE, Zuraw MJ. Confinement control in solid-state photochemistry. J Am Chem Soc 1989; 111:2358–2361.

60. Lazare S, de Mayo P, Ware WR. Biphasic photochemistry: the first actinometric reaction on a solid support. Photochem Photobiol 1981; 34:187–190.

Radiometry/Photometry

Robert Angelo
Gigahertz-Optik, Inc., Newburyport, Massachusetts, U.S.A.

ELECTROMAGNETIC SPECTRUM

Electromagnetic energy radiated in the wavelength range from about 200 nm to 10.6 μ is commonly but technically incorrectly referred to as "light." This spectral range encompasses ultraviolet (UV), visible (VIS), and infrared (IR) electromagnetic radiation. A more correct term, "optical radiation," is also used to describe this energy because it behaves according to the laws and principles of geometric optics. Optical radiation may be characterized by the related quantities of wavelength, frequency, and photon energy (1). Below 200 nm, the atmosphere is a poor transmitter of optical radiation and above 10,600 nm (10.6 μ), this energy is known as heat and radio waves.

Radiometry deals with the metrology of optical radiation throughout the entire spectrum. Technically, the term "light" should only be used to describe optical radiation within the human eye's spectral range of sensitivity. Photometry is the measurement of this particular VIS portion of the spectrum from approximately 380 to 780 nm.

Through testing of a sample of the population, the average human eye spectral response was first defined by the Commission Internationale de l'Éclairage (CIE), in 1924 (2). The "standard observer" photopic eye response, the CIE V(λ) function, resulted from testing of the light-adapted eye. The eye has a different response in its dark-adapted state, referred to as scotopic vision [CIE V'(λ)], where the eye can no longer determine color. Both of these responses are plotted in Figure 1.

The CIE next defined a standard source to gauge intensity using a specific type of candle, giving rise to the terms foot-candle (fc) and candlepower. Optical radiation wavelengths have been divided into spectral bands and again, for consistency, defined in a "CIE Vocabulary for Spectral Regions" (3). The various band assignments given in this document are given in Table 1.

The UV region spectrally ranges from about 180 to 400 nm, the VIS spectrum from 380 to 780 nm, and the IR from 780 nm to 1 mm. The actual point of separation between the (UV)-A and VIS region frequently overlaps.

RADIOMETRIC AND PHOTOMETRIC CONCEPTS AND UNITS

Radiant Flux (Power) and Luminous Flux (Lumen)

Figure 2 gives some of the basic terms of optical power. These terms refer to the total radiant output of a light source expressed in watts (radiometric) and total luminous output in lumens (photometric), the most basic units of optical flux.

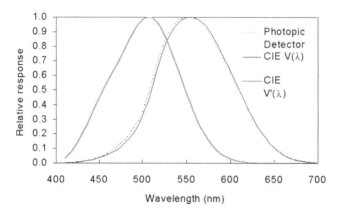

Figure 1 Photometry: visible "light" measurement. *Abbreviations*: CIE, Commission Internationale de l'Éclairage, CIE V(λ), Photopic – Light Adapted Eye Response, CIE V'(λ), Scotopic – Dark Adapted Eye (no color) Response.

The rate of flow of flux is radiant or luminous energy expressed in watt·seconds or lumen·seconds, respectively. In measuring power, the total flux must be captured by the photodetector effectively, underfilling it, as illustrated in Figure 3.

When the beam or source signal is larger than the active detection area of the photodetector, an integrating sphere can be used to trap and reflect signal evenly to the sensor, as shown in Figure 4.

Radiant and Luminous Intensity

As shown in Figure 5, intensity refers to the ratio of flux emitted by a source to the solid angle (steradian) of emittance of that source. Units are expressed in watts per steradian ($W sr^{-1}$) or lumens per steradian ($lm sr^{-1}$ = candela). Typically, beam intensity is measured with the photodetector overfilled and positioned to sample the peak output of the beam. The source to detector distance is factored out making

Table 1 Commission Internationale de l'Éclairage Vocabulary for Spectral Regions

Name	Wavelength range
UV-C	100 to 280 nm
UV-B	280 to 315 nm
UV-A	315 to 400 nm
Visible	360–400 to 760–800 nm
IR-A*	780 to 1400 nm
IR-B	1.4 to 3.0 micron (μm)
IR-C**	3.0 μ to 1 mm

Note: 1 micron = 1000 nm.
*Also called "near IR or NIR."
**Also called "far IR."
Abbreviations: UV, ultra violet; IR, infra red.

- Basic unit of optical power
- Radiometric: watts
- Energy: 1 joule = 1 watt • second
- Photometric: lumens

Figure 2   Radiant and luminous flux.

the measurement a function of the source itself, independent of distance. Mean spherical candlepower is a photopic measurement of the source in all directions over 4π steradian.

Irradiance and Illuminance

As shown in Figure 6, irradiance and illuminance or flux density refers to the power per unit area, either radiant or luminous. As a function of optical radiation falling on a surface, the photodetector area is overfilled, as shown in Figure 7, and positioned at the point of interest. This type of measurement is distance dependent.

The International System of Units (SI units) for irradiance are watts per square meter ($W\,m^{-2}$) and for illuminance, lumens per square meter ($lm\,m^{-2}$) or Lux. The illuminance unit, fcs or lumens per square foot ($lm\,ft^{-2}$), is also commonly used as a result of the early work by the CIE, as stated previously. Another common unit for irradiance is watts per square centimeter ($W\,cm^{-2}$).

Radiance and Luminance

These terms refer to the radiant or luminous intensity of source projected onto a surface, as shown in Figure 8. The intensity is divided by the area of a source projection on the surface. The radiometric SI units of measurements are watts per steradian per square meter ($W\,sr^{-1}\,m^{-2}$) or watts per steradian per square centimeter ($W\,sr^{-1}\,cm^{-2}$). Photometric SI units are candela per square meter ($cd\,m^{-2} = lm\,sr^{-1}\,m^{-2}$) but foot-Lamberts, ($lm\,sr^{-1}\,ft^{-2}$) are also commonly used terms.

- Total power emitted is measured
- Source underfills detector active area

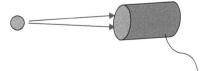

Figure 3   Optical power measurement.

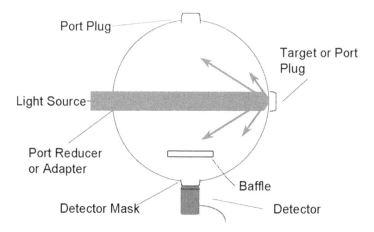

Figure 4   Integrating sphere.

A photodetector with a narrow field of view is used in this measurement to sample a small surface area. Ideally, the surface is uniformly illuminated, as illustrated in Figure 9.

For profiling or checking a large area source for uniformity, the restricted viewing angle radiance or luminance detector is used. Radiance and luminance (photometric brightness) are a function of the source. Because the detection area increases with distance, conserving signal, this measurement is independent of distance.

QUALITY OF MEASUREMENT

The uncertainty of optical radiation metrology is high compared to electrical measurements, which are accurate to many decimal places. But with electrical quantities, there is no distribution of signal, except perhaps for time. Because either voltage

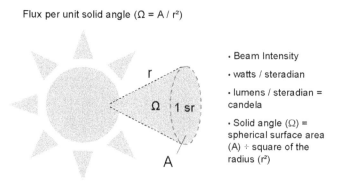

Figure 5   Radiant and luminous intensity.

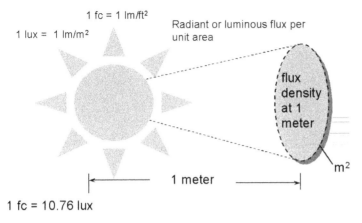

1 fc = 10.76 lux

Figure 6 Irradiance and illuminance.

or current is read from the photodetector, the electronic measurement part of the radiometer/photometer system contributes very little error to the overall optical radiation measurement, less than 0.5% measurement accuracy is easily achieved for a precision meter at its least sensitive ranges.

Optical radiation is distributed over time, position, direction, wavelength, and polarization, so field measurement accuracies within ±10% are normal. There are over 30 potential sources of error in light measurement (D. Ryer, personal communication, 1983–1999), six of these sources involving optical error are considered foremost.

Spatial Response
This term refers to the change in responsivity as a function of translational and angular displacements. A Lambertian or cosine spatial response refers to a particular angular response proportional to the cosine of the angle of the signal. This is the spatial response, which matches that of a perfectly absorbing flat surface. Because irradiance and illuminance are defined as the amount of light falling on a surface, the photodetector is provided with a diffuser so that its spatial response emulates that of a flat surface.

Source overfills detector active area

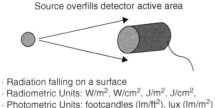

· Radiation falling on a surface
· Radiometric Units: W/m², W/cm², J/m², J/cm²,
· Photometric Units: footcandles (lm/ft²), lux (lm/m²)

Figure 7 Irradiance and illuminance measurement.

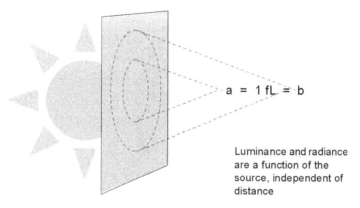

$a = 1\ fL = b$

Luminance and radiance
are a function of the
source, independent of
distance

Figure 8   Radiant or luminous intensity per area of source projection.

Diffusers are made from machined quartz, Teflon, or opal. The type of material is chosen based on the wavelengths of interest. Other applications, like radiance and luminance, call for a narrow field of view so an occluding hood or lensed barrel is used on the photodetector to limit the spatial response to a small, well-defined area.

Graphing the two responses on a polar plot provides an easy way to show how well a photodetector's spatial response matches the cosine response function.

As shown in Figure 10, with the photodetector normal to the signal at zero degrees, the cosine of the angle is one or 100%. As the signal angle changes to become parallel with the photodetector, the signal goes to zero, as does the cosine of the same angle ($\cos 90° = 0$). At 45°, a photodetector with a perfect cosine spatial response would read 70.7% of the signal ($\cos 45° = 0.707$) and so forth. As the angle increases, the receiving surface of the photodetector decreases, as illustrated in Figure 11.

Spectral Response
As shown in Figure 12, the spectral response refers to how the signal out of the photodetector changes as a function of the wavelength of optical radiation. Broadband

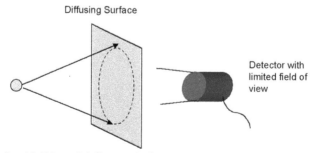

Diffusing Surface

Detector with
limited field of
view

• Radiometric Units: watts/m²/sr, watts/cm²/sr

• Photometric Units: candela/m², footlamberts (fL)

Figure 9   Radiance and luminance measurement. *Abbreviation*: Fl, foot Lamberts.

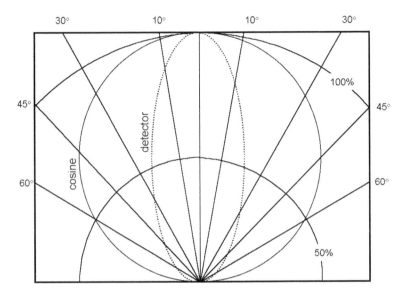

Figure 10 Polar spatial response.

radiometers and photometers employ bandpass filters to limit the bare detector response to the desired wavelength range. For absolute radiometry, a broadband detector with a "flat" (quantum) spectral response would be needed. This detector would, ideally, read 100% of the incident signal at every wavelength within the desired bandpass, plus have no response outside the range of interest. Unfortunately, perfectly spectrally flat broadband photodetectors with precisely square spectral responses do not exist. However, if the relative spectral output of the source is known, a calibration factor can be calculated and applied to the broadband photodetector reading to correct it to within a few percent of the absolute value. Otherwise, for absolute data, a true spectroradiometer (cosine correction may be necessary), which has the capability of wavelength selection by employing moving grating(s) to separate and isolate discrete wavelengths, is required. Because

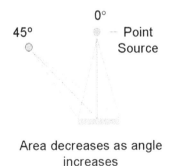

- $E_\theta = E\cos(\theta)$
- Cosine is 1 or 100% normal to detector at 0°. At 45°, cosine = 0.707, detector should read signal with 70.7% of value produced by same signal normal to detector
- cosine = 0 @ 90°

Figure 11 Cosine response.

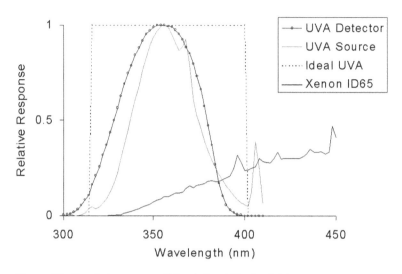

Figure 12 Spectral response. *Abbreviation*: UV, ultraviolet.

a high-quality spectroradiometer significantly escalates complexity and cost, for some, this approach may be impractical.

For many applications, a broadband detector can be made to maximize the spectral response in a wavelength range that covers both the intended spectral region of interest and the source being used while minimizing its response outside this range. This is an effective measurement, not an absolute measurement. Photometry is a good example of an effective measurement.

Without standardization, this approach, however, creates nonuniversal detectors with different characteristics, which can result in wide variances in readings, even when measuring the same source. In some cases, like photometry, this approach may work. The spectral response of the receiver (human eye) and the effective units of illuminance, Lux and fcs, are defined by the CIE (4). It must be pointed out, however, that not all photometers are created equal. Refer back to Figure 1, where a photopic detector response is shown plotted against the CIE $V(\lambda)$ spectral function. As you can see, this particular detector matches the CIE function very well. The CIE defines a methodology for assessing a photopic detector's relative spectral responsivity deviation from the $V(\lambda)$ function called the f1' error (5). The detector shown has an $f_{1'} = 3$, meaning it deviates by only 3% from the reference CIE function. A close match to the CIE function ensures better photometric accuracy when measuring different types of sources, other than the calibration source, especially monochromatic and near monochromatic sources.

CIE Publication 53-1982 (Methods of characterizing the performance of radiometers and photometers) and CIE Publication 69-1987 (Methods of characterizing illuminance meters and luminance meters: Performance, characteristics and specifications) provide detailed information on this error source and much more. CIE Division 2 TC 2-40 is currently working on an updated revision of these publications. Presently no radiometric applications have reached this level of standardization. However, both the European Thematic Network for UV Measurements (6) and the CIE TC 2-47 (Characterization and calibration methods of UV radiometers), established around 1998, (To prepare a CIE recommendation on methods of characterization and calibration of broadband UV radiometers in the UV-A and UV-B for industrial applications), are working toward this goal.

Another spectral consideration, in choosing a detector, is the photodetector's stray light rejection capability, or how well it can measure "in band" signal while excluding "out of band" signal. Stray light or "noise" can be many times larger than the signal of interest, to be measured. Extracting the UV-B content from sunlight is an example. "Solar blind" photodetectors with bandpass filters employing long wavelength blocking filters are required for this application, otherwise, gross errors due to stray light contamination will occur.

Distance

The source to photodetector distance should always be noted as part of any illuminance or irradiance measurement, as demonstrated in Figure 13. The inverse square law states that as distance increases, the signal decreases inversely proportional to the square of the distance. This is true for a point source relationship where the source to detector distance is at least five times greater than the largest source dimension. So in effect, if the distance is halved, the signal strength per unit area is quadrupled and vice versa. Luminance and radiance measurements are independent of distance because the detection area and source intensity vary in direct proportion with distance.

Absolute Calibration

Uncertainties in the absolute optical calibration of a photodetector by a U.S. National Institute of Standards and Technology (NIST) traceable laboratory are attributable to a combination of inherent uncertainties found in the primary NIST standard plus the uncertainties introduced by the calibration facility during transfer procedures.

Uncertainties are greater in the UV and IR spectral regions than in the VIS. At 254nm, the NIST uncertainty is ±1% (as last checked) added to the estimated transfer uncertainty from a qualified calibration facility of as much as ±10% yields a total estimated uncertainty of calibration from NIST to customer of ±11%.

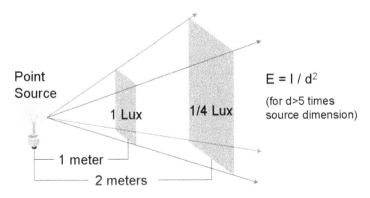

Point Source

$E = I / d^2$

(for d>5 times source dimension)

1 Lux   1/4 Lux

1 meter

2 meters

Figure 13 Inverse square law. Intensity per unit area varies inversely proportional to the square of the distance for a point source.

Temperature

The absolute sensitivity of a photodetector will change with temperature. This temperature coefficient, stated as a plus or minus percent per degree Centigrade, can cause thermal drift in readings, as the source under test heats the detector due to prolonged exposure. With a temperature coefficient of $-0.2\%/°C$, if the detector reaches a temperature that is 75°C higher than the temperature at calibration, the measured signal will be 15% lower than the actual signal. It is important to note that this pertains to the temperature of the active element of the photodetector, not its housing temperature but rather that of the sensing element. Photodiode temperature coefficients can be positive or negative and will vary with wavelength, making it very difficult to quantify the drift for multiwavelength sources.

Temporal Response

System time response is an important consideration especially for pulsed source and integrated energy or dose measurements. The meter/detector response rise times must be fast enough for short-term integration to follow pulse times faster than one cycle second. The system must also be stable enough for long-term integration over extended time periods. A hidden temporal problem arises when measuring average power from a fluorescent lamp, which is actually pulsing at 60 cycles per second or some other much higher frequency (e.g., 20,000 cycles), depending on the type of ballast used (J. Piechocki, personal communications, 2000). Here the radiometer/photometer must hold the reading from the individual pulses long enough to give a good average reading. More sophisticated, newer, systems are capable of displaying and analyzing these pulsed signals.

PHOTOSTABILITY

The current International Conference on Harmonization (ICH) guidelines (7) specify that drug substances and products under photostability testing receive a measured dose of both UVA and VIS optical radiation. The specified units of dosage measurement are to be in the terms of Lux hours and watt-hours per square meter. This guideline requires both radiometric and photometric measurements in terms of illuminance in Lux for the VIS photopic range and UVA (315–400 nm) irradiance in $W\,m^{-2}$ multiplied by exposure time in hours. It is important to remember that total or absolute UVA is implied. No effective UVA spectral function is specified. Referring to Figure 14, a perfect broadband UVA detector would have a flat square-wave spectral shape starting at 315 to 400 nm for 100% response at each wavelength across the UVA spectrum with no response outside this bandpass. Most currently available UVA detectors have a "bell" shaped spectral response which, if uncorrected through calibration, will read greater than 25% low the UVA fluorescent source as shown, and greater than 40% low for xenon glass ID65 type light sources. Note that, UVA fluorescent, xenon and metal halide lamps are the sources specified in the ICH guideline. A closer approximation to an ideal UVA broadband detector has recently been developed for photobiological applications, as shown in Figure 14. The typical total area error for this "flat" UVA detector is 14%, while that of a typical "bell-shape" is 34% as compared to the ideal UVA square function.

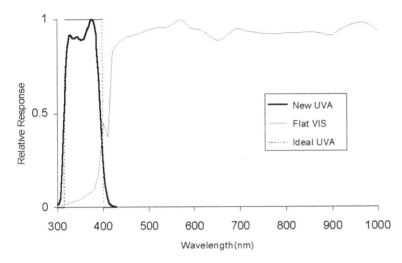

Figure 14   Ultraviolet-A and visible "flat" detectors. *Abbreviation:* UV, ultraviolet.

The ICH guidelines also state that to ensure spectral conformity of the source(s), a phototester "may rely on the spectral distribution specifications of the light source manufacturer." Often times, in actual practice, either the spectral data is not available or is not reliable. It is not that of a certified lamp but rather that of a typical lamp. As mentioned earlier, a spectroradiometer will be required to measure the actual spectral distribution of a certified lamp or an unspecified source.

The guideline offers no clear-cut answer as to why illuminance (Lux) measurement is specified in the ICH guidelines. It may be speculated that this was adopted from a similar application involving degradation of materials, specifically museum artifact conservation. In this particular application, typically, both UVA and Lux are measured, the readings divided, and the resulting ratio is to be kept within stated values for minimal degradation effect. In other terms, $W\,m^{-2}$ (UVA) divided by lumens m^{-2} (Lux) equals W/lumen or the ratio of the UVA content to the photopic, ambient, radiation.

In testing the photostability of drugs and drug products, where the effects of VIS radiation are much more uncertain, the relevance of measuring the VIS content of the test source based on the human eye [CIE $V(\lambda)$] response is questionable and potentially dangerous. Spectrally, "flat" and quantum broadband radiometers or spectroradiometers are available to measure the VIS spectrum in more absolute rather than effective terms, which would be more applicable, unless an action spectra is defined. Referring again to Figure 14, examples of spectrally "flat" UVA and VIS photodetectors are shown. As long as the current ICH guidelines stay in effect, as is, and no standardization of the instrumentation used in this application exists, the phototester faces a real dilemma. He/she is charged with measuring potentially unaccredited sources, using techniques associated with high levels of uncertainty and then adding the extra experimental error created by having to measure inside the test apparatus, normally a box or chamber, rather than in a classical radiometric setup.

SUMMARY

A photostability tester should always:

- Know the lamp type(s) and actual spectral power distribution
- Know measurement uncertainties of the metering devices to be used
- Have current calibration certification
- Have calibration correction factors for the specific lamp(s)

A good understanding of the measuring devices and their limitations is important.

REFERENCES

1. Hitchcock RT. Ultraviolet Radiation Nonionizing Radiation Guide Series. Akron: American Industrial Hygiene Association, 1991.
2. CIE Compte Rendu. 1924:67.
3. CIE International Lighting Vocabulary. Publication No. 17.4, 1987.
4. Kelley J. UV Curing Radiometry. New Orleans, LA: Radtech, 1986.
5. CIE. Methods of Characterizing Illuminance and Luminance Meters. Publication No. 69. Vienna: Central Bureau of the CIE, 1987.
6. Gugg-Helminger A. Characterizing the Performance of Integral Measuring UV-Meters. UV News 2000; 6:A-1–A-36.
7. International Conference on Harmonisation of Technical Requirements for Registration of Pharmaceuticals for Human Use. Stability Testing: Photostability Testing of New Drug Substances and Products. ICH Tripartite Guideline Recommended for Adoption at Step 4 of the ICH Process, 1996.

10 Sample Presentation for Photostability Studies: Problems and Solutions

Steven W. Baertschi
Eli Lilly and Company, Analytical Sciences Research and Development, Lilly Research Laboratories, Indianapolis, Indiana, U.S.A.

Scott R. Thatcher
Analytical Services, The Chao Center for Industrial Pharmacy & Contract Manufacturing, West Lafayette, Indiana, U.S.A.

INTRODUCTION

Many pharmaceutical drug substances and drug products are photosensitive and will undergo photodegradation upon exposure to outdoor or indoor photon sources. The topic of photostability has been dealt with by the International Conference on Harmonization (ICH) as a special case of stability (ICH Topic Q1A), and guidelines for photostability testing are presented as an annex (ICH Topic Q1B) to the parent stability guideline. Photostability testing, however, presents many unique difficulties when compared with thermal and humidity testing outlined by ICH Q1A. Issues such as the choice of photon source, irradiation homogeneity, containers/sample holders, sample thickness, and orientation are just a few of the parameters that are critical to meaningful evaluation of the photostability of a drug substance or product. Furthermore, careful consideration should be given to the type of analytical testing performed as part of the control strategy for evaluation of photostability postexposure. Sample presentation, the manner in which the samples are prepared, packaged, and aligned relative to the photolysis source for photostability testing, can have a significant effect on the outcome of the photostability experiment. This chapter will provide some practical information about the critical parameters involved in photostability testing with respect to "presentation" of the sample to the photon source.

TYPES OF STUDIES

Forced Degradation

Two types of studies are presented under the current ICH Q1B guideline (referred to as the ICH guideline hereafter) (1): forced degradation studies and confirmatory studies. Forced degradation studies are intended to provide information regarding the intrinsic photostability of the molecule, and compatibility of the drug substance with excipients, potential formulations, and dosage forms. Thus, forced degradation studies may be regarded as a stress test (2). Exposure requirements are not specified and are typically conducted as part of analytical method development, formulation development, validation for specificity, and evaluation of photodegradation pathways (3). Early versions of the ICH guideline suggested that forced degradation studies should use exposures that are up to 5 to 10 times the exposure level used in confirmatory studies.

The use of different photon sources (i.e., Q1B Option 1 vs. Option 2) for forced degradation studies and confirmatory studies is not specifically discussed in the ICH guideline; however, the use of different photon sources can lead to differences in the photostability results (i.e., different photodegradation rates and potentially different photodegradation profiles). Thus, it may be prudent to use the same photon source for both studies, as was shown in the work of Sequeira and Vozone (4–6).

Confirmatory

Confirmatory studies should be conducted on both the drug substance and the final formulation as part of the formal stability studies for registration, and may be regarded as an accelerated stability test (2). It is very important that the physical state of the sample be in its manufactured and/or marketed form (i.e., final crystal form, particle size, and hydration state) during exposure. Techniques used to alter the physical state of the sample (e.g., such as grinding to reduce particle size) as recommended in some publications[a] (7) should be avoided.

Confirmatory studies are used to determine if precautionary measures are needed during formulation or manufacturing and whether special precautionary labeling and/or packaging is/are needed to mitigate photoexposure. Validated methods are used to analyze the exposed drug substance or drug product. If an "unacceptable change" is observed while directly exposing (no protective packaging) the drug substance or drug product, photostability testing is to be repeated on the materials in their immediate packaging or final packaging. This sequential testing is to continue through dosage alteration, packaging, or formulation changes until photostability is demonstrated, as described in the flow diagram in the ICH guideline (1).

Confirmatory studies can be viewed as a limits test given that the final results, qualitative and quantitative, have to be judged as "acceptable" or "unacceptable." An acceptable change as described in the ICH guideline is a "change in limits as justified by the applicant." This definition is broad and nondefinitive. A more practical definition of "acceptable change" might be the change within the justified limits that does not compromise quality, safety, and/or effectiveness of the drug substance or drug product. Conversely, an unacceptable change is a change outside the justified limits that compromises quality, safety, and/or effectiveness (8).

PHOTOLYSIS SOURCES OR CHAMBERS

Sources

Two options are proposed in the ICH guideline for the photostability testing of pharmaceuticals. Option 1 sources are to have outputs that are "similar" to either the D65 or the ID65 emission standard listed in International Standards Organization (ISO) 10977 (9). The D65 emission standard pertains to outdoor daylight radiation (unfiltered sunlight) and the ID65 standard pertains to indoor-indirect daylight (i.e., sunlight filtered through window glass) (10). The D65 and ID65 emission standards are not identical in that the ID65 standard cuts off a significant amount of

---

[a]Grinding of the drug substance powders to reduce particle size was suggested in the Pharmacopeial Forum (PF). The authors of the present chapter strongly feel that this suggestion is in direct contrast to the ICH Q1B guideline, and therefore should not be followed.

high-energy radiation below 320 nm. Either standard D65 or ID65 is acceptable according to the ICH guideline (1).

Option 2 photolysis sources attempt to mimic "indoor lighting conditions" (a combination of window-glass–filtered daylight and artificial radiation, although "indoor lighting" varies greatly and has no "standard") as well as those commonly encountered in manufacturing facilities, warehouses, pharmacies, and homes, with an additionally added ultraviolet (UV) component to account for potential indirect daylight exposure. The cool white fluorescent sources are to have output similar to the emission standards as listed in the ISO 10977 for the visible (VIS) region. The UV fluorescent source (11,12) must have an emission similar to the spectral distribution requirements stated in the ICH guideline. The sample is to be exposed to both sources (either sequentially or simultaneously) so that exposure to both near-UV and VIS radiation is obtained.

Each option in the ICH guideline presents its own advantages and disadvantages (8). One distinguishing feature of Option 1 versus Option 2 is that Option 1 involves use of a single lamp, and therefore samples are exposed to near-UV and VIS radiation simultaneously. Option 2 utilizes a two-lamp system (i.e., a "cool white" fluorescent lamp for VIS light emission and a UVA lamp for near-UV emission). The ICH guideline implies that samples are to be exposed to the sources sequentially (i.e., the same sample is exposed to the UVA lamp and the fluorescent lamp at different times). Because no guidance is given as to which lamp should be used first, it is apparently acceptable to expose the sample in either order. The wording of the ICH guideline also makes it acceptable to expose the sample to both sources at the same time. Because it is possible that different results can occur as a function of which lamp the sample is exposed to first, simultaneous exposure is recommended (2,7).

The spectral power distribution (SPD) of a source, referred to as the "output" in the ICH guideline, is a plot of wavelength versus calibrated intensity across a spectral range (e.g., 320–800 nm). An SPD provides both qualitative and quantitative data regarding a source, and is considered the gold standard for source calibration. SPD data can be used for calculating dosages for a particular source in both the near-UV and the VIS regions. SPDs supplied by photolysis source manufacturers will often be an average of several lamps and not an individual lamp (13). The rate of change in the SPD of a source will depend on the frequency of use and the type of source. The topic of lamp or source selection is dealt with in more detail in Chapter 6 of this book.

Chambers

Most marketed test chambers are designed for weathering simulation, where simulated daylight and/or indoor lighting conditions are used for the stress testing of cosmetics, paints, textiles, and building materials. Only recently have chambers been advertised as designed specifically for the photostability testing of pharmaceuticals. These chambers are typically adaptations of previously marketed weather simulator designs and therefore are not always well suited for pharmaceutical needs. Consequently, a researcher must keep in mind the type of samples tested (e.g., powder and liquid), size, orientation of the photon source relative to the samples environmental control (i.e., temperature and humidity), and number of test samples that need to be tested at one time.

One important aspect of chamber selection is the overall size of the apparatus. Chambers range in size from benchtop models to whole room models. This constraint alone may narrow the field of choice depending on available space for the unit or

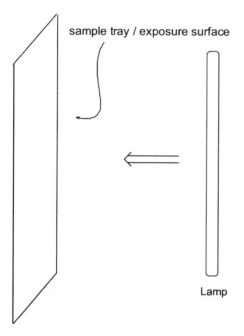

sample tray / exposure surface

Lamp

Figure 1   Vertical exposure design.

the total throughput of samples that need to be tested. Benchtop and freestanding models will vary with the available space for samples (i.e., footprint). This varies due to number and size of source(s), location of the sources(s) in the chamber, and efficiency of design.

Some chambers are designed to present samples to the source in a vertical orientation (Fig. 1) while others are designed to present samples lying flat (horizontally) (Fig. 2). The chambers designed to present samples vertically are well suited for paints, textiles (which can easily be hung vertically), and large bottles, but the presentation of pharmaceutical samples (such as powders) in a vertical orientation is more challenging.

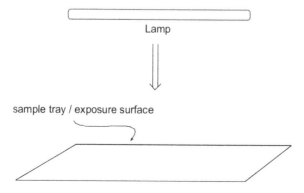

Lamp

sample tray / exposure surface

Figure 2   Horizontal exposure design.

Figure 3   Illustration of a rotating horizontal sample presentation tray. *Source*: Courtesy of Nagano Science, Japan.

A simple but effective apparatus for such a powder sample holding device is discussed later on in this chapter under the heading Containers.

The sample exposure area inside the chamber can vary from a simplistic flat horizontal tray, to a rotating horizontal tray, to a vertical rotating design (an example of a rotating horizontal tray is shown in Fig. 3).

Rotating trays are designed to aid in the uniform exposure of samples to the irradiation, because irradiation intensity can vary significantly within a chamber. With a vertical rotating design (i.e., the samples oriented vertically from a center lamp), samples are placed "upright" and can be distributed 360° around the center lamp in a carousel type design. A circular rotation of the carousel ensures uniform exposure for all the samples. With a "flat" or horizontal rotating tray (with the lamp oriented above the samples), samples placed at the same distance from the center of the tray receive the same exposure. Conversely, samples placed at different distances from the center of the tray may receive different irradiation intensities dependent on the spatial homogeneity of the lamp and the chamber. One attempt to overcome this limitation of the rotating tray design is a "double rotating tray" design (Fig. 4). This design does provide an averaging effect on sample irradiation exposure for the various locations on the trays, but accurate measurement of the irradiation intensity by a physical actinometer requires the radiometric sensors to be built into the moving trays, as shown in the Figure 4, which provide only single point measurements.

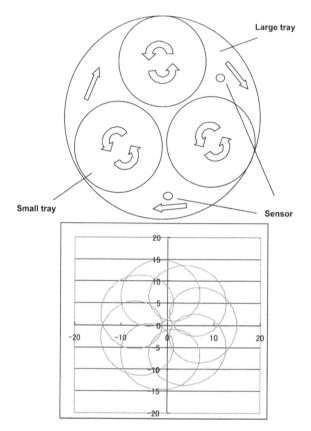

Figure 4 Double rotating tray design (*upper*) Paths of samples on outermost peripheries of small trays (*lower*). *Source*: Courtesy of Nagano Science, Japan.

The horizontal, flat exposure, nonrotating design is the more common design for pharmaceutical photostability testing apparatus in Europe and America. The horizontal rotating sample surface design is more common in Japan. Consideration of the irradiation homogeneity over the surface of the exposure area is important to the accurate determination of the irradiation intensity a sample will receive when placed on the exposure area.

Mapping

The process of determining the irradiation homogeneity as a function of a location on the exposure area is often referred to as "mapping," a term commonly used in the pharmaceutical stability field when referring to temperature homogeneity within a thermal stability chamber. It should be noted that such measurements do not provide any information about the SPD of the radiation impinging upon the samples.

Chamber irradiation intensity mapping involves measurement of the irradiance (measured in Watts/m², near UV region, 300–400 nm) and illuminance (measured as lumen/m² or Lux, VIS intensity) at different locations of the sample exposure surface

(14). This does not need to be a complex procedure. Measurements (using broadband radiometers, quantum detectors, spectroradiometers, or chemical actinometers) simply need to be made at several locations, including the corners (near the walls) and the center of the chamber. The more locations measured, the higher the resolution of he resulting intensity map. The average of the measurements can be considered to represent the average irradiance or illuminance of the chamber. Areas of the exposure surface representing lower intensities can either be "marked off" to avoid sample placement in these areas, or the exposure times for samples placed in previously calibrated spots can be increased (relative to the higher intensity areas) to compensate for their lower intensities.

Alternatively, one can simply adjust the exposure times to ensure that all areas of the chamber receive at least the ICH minimum recommended exposure for confirmatory studies (i.e., 200 W h/m^2 near-UV and 1.2×10^6 Lux hours) (1). This approach will lead to what some have called "overexposure" of the samples in the more intense regions of the chamber with higher intensities, a practice deemed acceptable under the ICH guideline.

"Overexposure" is not really an appropriate term, however, because ICH confirmatory photostability testing involves meeting a minimum recommended exposure, and *not* a maximum. Strictly limiting the exposure to the minimum recommended exposure, while acceptable, is not a requirement and is probably less predictive of what would happen in a real-world exposure to outdoor sunlight.

Several examples of the mapping of chambers to determine the homogeneity of irradiation intensity have been presented at photostability conferences (15–17). An example of irradiance and illuminance intensity mapping of a photostability chamber is shown in Figure 5.

As seen in the Figure, the irradiance (UV) and the illuminance (VIS) maps can look very different, indicating that the ratio of intensities, i.e., UV/VIS ratio, can vary across the sample exposure area. Although in the example shown in the figure there are multiple lamps, differences in irradiance and illuminance maps can be expected even when the chamber consists of a single lamp. Thus, intensity mapping should include separate measurements of both irradiance and illuminance, unless a spectroradiometers is used, because this instrument measures all wavelengths in a "single" measurement, irradiance.

Mapping for the irradiance and illuminance levels is conducted separately, unless a spectroradiometer is used, because near-UV and VIS radiation are measured using different detectors. In the end, there will be two maps for the chamber: one for the near-UV region and one for VIS region. Mapping is conducted using calibrated physical or chemical actinometers, placing the detector (or solution) in several different locations throughout the chamber. (Note: detectors should be "cosine corrected" to correct for the dependence of a radiometer's response on the angle between the direction of the incident radiation and the normal of the collector for the radiometer.)" Remapping should be done periodically and should be considered each time the lamp is replaced.

For chambers that have the capability for different power settings for the source, changes in the power setting will almost certainly affect the intensity map, both in the UV and in the VIS regions. This is illustrated in Figure 2A and 2B (Chapter 2), which shows an irradiance map at a low and high lamp power setting for a xenon long-arc photostability chamber. Thus, if photostability studies are to be performed at different lamp power settings, intensity mapping should be performed at the relevant power setting.

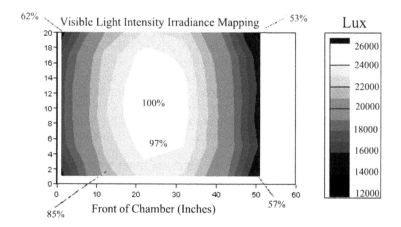

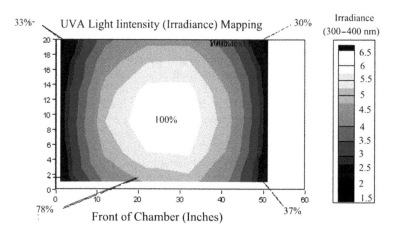

Figure 5   Illuminance (*upper*) and irradiance (*lower*) photostability chamber mapping of a single photostability chamber configured with both cool white fluorescent lamps and UVA lamps. (Chamber configured with six cool white fluorescent and two UVA fluorescent bulbs, symmetrically arranged—2 cool white, 1 UVA, 2 cool white, 1 UVA, 2 cool white.

An alternative to using chamber mapping as a means to facilitate accurate photoexposure is to make radiometric measurements at the site of placement of the sample at the start and end of the photoexposure. Thus, for example, if several samples are to be photoexposed, the site of placement of each sample can be measured, and the duration of the photostability study can be adjusted as needed for each sample location.

It is important to understand that the differences in irradiance or illuminance intensities observed during chamber mapping are not simply artifacts of broadband radiometric measurements; that is, the differences measured should translate into significant corresponding differences in the rates of drug photodegradation as a function of location in the chamber. This topic has been studied by both Baertschi (16) and Clapham and Sanderson (17).

As shown by Baertschi in Figure 6, the amount of photodegradation of a UVA-photounstable drug after irradiation for a specified time does vary as a function

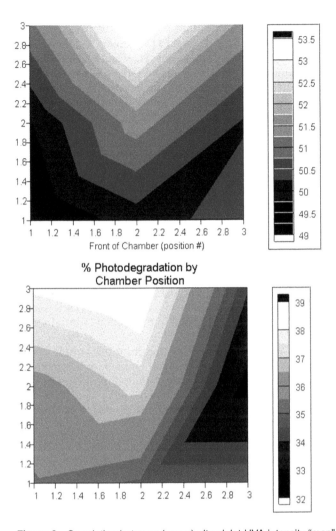

% Photodegradation by Chamber Position

Figure 6 Correlation between (*upper*) ultraviolet UVA intensity "map" (measured by an IL 17000 radiometer with a UVA detector) and (*lower*) the amount of photodegradation of a UVA photosensitive drug as a function of sample position in the chamber. All measurements made in a Suntest CPS+ photostability chamber equipped with a Xenon-arc lamp, Atlas Materials Testing Technologies, Chicago, IL, U.S.A.

of sample position, and the resulting "photodegradation map" is quite similar (though not identical) to the UVA intensity map determined using a broadband radiometer (16). In this case, the drug was dissolved in a solution and presented to the photon source in scintillation vials in the upright position, covered with a clear polyethylene film. Differences between the radiometer-derived map and the drug degradation map are likely due to the effects of "shading," i.e., how effectively the irradiation actually reaches the drug solution from all angles by which the container is transparent. An intriguing example of this concept was demonstrated by Clapham and Sanderson (17) and is discussed in the section on "Shading" in this chapter.

Another important concept to be kept in mind is the fact that irradiance or illuminance measurements are inversely proportional to the square of the distance from the source. This is known as the inverse square law (13), and applies strictly only when the ratio of the distance from the lamp to the lamp diameter ratio is greater than 5:1. For most photostability chambers, this rule does apply because the "lamp diameter" does not refer to lamp "length," but rather the effective "width" of the source of the radiation. Therefore, it is important to make radiometric measurements *at the level* of the sample to be exposed.

Chamber Temperature and Lamp Intensity

Consideration must be given to the temperature of the chamber during photo-exposure, as well as to potential sample surface heating effects (18). Surface heating can be significant with all lamps, and particularly with Option 1 sources such as xenon and metal halide lamps, because these lamps emit significant radiation in the infrared (IR) region. The effect of sample color on surface heating is illustrated in Figure 7, which compares surface heating of different colors to a standard "black panel temperature."

Such surface heating effects, coupled with elevated chamber temperatures, can be significant variables to consider during photostability testing. Substantial surface heating effects have been observed in photostability testing of tablets and capsules, and the importance of this topic is discussed in the section "Containers," later in this chapter.

High chamber temperature or high surface temperature of the sample being exposed can also lead to chemical degradation of the drug molecule if it is susceptible to thermal degradation. The apparent photodegradation profile (i.e., the profile of products observed upon analysis after photoexposure) of a drug can change depending on the temperature and duration of the photoexposure. It is

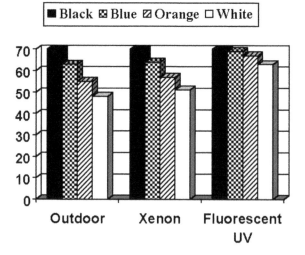

Figure 7 Typical surface temperature profile for colored samples exposed to outdoor sunlight (45° south facing-backed exposure), xenon long-arc radiation (intensity set to 0.55 W/m2 at 340 nm, filtered to mimic outdoor sunlight), and a fluorescent UVA-340 lamp. *Abbreviation*: UV, ultraviolet. *Source*: Courtesy of Atlas Materials Testing Technology, LLC, Chicago, IL, U.S.A.

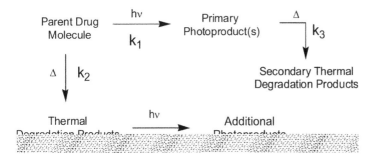

Figure 8 The effects of thermal degradation on the photodegradation profile obtained during photostability testing.

equally important, however, to consider the effect of lamp intensity. This can be illustrated by consideration of the interplay of lamp intensity and temperature as shown in Figure 8.

As shown in the Figure 8, a drug molecule and / or its photodegradation products can degrade by both thermal and photolytic pathways. If the rate of thermal degradation is fast enough (k_2), some thermal products can accumulate during the photoexposure time; these thermal products can either accumulate or further photodegrade (if they are photoactive). Additionally, if the primary photoproducts can degrade via thermally catalyzed pathways and if the rate of thermal degradation (k_3) is comparable to the primary photodegradation rate (k_1), secondary thermal degradation products can accumulate. The photodegradation profile can therefore be potentially different when photostability studies are conducted at different temperatures or at different lamp intensities.

It is worthwhile to consider in more detail the effect of lamp intensity on the rate of photodegradation (k_1) and the effect of temperature on the rate of thermal degradation $(k_2$ and $k_3)$ in order to better understand the implications of these two important variables. In general, it can be expected that

(1) Any change to the k_1/k_2 or k_1/k_3 ratios can affect the observed degradation profile
(2) If $k_1 >> k_2$ or k_3, mostly primary photoproducts will be observed, especially at early time points
(3) If $k_3 >> k_1$, secondary thermal products may predominate.

The effect of lamp intensity on k_1 is straightforward: k_1 increases linearly with increasing lamp intensity. This assumes, of course, that two-photon events are negligible. Two-photon events are typically negligible for the lamp sources and power levels typically used in pharmaceutical photostability testing (2). If, however, high-power laser or pulsed photon sources are used, the possibility of two-photon events should be considered. Thus, if the intensity of the photon source is doubled, the photodegradation theoretically rate should double. In contrast, the relationship of temperature to the thermal degradation rate (k_2) is not linear, but typically follows Arrhenius kinetics (19), where $k_2 = A^* \exp -E_a/RT$, where k_2 = the deg. rate constant, A = the "preexponential" constant, E_a = the energy of activation, R = the universal gas constant, and T = degrees in Kelvin. (Arrhenius kinetics requires that the thermal degradation pathway(s) remain(s) the same for the temperature ranges considered.

Table 1 Effect of Temperature on Thermal Degradation Rate (k_2) (Assuming Arrhenius Kinetics)

	Relative rate with different energies of activation		
Temperature (°C)	$E_a = 12$ kcal/mol	$E_a = 15$ kcal/mol	$E_a = 20$ kcal/mol
25	1.00	1.00	1.00
30	1.39	1.52	1.75
40	2.62	3.37	5.04
50	4.78	7.10	13.65
60	8.36	14.33	34.79
70	14.20	27.74	83.98

This is usually the case in the range of room temperature to 60°C for typical small molecule drug compounds.)

If we make the reasonable assumption that the thermal degradation follows Arrhenius kinetics, we can estimate the effect of temperature on k_2 by assuming the E_a for the degradation process. It has been empirically determined that the energy of activation for most small molecule drugs falls in the range of 12 to 24 kcal/mol (19,20). It is possible, therefore, to estimate the effect of temperature on the thermal degradation rate, as shown in Table 1. If we assume an E_a of 20 kcal/mol, we see that k_2 increases about fivefold if the temperature experienced by a sample during a photostability study is increased from 25°C to 40°C. From the information presented above, some important conclusions can be made regarding the two critical variables of temperature and lamp intensity. Consider a photostability test performed with a lamp intensity that requires seven days (i.e., 168 hours) to meet the required ICH minimum photoexposure (as is often the case for Option 2 lamp combinations and other fluorescent lamps) (8) and the temperature is maintained at 25°C. Compare this to a study performed using a much higher intensity lamp (e.g., as is often the case with Option 1 lamps such as xenon or metal halide) where the required photoexposure can be met in as little as six to eight hours but at a chamber temperature of about 40°C. The thermal degradation rate at 40°C should be 2.6 to 5 times that of the rate at 25°C (assuming E_a's of 12–20 kcal/mol). The duration of the photoexposure at 25°C is 21 to 28 times longer than the 40°C exposure. Therefore, more thermal degradation will occur during the photoexposure at the *lower* temperature of 25°C! The use of a higher intensity lamp, therefore, even though it causes the photostability chamber to operate at a higher temperature, may create less thermal degradation than a photostability study using a low intensity lamp at a lower temperature. Table 1 can be used as a guide to help in estimating the effects of thermal stress during a photostability study.

Temperature Control
The control of temperature in many chambers (especially some of those with Option 1 sources such as xenon or metal halide lamps) is often accomplished using a combination of both IR (dichroic) filters and refrigerated air exchange, using fans to circulate the air. Consideration must be given to the development of cooling gradients and also the possibility of "blowing" of samples (e.g., dispersal of powders and toppling of containers) during photoexposure. When using fans, samples

may need to be secured with adhesive tape or by other means and/or covered to prevent sample loss, cross-contamination, and safety problems.

CONTAINERS

Sample presentation can have a significant effect on the outcome of the photostability experiment. Awareness that different dosage forms have varying levels of susceptibility needs to be incorporated into experimental design. Table 2 illustrates this point. Experiments should be designed such that the physical characteristics of the sample being tested are conserved. As outlined in the ICH guideline, the control of temperature should be used to minimize the effects of physical state changes such as sublimation, evaporation, or melting (1). Containers used in photostability studies should have known (i.e., "suitable") transmittance properties and be chemically inert (21). The transmittance properties of a container/cover can be easily determined using a UV/VIS spectrophotometer. The sample container or cover, or a representative piece thereof, can be placed in the beam path and scanned from at least 300 to 800 nm in the transmission mode against an air blank. One should remember that during photostability studies the incident radiation passes only through one wall of the container during testing before reaching the sample. So to accurately measure the transmittance of a container, only one wall of the container should be placed in the beam path. Because that may not be practical when using, e.g., a quartz cell (without breaking the cell), it may be simpler to just place the container in the beam path such that it passes through both walls. This will lead to lower than actual transmittance data, and would result in a slightly higher photoexposure (a conservative approach). Alternatively, the data could be mathematically corrected for transmittance through one wall.

For containers that are relatively "inexpensive," one can literally break the container and obtain measurements on a single wall. Regardless of the approach taken, reflective or absorptive losses due to the container should be included in the final calculation of the dose.

It cannot be assumed that a visibly transparent container is completely transparent to radiation from the UV through the VIS region. Even some clear plastic containers, such as those composed of polypropylene, may contain plasticizers, stabilizers, or mold release agents that may absorb strongly in this region. Some glass or plastic containers, such as Pyrex® glass, may contain impurities that can absorb significant radiation in the near-UV region (Chapter 7).

Scintillation vials appear to be a suitable container of choice for photostability studies, because they are manufactured of good optical quality glass and have reasonable transmission properties in the near-UV (>320 nm) and VIS

Table 2 Susceptibility of Different Dosage Forms to Photodegradation

Very high	High	Low
Parenterals	Soft gelatin capsules	Tablets
Infusions	Ointments	Capsules
Ophthalmic drops	Creams	Powders
Nasal solutions	Lotions	Coated tablets

regions, even though they do absorb significant radiation below about 320 nm. (According to the ICH guideline, it is acceptable to filter significant radiation below 320 nm). As discussed above, petri dishes with an appropriate cover (e.g., polyethylene film or quartz glass) are also suitable. In addition, quartz cells fitted with Teflon-lined screw caps are available and well suited for photostability testing.

Cross-Contamination of Samples

Depending on the instrument/chamber being used, a solid drug form may have to be covered to protect the sample in the container and other samples (if present in the chamber) from cross-contamination. For example, some photostability chambers equipped with a cooling fan, could disperse a drug substance powder throughout the chamber if the sample is not covered, causing not only cross-contamination but also potential safety problems.

The cover selected, based upon the foretasted principles, must be reasonably transparent. A quartz cover provides excellent transparency throughout the UV and VIS region. The Pyrex® glass cover of a typical petri dish absorbs a significant amount of radiation in the <335 nm range and is not as transparent as a typical scintillation vial of comparable thickness (Fig. 9).

Plastic wraps, thin films of polyethylene (e.g., Handi-Wrap™ and Glad® Wrap) have been shown to have acceptable transmittances in the near-UV and VIS regions, with Handi-Wrap™ being the most transparent and stable during photoexposure (22). For reference, the UV transmittances of three commonly used plastic wraps and a borosilicate scintillation vial are presented in Figure 10. It should be noted that the composition of Saran Wrap™ was changed in 2004 from polyvinylidine chloride (Saran Original) to low density polyethylene (Saran Premium). The transmittance data provided in Figure 10 were obtained prior to 2004.

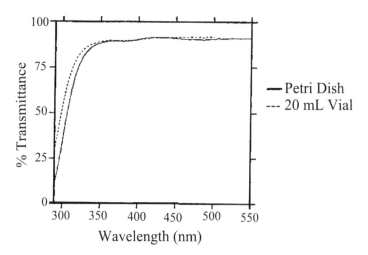

Figure 9 Transmittance of a Pyrex® Petri Dish cover compared to a 20 mL borosilicate glass scintillation vial.

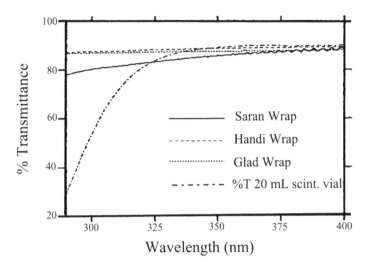

Figure 10 Transmittance of plastic wraps and a 20 mL borosilicate glass scintillation vial. *Source*: Courtesy of Kimble Glass, Vineland, NJ, U.S.A.

Greenhouse Effect

Samples covered during photostability studies may experience surface-heating effects as previously mentioned especially, for example, with Option 1 sources that emit high levels of IR radiation (such as xenon and metal halide lamps). Surface heating effects can be significantly accentuated by the use of plastic wrap covering, especially if the samples are colored and the wrap provides a tight seal. Elevated surface temperatures can greatly affect the results of photostability testing.

As shown in Figure 11, capsules exposed for 10 hours to xenon lamp radiation (Suntest CPS™ + chamber with UV and IR filters and SunCool™ refrigerated-air cooling in place, Atlas Electric Devices, Chicago, IL, U.S.) in a petri dish covered with polyethylene film showed significant visual signs of excessive heating (e.g., browning of pellets and melting of gelatin). An investigation of this problem determined that the maximum temperature of some of the capsules reached were in excess of 70°C during photoexposure (16)! A solution to this excessive surface-heating problem was simply to make a few small holes in the polyethylene film, which allowed the excess heat to dissipate, thereby reducing the "greenhouse effect" caused by the film. Using this simple solution, the surface temperatures were reduced to around 30°C to 35°C, just a few degrees above the chamber temperature, eliminating the melting and browning problem.

This same surface-heating problem was also observed during a tablet photostability study, where the coated tablets being photoexposed were reddish-brown in color. After 10 hours of photoexposure to a xenon lamp under the same conditions as the capsules (see above), the surface of several of the tablets showed obvious signs of "melting," and several of the tablets "fused" together. Subsequent investigations revealed that the surface temperature of some of these tablets exceeded 80°C. Again, either removing the polyethylene film cover from the petri dish or cutting a few small holes in the film alleviated the excessive surface-heating problem.

In addition to potential surface heating problems, consideration should be given to the potential for relative humidity changes during photostability testing. If

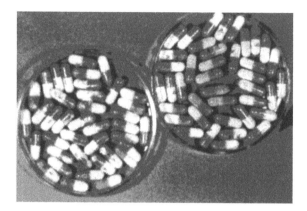

Figure 11 Photograph of capsules after photoexposure to a Xenon long-arc lamp (intensity of 765 W/m2, 300–800 nm) when tightly covered with polyethylene film (Glad® Wrap). Note the melted gelatin in the left petri dish and the darkened pellets in the right petri dish.

a sample is experiencing significant heating during the testing, the water present in the sample (from either the drug substance or, in the case of formulated products, the excipients) can be driven off into the atmosphere in the enclosed space of the sample. This phenomenon could result in polymorphic and other physical changes.

Another potential problem one should be cognizant of is the potential for condensation of water vapor, when using refrigerated air cooling, after the photon source is turned off (or the sample is removed from the chamber and cooled). The potential for such condensation is also present when samples are directly exposed (i.e., with no cover) to the photon source. For example, consider the measurements shown in Table 3. These measurements were made during and after a photostability test in an air-cooled xenon lamp chamber that utilizes a chiller to cool the air brought into the chamber to help with temperature control. Unfortunately, the cooling continued well after the xenon lamp was automatically turned off, and the chamber temperature subsequently dropped from 33°C to 14°C. The tablets undergoing the photostability testing showed evidence of water vapor condensation, and exhibited a dramatic (and unacceptable) loss of hardness. Redesigning the photostability test to prevent the inadvertent cooling period after the lamp is turned off eliminated the loss of hardness, showing that the tablets were not susceptible to photostability problems as had been first concluded.

For the presentation of powdered samples in either a horizontal or a vertical position, Stahl (23) designed a special apparatus shown in Figure 12. As shown in this figure, the holder is a metal (e.g., polished aluminum or stainless

Table 3 Changes in Chamber Temperature and Relative Humidity During a Photostability Test (Lamp "On") and After (Lamp "Off")

	During test (photon source "on")	After test (photon source "off," 2 hr later)
Chamber temperature	33°C	14°C
Relative humidity	43%	70%

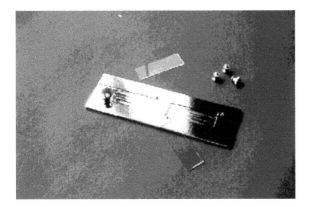

Figure 12   Sample holder for powder samples for photostability testing.

steel) plate with a trough (i.e., a hollowed well) machined into the surface to a specified depth; different depths can be designed as needed. The powder is placed in the indented region and covered by a quartz (or glass, if desired) slide held in place by thumbscrews, providing a powder in a contained area with a controlled thickness.

Shading

As documented in the ICH guideline under "Drug Product: Presentation of Samples (1)," "all precautions should be chosen to provide a minimal interference with the irradiation of samples under test." One of the potential interferences with the irradiation is the inadvertent shading (i.e., partial blocking of irradiance) of a sample by the presence of other samples positioned in close proximity. While it is difficult to prescribe precise recommendations for the "closeness" of neighboring samples, a rule-of-thumb that some in the industry have adopted is to place neighboring samples no closer than a distance of 1 to 1.5 times the height of the sample container. Such a practice will minimize the effects of shading from neighboring sample containers. The orientation of the container relative to the photon source is also a potential source of shading.

A dramatic example of this was presented by Clapham and Sanderson (17). Clapham examined the amount of photodegradation of a photounstable compound as a function of its position in a sunlight simulation chamber; the compound was presented to the photon source after dissolving in an appropriate solvent and placed in vial containers in an upright position.

As shown in Figure 13, the amount of photodegradation observed in the vials was lowest in the vials placed in the center of the chamber (where the irradiance measured was the highest), and highest in the vials placed near the walls and corners of the chamber (where the irradiance measured was the lowest). These unexpected results were eventually attributable to the shading effects of the opaque caps of the vials.

As can be seen in Figure 14, the solutions in the vials in the center experienced the greatest amount of "shading" from the opaque caps with the photon source directly "overhead." It is important therefore to consider the orientation of the sample

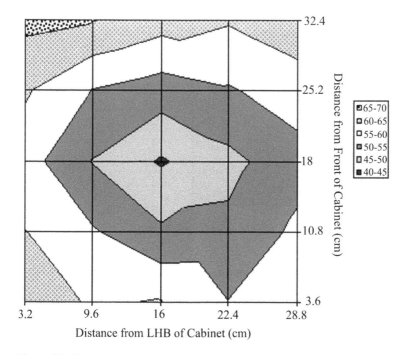

Figure 13 Percentage loss of compound Y in capped vials placed at different positions in a SOL 2 light cabinet. *Abbreviation*: LHB, Left hand base. *Source*: Courtesy of Dr. Hönle AG, Gräfelfing, Germany.

containers to the photon source and the potential shading effects of caps, labels, etc. According the ICH guideline, when performing confirmatory studies, the samples should be presented "horizontally or transversely to the light source, whichever provides for the maximum exposure area of the sample."

DRUG SUBSTANCE

For forced photodegradation studies, we recommend that samples of solid drug substances, mixtures with potential excipients (compatibility studies), and potential formulations be evaluated using a very thin layer (e.g., 1 mm or less) to maximize the photodegradation of the drug because photodegradation of solids is nonuniform and occurs primarily at the surface (24). Drug substances prepared as films (i.e., residue from evaporation of a drug substance dissolved in a solvent) will typically degrade more rapidly than the drug as a solid dry powder (25). However, because photodegradation rates and pathways can change depending on (*i*) the physical form of the drug (i.e., amorphous, crystalline, or polymorphic state) (25,26), (*ii*) the particle size, and (*iii*) the chemical form (i.e., salt form) of the drug substance used for manufacturing, the drug product and the final dosage form to be marketed should always be part of the forced degradation study (27).

It is recommended that drug substances be evaluated as a solid in a solid-state mixture with potential excipients as part of normal excipient compatibility studies.

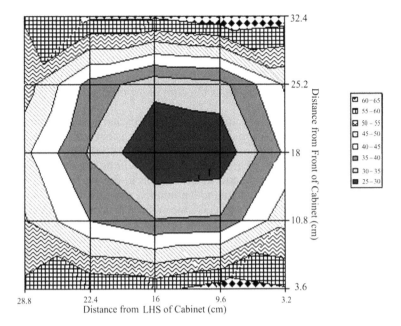

Figure 14 Percentage of maximum available radiation transmitted through a 14 mL clear glass screw capped vial.

An excellent example of the utility of excipient screening has been documented for sorivudine, where excipients (iron oxides) were selected and shown to markedly improve the photostability of the drug (28).

Drug substances should also be evaluated in simple aqueous solutions, in appropriate buffers, or in the presence of cosolvents for insoluble drugs. The dissolved drug substance should be exposed to pH values one to two units above and below the $pK_a(s)$ of the molecule to determine the photoreactivity of the ionized and unionized species. More detailed pH-photostability studies should be conducted for drug substances intended for formulation as solutions such as a parenterals or ophthalmic drops.

To conduct these kinds of studies, an appropriate container must be chosen. The only guidance provided by the ICH guideline on the choice of containers is that the container be "chemically inert and transparent." Some of the practical test container choices include glass Erlenmeyer or volumetric flasks, scintillation vials, quartz UV cuvettes and storage vessels, and final product immediate packages. It is important to remember, as previously mentioned, that the shape, size, wall thickness and transmittance of the container, the concentration of the drug substance, the volume or depth of the solution, and the orientation of the container relative to the radiation source are all variables that will affect the rate of photodegradation observed in the study. As an example, consider the relatively simple case of a scintillation vial containing a solution of a photoactive compound presented to the radiation source (with identical exposure times) as shown in Figure 15. The results of this experiment are summarized in Table 4. Some empirical conclusions can be made from a simple evaluation of these results. It is observed that the opaque cap, as would be expected, reduced the amount of photodegradation (by 5%), presumably by blocking some of the radiation. Doubling the sample volume greatly reduced (by 16%) the amount of photodegradation, as would be expected from any change in volume that results in a reduction of the ratio of exposed area to volume. The greatest amount of photodegradation was observed in the horizontal orientation (45% loss), as would be expected from an increase in the ratio of exposed surface area to volume.

This experiment is a rudimentary example of the differences that can be expected to occur with the changes described—a more detailed analysis is beyond the point of this example. The practical learning here is that the amount of photodegradation that will occur in solution in a study of a defined duration will be highly dependent on the container geometry, volume, and its orientation to the source (among other variables). Therefore, precise rates of photodegradation obtained in solutions for forced degradation studies may not be especially meaningful in an absolute sense (i.e., photo degradation rates obtained using the same container to be marketed at the relevant solution concentration(s) are more meaningful) (29,30). In a relative sense, however, such studies can provide excellent information on the effects of variables such as pH. Certainly, a great deal of valuable information can be obtained from the profile of photodegradation products and such information is used for method development purposes.

The choice of solvent(s) or cosolvent(s) to use for solution forced degradation studies is an important consideration. Photostability testing in water should be performed when solubility is not a problem. When the drug has a very low solubility in water, buffers or a cosolvent can be considered to help solubilize the drug substance.

The cosolvent or solvating agent should ideally be UV and VIS radiation transparent and photochemically inert. A common driving force for the choice of a cosolvent is the solvent used for the analytical methods (to help assure photostability of the drug during analytical workup). Thus, acetonitrile and methanol (the most frequently used reversed phase high performance liquid chromatography solvents) are frequently

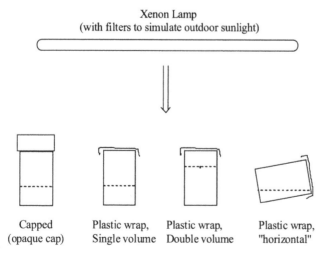

Figure 15 Illustration of a 20 mL borosilicate glass scintillation vial with different sample volume, cover, and orientation used during a photostability study.

Table 4 Effect of Volume, Cover, and Orientation on Photodegradation

Container	Percentage photodegraded
Capped, upright position, 1 volume	32
Plastic wrap cover, 1 volume	37
Plastic wrap cover, 2 volume	21
"Horizontal" exposure, 1 volume	45

used for forced degradation and photostability studies in the pharmaceutical industry. These studies, not specifically discussed in the ICH guideline, are an essential and regulatory-mandated part of the methods development and validation process.

It should be remembered when comparing photostability data that the selection of the solvent (cosolvent, buffer, and solvating agent) may affect the results obtained. For example, the solubility of oxygen in methanol, acetonitrile and water are 10.3, 9.1, and 1.39 mol/L, respectively (31). Thus the choice of solvent may significantly affect the results of photodegradation processes that are sensitive to oxygen.

For mechanistic studies of a photochemical reaction pathway, studies are often carried out in dilute solutions in an organic solvent under defined conditions (e.g., with and without oxygen present; in the presence of oxygen triplet sensitizers or quenchers). The photodegradation of the molecule can be measured kinetically, if needed, to determine the overall rate of photodegradation for the source being used (29,30). Kinetic studies are most applicable to solution products. Partially degraded drug substance solutions (e.g., <30% degradation) can be used for specificity testing during method development and validation. Preliminary formulations can be photoexposed as part of formulation development.

The number of experiments and the experimental design of photostability studies conducted under the "forced degradation studies" are left completely to the judgment of the applicant. Careful studies designed to discover photostability problems early in development could provide significant savings by avoiding costly surprises later on.

For confirmatory studies, according to the ICH guideline solid drug substances are to be placed in a suitable container and spread as to provide a layer that is no greater than 3 mm. This is supposed to minimize shielding that occurs using thicker samples because photodegradation of solid is typically a surface phenomenon. However, for most powders 3 mm is significantly thicker than the incident UV–VIS radiation will effectively penetrate, thus, significant shielding will occur. Because of this shielding, changes in the powder thickness can dramatically affect the results obtained during a confirmatory study. Figure 6 in chapter 16 for a graphical illustration of the problem.

Liquid drug substances can be placed in chemically inert transparent containers (e.g., quartz cells or scintillation vials). The ICH guideline recommends that such samples be exposed along their greatest axis, i.e., placed on their sides.

Only one batch of drug substance or drug product needs to be tested during the development phase, and then the photostability characteristics (i.e., either photostable or photolabile) should be confirmed on a single final batch. This batch should be one of the three final batches used for regulatory submission. Batch selection should be based on the criteria outlined in the parent guideline for selection of batches placed on definitive stability (32).

DRUG PRODUCTS

For confirmatory photostability studies, solid oral dosage forms should be directly exposed in the form the product will be received and used by the patient. The number of samples tested is dictated by analytical requirements (e.g., the guideline suggests an appropriate amount of 20 tablets or capsules). Photodegradation of solid oral dosage forms will most likely occur on the surface of the product (17,18,33). Therefore, the intact tablets and capsules should be tested using a single layer of the product in an open petri dish, or any other suitable container, either open or covered with a material of known transmittance.

A configuration for presenting tablets or capsules in their immediate pack for photostability testing is shown in Figure 16.

The sample configuration chosen can produce additional stress conditions on the product, especially when using some Option 1 photolysis sources that can generate heat. As discussed previously, there could be a "greenhouse" effect in the sample bottle where the temperature in the container could be significantly raised (e.g., >50°C). This heat could increase the relative humidity inside of the container if water-containing excipients or drugs are used.

These heat effects may not be captured by use of a dark control sample, if the container is covered with aluminum foil, which blocks the "greenhouse" effect. In cases where significant heating is suspected to be contributing to thermal degradation processes, the container can be left open or vented, or one can use a reduced irradiance level to minimize this effect. Another approach would be to include the dark control in the same container after wrapping the control samples in aluminum foil prior to their placement in the same bottle as the unwrapped test samples. In this case, the environment that the dark controls experience would be much closer to the environment the test samples experience, save for any surface heating caused by the sample color.

Placebo formulations can also be exposed to the photon source to aid in determining if the origin of a photodegradation product is from an inactive component;

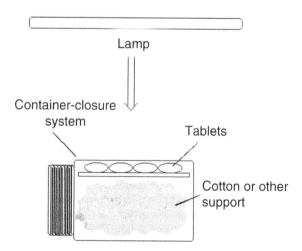

Figure 16   Illustration of a 20 mL borosilicate glass scintillation vial with different sample volume, cover, and orientation used during a photostability study.

however, if one of the excipients were acting as a photosensitizer (leading subsequently to photosensitized degradation of the active), this would not be detected. The presentation of drug products that are solid powders should be done in the same manner as solid drug substances (i.e., <3 mm thick layer, homogeneous sample).

For liquid or semisolid drug products such as parenterals, emulsions, suspensions, dermal creams, etc., the issue of "direct exposure" is not clearly addressed and can be somewhat confusing. The ICH guideline states, "Where practicable when testing samples of the drug product outside the primary pack, these should be presented in a way similar to the conditions mentioned for the drug substance."

For some types of products, it does not seem "practicable" to do direct exposure outside the "primary pack," in that the "primary pack" may represent the form of the product that will be received and used by the patient (e.g., unit dose inhalants, drugs delivered by personal injection devices, and intravenous bags or vials). In these cases, it seems logical that exposure in the immediate package effectively represents direct exposure for confirmatory studies. Because this is not clearly defined in the ICH guideline, it may be prudent to consult with the FDA prior to designing such a study. If a liquid or semisolid is to be exposed outside the immediate package, the details of how to perform such an exposure are again not defined in the ICH guideline.

The use of quartz cells as actinometric and sample containers are mentioned in the ICH guideline, but no mention of cell dimensions is made. It may be presumed that a standard 1 cm path length cell would be appropriate for semi-solids, emulsions, and creams but this is not analogous to the less than 3 mm thickness requirement for solid drug substances, and sample thickness will greatly affect the results. There are quartz cells available with 3 mm and less path lengths, and use of a cell with less than a 3 mm path length seems appropriate. Alternatively, a semisolid product could be dispensed into an open petri dish and covered with a transparent cover (e.g., quartz slide or polyethylene plastic wrap). A cover can help prevent unwanted physical state changes (14).

GENERIC DRUGS

For generic products, the guideline states "Photostability studies need not be conducted for products that duplicate a commercially available listed drug product provided that the packaging (immediate container/closure and market pack) and labeling storage statements regarding light duplicate those of the reference list product" (1). If there are any significant changes (e.g., formulation or packaging differences) from the commercially available product, confirmatory photostability testing may need to be conducted to ensure equivalence. Based on published and publicly presented data, even for the same ingredients, equivalence is not a given. This point was clearly illustrated in a presentation by Reed et al. (34) where photostability differences were attributed to trace levels of iron.

CONCLUSION

A critical aspect of the design and execution of meaningful (and reproducible) pharmaceutical photostability studies involves the presentation of the samples being tested to the photon source. Different considerations need to be made for forced photodegradation studies and confirmatory photostability studies. The choice of photon source (e.g., Option 1 vs. Option 2) is a very important part of any study. It is recommended that the same photon source be used for both forced degradation and confirmatory photostability studies.

Irradiation intensity mapping of the photostability chamber is needed in order to understand the ramifications of placing samples in different locations in the chamber. The potential effects of thermal degradation during photostability studies (i.e., the potential for different photodegradation profiles as a function of chamber temperature and duration of the study) should be appreciated. To minimize thermal degradation during a photostability study, the use of a higher intensity lamp with a shorter exposure time may be more effective than reduction of the chamber temperature (with a longer exposure time) during the study.

Containers used to present samples to the photon source can greatly affect the photodegradation rates, and therefore the transmittance characteristics of the containers should be known in order to determine the amount of radiation (UV and VIS) that will reach the sample. Orientation of the samples relative to the photon source should maximize the amount of surface area exposure, especially when performing ICH confirmatory photostability studies.

REFERENCES

1. International Conference on Harmonization. Guidelines for the Photostability Testing of New Drug Substances and Products. FR 1997; 62:27115–27122.
2. Riehl J, Maupin C, Layloff T. On the choice of a photolysis source for the photostability testing of pharmaceuticals. PF 1995; 21:1654–1663.
3. ICH. Guidance on the validation of analytical procedures: methodology. FR 1997; 62(96): 27463–27467.
4. Sequeira F, Vozone C. Photostability studies of drug substances and products. Pharm Tech 2000; 24(8):30–35.
5. Thoma K, Kerker R. Photoinstability of drugs. Part 1. The behavior of UV-absorbing drugs in simulated daylight. Pharm Ind 1992; 54(2):169–177.
6. Thoma K, Kerker R. Photoinstability of drugs. Part 2. The behavior of visible light-absorbing drugs in simulated daylight. Pharm Ind 1992; 54(3):287–293.
7. Photostability Testing (1172), Pharmacopeial Forum 2000; 26:384–387.
8. Thatcher SR, Mansfield RK, Miller RB, Davis CW, Baertschi SW. Pharmaceutical photostability: a technical guide and practical interpretation of the ich guideline and its application to pharmaceutical stability—Part I. Pharm Tech 2001; 25(3):98–110.
9. International Organization for Standardization (ISO). Photography—Processed Photographic Colour Films and Paper Prints—Methods for Measuring Image Stability. ISO 10977:1993, Geneve, Switzerland: 1993.
10. A Method for Assessing the Quality of Daylight Simulators for Colorimetry. Commission Internationale De L'Eclairage (CIE), 1999: 51(2); ISGN 92 9034 051 7.
11. Standard Practice for Operating Fluorescent Light Apparatus for V Exposure of Non-metallic Materials. American Society of Test Materials (ASTM) Designation G154-00ae1.
12. Standard Practice for Accelerated Testing for Color Stability of Plastics Exposed to Indoor Fluorescent Lighting and Widow-Filtered Daylight, ASTM Designation D 4674–4689.
13. Piechocki J. Selecting the right source for pharmaceutical photostability testing. In: Albini A, Fasani E, eds. Drugs: Photochemistry and Photostability. The Royal Society of Chemistry, 1998:247–271.
14. Thatcher SR, Mansfield RK, Miller RB, Davis CW, Baertschi. Pharmaceutical photostability: a technical guide and practical interpretation of the ICH guideline and its application to pharmaceutical stability—Part II. Pharm Tech 2001; 25(4):50–60.

15. Brumfield JC. The Pharmacia and Upjohn experience in selecting a photostability guideline option. Photostability 99: 3rd International Conference on the Photostability of Drug Substances and Drug Products, Washington, DC, July 10–14, 1999.
16. Baertschi SW. Practical aspects of pharmaceutical photostability testing. Photostability 99: 3rd International Conference on the Photostability of Drug Substances and Drug Products, Washington, DC, July 10–14, 1999.
17. Clapham D, Sanderson D. The importance of sample positioning and sample presentation within a sunlight simulation cabinet. Photostability 99: 3rd International Conference on the Photostability of Drug Substances and Drug Products, Washington, DC, July 10–14, 1999.
18. Boxhammer J. Technical requirements and equipment for photostability testing. In: Tønnesen, ed. Photostability of Drugs and Drug Formulations. Taylor & Francis, 1996:39–62.
19. Kennon L. Use of models in determining chemical pharmaceutical stability. J Pharm Sci 1964; 53(7):815–818.
20. Connors KA, Gordon AL, Valentino SJ. Chemical Stability of Pharmaceuticals. A Handbook for Pharmacists, 2nd ed. Wiley Interscience, 1986:19.
21. Yoshioka S, Ishihara Y, Terazono T, et al. Quinine actinometry as a method for calibrating ultraviolet radiation intensity in light-stability testing of pharmaceuticals. Drug Dev Ind Pharm 1994; 20:2049–2062.
22. Joshi HN, Formicola S, Pazadan JL, Carpenter P. Photostability and transmittance of four commercially available plastic wraps. Pharm Tech 2000; 24:74–82, 129.
23. Stahl PH. Stabilisierungstechnologie. Wege zur haltbaren Arzneiform. In: Essig D, Hofer J, Schmidt PC, Stumpf H, eds. APV Paperback Vol. 15, Wissenschaftliche Verlagsgesellschaft Stuttgart; 15(APV Paperback):15–32.
24. Teraoka R, Otsuka M, Matsuda Y. Evaluation of photostability of solid-state dimethyl 1,4-dihydro-2,6-dimethyl-4-(2_nitrophenyl)-3,5-pyridinedicarboxylate by using Fourier-transformed reflection-absorption infrared spectroscopy. Int J Pharm 1999; 184(1):35–43.
25. DeAngelis N. Photostability—industry practice and experiences. Photostability 99: 3rd International Conference on the Photostability of Drug Substances and Drug Products, Washington, DC, July 10–14, 1999.
26. Chongprasert S, Griesser UJ, Bottorf AT, Adeyinka WN, Byrn SR, Nail SL. Effects of processing on the crystallization of pentamidine isethionate. J Pharm Sci 1998; 87(9):1155–1160.
27. Baertschi SW and Jansen PJ, "Stress Testing: A Predictive Tool", in *Pharmaceutical Stress Testing: Predicting Drug Degradation*, Baertschi SW, Editor, Taylor & Francis, New York, p.22 (2005).
28. Serajuddin A, Thakur A, Ghoshal R, et al. Selection of solid dosage form composition through drug-excipient compatibility testing. J Pharm Sci 1999; 88:696–704.
29. Moore DE. Kinetic treatment of photochemical reactions. Int J Pharm 1990; 63:R5–R7.
30. Moore D. Standardization of photodegradation studies and kinetic treatment of photochemical reactions. In: Tønnesen H, ed. Photostability of Drugs and Drug Formulations. Taylor & Francis, 1996:63–82.
31. Muriv L, Carmichael I, Hug GL, eds. Solvent Properties, Handbook of Photochemistry, 2nd ed. New York: Marcel Dekker, 1993:289–293.
32. International Conference on Harmonization. Draft guidance for industry: stability testing on new drug substances and products. FR 2000; 65(78):21446–21453.
33. Teraoka R, Otsuka M, Matsuda Y. Evaluation of photostability of solid-state nifedipine by using Fourier transformed reflection-absorption infrared spectroscopy. Photostability 99: 3rd International Conference on the Photostability of Drug Substances and Drug Products, Washington, DC, July 10–14, 1999.
34. Reed RA, Harmon P, Manas D, et al. The role of excipients and package components in the photostability of liquid formulations. PDA J Pharm Sci Technol 2003; 57(5):351–368.

11 Determining the Kinetics and Mechanism of a Photochemical Reaction

Douglas Moore
*Department of Pharmacy, The University of Sydney,
Sydney, Australia*

REACTION KINETICS AND PHOTOSTABILITY

Testing of the photostability of a drug substance at the preformulation stage often involves a study of the rate of degradation of the drug in solution when exposed for a period of time to a source of ultraviolet and visible (UV-vis) radiation. If aliquots of the solution are analyzed after varying times of irradiation, it is possible to determine the rate of photodegradation of the drug under the experimental conditions as set up. A testing question is: "How is the data to be treated—is the reaction expected to follow zero-, first-, or second-order kinetics?"

The kinetic relationship most frequently applying to stability studies in solution is that of first-order kinetics in which the rate of degradation of a drug substance depends on the concentration raised to the first power, as in Eq. 1,

$$\text{Rate of degradation} = -d[\text{Drug}]/dt = k_1[\text{Drug}] \tag{1}$$

where k_1 is a proportionality constant called the specific reaction rate constant for the degradation. The integrated form of the first-order rate equation can be expressed in either exponential form (Eq. 2) or logarithmic form (Eq. 3).

$$[\text{Drug}]_t = [\text{Drug}]_0 e^{-k_1 t} \tag{2}$$

$$\ln [\text{Drug}]_t = \ln [\text{Drug}]_0^{-k_1 t} \tag{3}$$

where $[\text{Drug}]_0$ and $[\text{Drug}]_t$ represent the drug concentration at times zero and t respectively.

Verification of degradation by first-order kinetics can be obtained when, according to Eq. 3, a plot of the logarithm of the remaining drug concentration versus time is linear. As an example, Figure 1 displays the photodegradation of benzydamine hydrochloride at two pH values of the solution (1). The straight-line plots confirm the first-order kinetics, and a value of the apparent rate constant for each pH can be obtained from the slope of the respective line.

The alternative kinetic situation that is often encountered experimentally in photochemical reactions is zero-order kinetics, corresponding to the situation where the rate of degradation of the drug is a constant, independent of concentration, as shown in Eq. 4

$$\text{Rate of degradation} = -d[\text{Drug}]/dt = k_0 \tag{4}$$

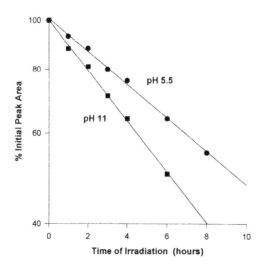

Figure 1   First-order (logarithmic) plot of the photodegradation of benzydamine hydrochloride (5×10^{-5} M) in air-saturated aqueous solution at pH 5.5 and 11.

The integrated form of Eq. 4 is given in Eq. 5 showing that the concentration of the irradiated drug will fall linearly with time, as demonstrated for the photodegradation of ketrolac tromethamine in (Figure 2) (2).

$$[\text{Drug}]_t = [\text{Drug}]_0^{-k_0 t} \qquad\qquad (5)$$

Other more complicated kinetic relationships, such as second- or fractional-order reactions, arise very infrequently from complex reaction mechanisms, and do not need to be considered for photodegradation reactions.

At first glance, it would seem that distinguishing between zero- and first-order kinetics is not a difficult task. However, a complication does arise because a reaction has to be taken to at least 50% conversion before an unequivocal

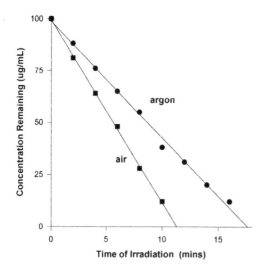

Figure 2   Zero-order (linear) plot for the photodegradation of ketrolac tromethamine (initial concentration 100 µg/mL) in ethanol solution under argon and air atmospheres.

delineation of the order can be made, even with good quality analytical data for the photodegradation. This point is demonstrated in Figure 3, where the early data of Figures 1 and 2 are replotted, according to the alternative kinetic equation, i.e., the benzydamine data is plotted on a linear scale in Figure 3A, and the ketrolac data is plotted on a logarithmic scale in Figure 3B. As can be seen, the linear plots show a reasonable agreement with the data in both cases, indicating that the benzydamine degradation could be zero order and the ketrolac degradation first order.

In the majority of cases, the drug will be relatively stable to irradiation and, unless very high radiation intensities are used, only a small extent of reaction is seen, so that the data will frequently fit any order kinetics. Intuitively, one expects that the photodegradation of a drug will depend on the amount of radiation absorbed by the drug, and that, in turn depends on the concentration of the drug. With that understanding, first-order kinetics should apply. Hence, many investigators will use a logarithmic plot, observe a straight line with a precision of 5% to 10%, and thereby obtain a first-order rate constant for the photodegradation process.

The value of the rate constant so determined, however, depends very much on the design of the experiment and is likely to vary greatly from one experimental setup to another due to a number of factors, of which only one is the variation

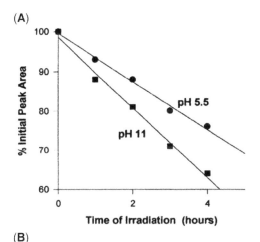

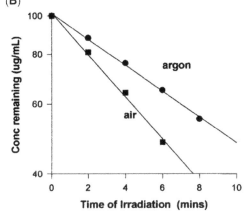

Figure 3 (A) Replot of early points for benzydamine hydrochloride photodegradation (Fig. 1) on a linear scale; (B) Replot of early points for benzydamine hydrochloride photodegradation (Fig. 2) on a logarithmic scale.

of the source of irradiation. Therefore, a rate constant value obtained experimentally for a photodegradation reaction should be designated as the "relative" or "apparent" rate constant for the process. Nonetheless, determination of the apparent rate constant for the photochemical reaction is useful for comparative studies using the same experimental arrangement, such as the variation of reaction rate with pH of the solution, as shown in Figure 1 for benzydamine.

The Quantum Yield

It must be emphasized that the first- or zero-order rate constant obtained as described above by analysis of a drug photodegradation system is applicable only to that experimental arrangement. For a "true" constant to describe the photochemical reaction of drug D degrading to products P1, P2, etc.,

$$D + h\nu \rightarrow P1 + P2 + \ldots \ldots \tag{R1}$$

the characteristic quantity, independent of the experimental arrangement used, is the "quantum yield" or "quantum efficiency." The quantum yield of a photoprocess, designated by the symbol φ, describes the rate or efficiency of the reaction, and is defined in Eq. 6:

$$\varphi = \frac{\text{Number of molecules transformed per unit volume per unit time}}{\text{Number of photons absorbed per unit volume per unit time}} \tag{6}$$

The quantum yield would normally be obtained by analyzing the amount of D remaining and expressed as φ_{-D}. When correctly measured, the quantum yield of a particular photochemical reaction should be the same, irrespective of whether it is determined in Washington, Sydney, or Oslo. When reporting the photochemical reactivity of a drug in a quantitative sense, the quantum yield is what should be quoted for each reaction (3,4). The reason for the interest of photochemists in the quantum yield of a photochemical reaction is that φ is the measure of the amount of reaction corresponding to the photons actually absorbed by the sample, and therefore is the true constant. Chemical actinometer systems have been widely used in basic photochemical studies to enable the determination of the quantum yield of a photochemical reaction (5).

Derivation of the Photochemical Rate Equation

The quantum yield is one of the two factors that determine the rate of a photochemical reaction, and is a characteristic of the compound involved. The other factor is the quantity of absorbed incident radiation that varies from one experimental arrangement to another, i.e., the extent to which the radiation is absorbed by the test sample (6). This is the denominator of Eq. 6, the number of photons, N, absorbed per unit volume per unit time. Thus, the rate of a photochemical reaction is defined as:

Rate = Number of molecules transformed per unit volume per unit time

= Number of photons absorbed per unit volume per unit time

$\times$ Efficiency of the photoreaction (φ)

$$= N\varphi \tag{7}$$

In the first instance, the rate can be determined for a homogeneous liquid sample in which the only photon absorption is due to the drug molecule undergoing transformation, with the restriction that the concentration is low, so that the drug does not absorb all of the available radiation in the wavelength range corresponding to its absorption spectrum. If one is using a monochromatic photon source, the number of photons absorbed depends upon the intensity of the photon source and the absorbance at that wavelength of the absorbing species.

The value of N can be derived at a particular wavelength λ and is given by:

$$N_\lambda = I_\lambda - I_t = I_\lambda(1 - 10^{-A}) \tag{8}$$

where I_λ and I_t are the incident and transmitted radiation intensities, respectively, and A is the absorbance of the sample at the wavelength of irradiation. The relation between these variables is illustrated schematically in Figure 4. The exponential expression in Eq. 8 can be expanded as a power series:

$$N_\lambda = 2.303\, I_\lambda(A - A^2/2 + A^3/6 - A^4/24 + \ldots\ldots) \tag{9}$$

When the absorbance is low ($A < 0.1$), the second- and higher-order terms can be considered negligible, and the expression simplifies to the first term in Eq. 9. Using the Beer-Lambert Law relation between absorbance, path length and concentration, together with the assumption that the system contains a single absorbing substance, N can be seen to be directly proportional to the concentration of that absorbing substance:

$$N_\lambda = 2.303 I_\lambda A = 2.303 I_\lambda \varepsilon_\lambda bC \tag{10}$$

where ε_λ is the molar absorptivity at wavelength λ, and C the molar concentration of the absorbing species, and b is the optical path length of the reaction vessel. Now I_λ and ε_λ vary with wavelength, so the expression has to be integrated over the relevant wavelength range where each has a nonzero value:

$$N = 2.303 bC \int I_\lambda \varepsilon_\lambda d\lambda \quad \text{integrated from } \lambda_1 \text{ to } \lambda_2$$

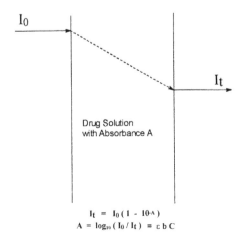

I₀

It

Drug Solution
with Absorbance A

$I_t = I_0(1 - 10^{-A})$

$A = \log_{10}(I_0/I_t) = \varepsilon bC$

Figure 4 Schematic representation of the fall in the intensity of radiation passing through a solution of an absorbing drug.

Thus,

$$\text{Rate} = 2.303bC\varphi \int (I_\lambda \varepsilon_\lambda) \, d\lambda \tag{11}$$

Now the overlap integral, $\int I_\lambda \varepsilon_\lambda d\lambda$, is a constant for a particular combination of photon source and absorbing substance, b is determined by the reaction vessel chosen, and φ is a characteristic of the reaction. By grouping the constant terms into an overall constant k, the expression is simplified to:

$$\text{Rate} = kC \tag{12}$$

According to Eq. 12, first-order kinetics apply, as was predicted intuitively. Thus a plot, according to Eq. 2, of the logarithm of (concentration of drug remaining) should be linear with slope equal to $(-k)$.

Experimental Examples

It is useful to use the apparent rate constant, determined from the slope, to compare results, where the only variable is a parameter, such as the pH of the reaction solution or the concentration of additives or excipients designed as inhibitors or promoters of the photoreaction. This is appropriate for purposes of mechanism determination and should be done under relatively low concentration conditions where first-order kinetics apply, as exemplified by the pH study with benzydamine as shown in Figure 1. On the other hand, one should be careful with the interpretation of apparent rate constants when the variable being studied is the concentration of the photon-absorbing component. Figure 5 is another example, showing the photodegradation of midazolam, in which the extent of reaction has been expressed as (Residual Drug as Percent of Initial Concentration) (7). As the initial drug concentration is increased, this presentation of the data shows that the slope of the plot (the apparent rate constant) diminishes, i.e., an inverse concentration relation appears to hold. The reason for this anomaly is that all the concentrations studied are so high that most of the UV radiation (UVR) in the absorption range of the drug is absorbed. If the data were to be replotted in terms of the (amount of drug reacted), it would be found that there is no dependence on drug concentration at all, i.e., zero-order kinetics, apply. In other words, the intensity of the photon source is having the dominant effect on the rate of reaction.

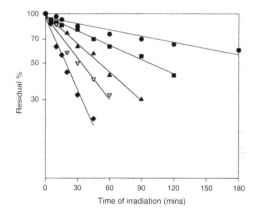

Figure 5 Photodegradation of Midazolam in ethanolic solutions plotted on a logarithmic scale. (●) 20 mM; (■) 15 mM; (▲) 10 mM; (▼) 7.5 mM; (■) 5 mM. *Source*: From Ref. 7.

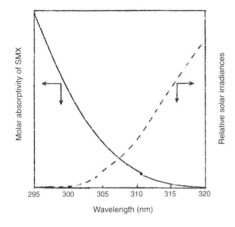

Figure 6 Overlap region of the absorption spectrum of sulfamethoxazole and the output spectrum of sunlight.

Strictly, the overlap integral determines the rate and therefore the rate constant for the reaction. When the substance being examined has only a relatively small absorption in the UVB and UVA regions, the overlap integral will be small and first-order kinetics are observed. This will be the result even though a large amount of the absorbing substance might be present in the system. Such is the case for the study of sulfamethoxazole photodegradation, under irradiation by a sunlight-simulating UVR source. In Figure 6, the very small overlap of sulfamethoxazole absorption and the sunlight spectrum is shown. The consequence of the small overlap is that an initial concentration of 1 mM sulfamethoxazole will yield a first-order kinetic plot. On the other hand, nifedipine has a more extensive overlap integral as shown in Figure 7, and a concentration of 5×10^{-5} M is the maximum that can be used to maintain first-order kinetics.

Deviations from First-Order Kinetics
It should be recognized that the derivation leading to Eq. 12 is an approximation, valid only at low values of the absorbance A in Eq. 9. At higher values of A, second,

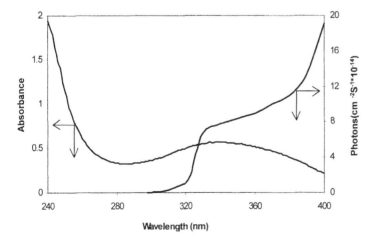

Figure 7 Overlap region of the absorption spectrum of nifedipine and the output spectrum of sunlight.

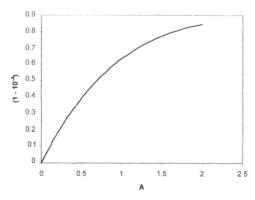

Figure 8 Theoretical plot showing the variation of the expression $(1 - 10^{-A})$ in Eq. 8 with increasing values of A.

third, and higher order terms of the power series become significant contributors, leading to a deviation from the first-order kinetic equation. Figure 8 is a theoretical plot that shows how the expression in Eq. 8 changes with increasing values of the absorbance A. Clearly, at low values of A, the relationship is linear, showing a direct dependence on A. This gradually changes so that when the absorbance approaches 2, the drug absorbs essentially all of the incident radiation. Here, the rate-limiting factor is the intensity of the incident radiation, and the reaction follows pseudo–zero-order kinetics. This type of kinetic order variation is referred to as saturation kinetics, commonly seen in reaction situations where one of the important factors is limiting, an example being simple enzyme-catalysed reactions. In between the two extremes, the reaction order is nonintegral, so that neither first nor zero-order kinetic plots will adequately describe the data when the reaction is taken to more than 50% conversion.

Experimentally, it is important that the irradiated solution is stirred effectively to ensure that a uniform concentration is maintained throughout the solution, and that products are being dispersed to minimise secondary degradation. In the high concentration situations, it is rare to follow the reaction to the extent (50% conversion) necessary to clearly decide which kinetic order is applicable, so most experimenters treat their data according to first-order equations. While this may be erroneous in principle, it is not important because the rate constants so determined only apply to the apparatus used. Nonetheless, it is recommended that preformulation studies of photodegradation be conducted with low solution concentrations of the drug so that first-order kinetics apply, and the reaction rate is limited by the drug concentration rather than the radiation intensity. The concentration limit can only be decided from a calculation of the overlap integral for the drug and photon source configuration. As a rule of thumb, Figure 8 can be used as a guide, for example, indicating that at an A value of 0.2, the kinetics have deviated from first order by about 10%. In general, it has been the practice of many experimenters to select a drug concentration that will give an absorbance of about 0.5 at the substances absorption maximum. This is done so that a sufficient sensitivity is available for the analysis of the photodegradation products. As pointed out above, it is the overlap integral value equivalent to an absorbance of 0.2 to 0.3 that should be considered when this selection is being made, if a good approximation to first-order kinetics is to be achieved. The corollary, ensuing from this discussion, is that there is little point in a study of the kinetics of photodegradation as a function of the concentration of the absorbing substance because the kinetic order will change from first to zero as the concentration increases.

Systems with More Than One Absorbing Species

The derivation of Eq. 12 would need to be modified if other components of the system absorb the incident radiation. Two possibilities arise, depending upon whether the other components(s) participate in subsequent reactions. If the other absorbing components do not react, the only effect is a reduction or filtering of the incident light. The fraction of the light absorbed by the drug can be expressed as, Eq. 13:

$$FC = \varepsilon_\lambda C / (\alpha_\lambda + \varepsilon_\lambda) \tag{13}$$

where α_λ is the attenuation coefficient of the medium at wavelength λ (i.e., the sum of the absorbances of components other than the drug of interest).

Where the other absorber(s) react and influence the degradation of the drug, for example, an impurity or additive which sensitizes the photoreaction, the overall kinetics will be a combination of both pathways. Thus, the reaction rate may depend on both the drug and the sensitizer concentrations. As a consequence of such considerations, the photochemical stability of a drug in a formulation is not necessarily predictable only from its absorption spectrum and stability studies in a pure solvent. The final form of the drug in the formulation should also be considered.

Application of the Derived Rate Constant Expression to Specific Examples of Photonic Exposure

If the output of a particular photon source is known in terms of intensity as a function of wavelength (its spectral power distribution), it is possible to use Eq. 11 to predict the rate at which a compound, of known absorptivity and degradation quantum yield, will degrade when irradiated by that source. In practice, the integration in Eq. 11 is replaced by a summation of the values measured for finite wavelength bands ($\Delta\lambda$) across the region of interest, so that the rate constant expression becomes:

$$K = 2.303\varphi b \sum I_\lambda \varepsilon_\lambda \Delta\lambda \qquad \text{summed from } \lambda 1 \text{ to } \lambda 2 \tag{14}$$

Using Eq. 14, the reaction rate constant for the photodegradation, of a particular compound, can be predicted from the quantum yield and the molar absorptivity at specific wavelengths, together with the radiation source intensity data. For example, average intensity data is known for sunlight as a function of latitude and season (8), enabling the calculation of an expected rate constant of degradation when a sample is exposed to direct sunlight. A satisfactory agreement was found between predicted and experimental first order rate constants for the photodegradation of sulfamethoxazole when exposed to sunlight at midday in different seasons (9). This is an example of how the determination of the rate constant, for a well-documented photodegradation occurring in a particular experimental arrangement, can provide information about the radiant intensity of the source used, in the region of absorption by the solute. In other words, it can act as an actinometer system for the wavelength region in which the reaction system absorbs.

KINETICS OF PHOTOSENSITIZED REACTIONS

When a photosensitizer is involved in the photoreaction, i.e., the absorbing substance transfers the energy to another nonabsorbing molecule that then undergoes reaction, the above principles still apply, with the analysis applied to the acceptor molecule. An example is photooxidation via the singlet oxygen mechanism, details of which are given in Section "Application of Kinetic Measurements to Reaction Mechanism Determination in Photostability."

A perfect photosensitizer will not be transformed in the photosensitizing process, although in reality, some degradation of the photosensitizer will occur in time. Nonetheless, the rate of the photosensitized reaction will depend directly on the sensitizer concentration at low values where not all of the relevant incident radiation is being absorbed. Most studies of this type of photoreaction use a high concentration of the acceptor A, in which case the transformation of A will follow apparent zero-order kinetics.

Experimental Considerations

Consistency (plenty of it!) and contamination (none at all!) are the key factors in achieving reliable outcomes from a photochemical kinetics study. The experimental arrangement must be, exactly, the same throughout so that no variation in the conditions of irradiation occurs. This is best achieved by the use of a radiometer to monitor radiation intensity, and a thermostatted sample vessel. A useful apparatus for preformulation kinetic studies in solutions is shown in Figure 9 (4). Solutions should be effectively stirred to maintain homogeneity, and the removal of samples should not affect the volume being irradiated. Contamination of the reaction solution must be avoided at all costs to prevent variable side reactions. The most destructive form of contamination is attributable to metal ions, such as iron and copper, because of their high level of reactivity by electron transfer mechanisms leading to free radical mediated degradation processes.

APPLICATION OF KINETIC MEASUREMENTS TO REACTION MECHANISM DETERMINATION IN PHOTOSTABILITY

Rate measurements are an integral component of mechanism determination in photochemical reactions. Although the values of rate constants are specific to the particular apparatus used, as emphasized in Section "Reaction Kinetics and Photostability," the comparison of rates obtained as a function of, for example,

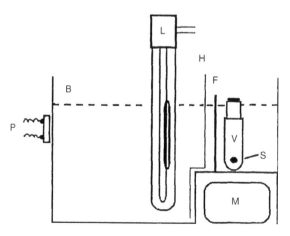

Figure 9 Apparatus for preformulation testing of the kinetics of photostability of drugs in solution. *Abbreviations*: V, reaction vessel; S, stirrer bar; M, magnetic stirrer; F, filter; H, shutter; L, arc lamp; B, thermostat bath; P, photocell. *Source*: From Ref. 4.

additive concentration or pH, provide mechanistic information. It is important to maintain concentration conditions such that first-order kinetics are always observed in these experiments.

There are a large number of different reaction types that can be initiated by photochemical means. These reaction types have been categorized for drug photostability by Greenhill and McLelland (10) and include oxidation, isomerization, hydrolysis, and the removal of various substituents, such as halogens or carboxyl groups. Knowledge of the mechanism by which the photodegradation occurs will be of benefit in designing ways to inhibit the reaction. Clearly, the prevention of UV-vis photon absorption is the primary means of inhibition, but unless the drug substance has a well-documented susceptibility to photodegradation, it is very likely that the product may be photodegraded at various stages of manufacture or in the hands of the consumer.

This section contains a description of the rate methods normally used in elucidating the mechanism of drug photodegradation, i.e., the nature and efficiency of the various steps, which make up the overall process. For a complete understanding, an array of physical techniques is required, ranging from laser flash photolysis to identify the primary photochemical species formed in the first nanoseconds of the reaction, to electron paramagnetic resonance (EPR) for the identification of radical intermediates. These techniques are required to supplement the skills of the organic chemist in identification of the end product(s) of photodegradation. This section will consider mainly the physical techniques.

Initial Mechanism Studies

In these experiments, only a small amount of drug is irradiated in solution at a concentration such that the absorbance is in the region 0.3 to 0.5. The temperature of the samples should be controlled to avoid complications due to heating by the arc source lamps. UV-vis absorption spectra and high performance liquid chromatography (HPLC) are used to monitor the reaction from the very beginning. It is most important to detect the first alterations to the spectrum and appearance of product(s). The use of a long irradiation time without intermittent monitoring may result in failure to see the primary chemical change.

The spectral changes are a clear indication of changes occurring in the chromophore of the absorbing molecule. If the rate of alteration in the spectrum can be related to the appearance of a single product by HPLC, then a simple process is clearly indicated. If a stable product is formed directly in the primary process, the quantum yield of the product is equal to that for the loss of the drug, and is a clear measure of the efficiency of the process. Unfortunately, the vast majority of photochemical processes are complex and require a plethora of detailed experimental information before one can begin to elucidate the mechanism. In some photochemical changes the reaction occurs at a site sufficiently removed from the chromophore that no substantial alteration can be seen in the absorption spectrum, e.g., the decarboxylation of certain anti-inflammatory drugs such as naproxen and benoxaprofen (11). In the photodechlorination of frusemide, the spectral shift is quite small as the chlorine is replaced by either H or OR from the solvent ROH (12). Many examples of drug photodegradation studies are documented with spectrophotometric changes and the corresponding appearance of products by HPLC analysis, e.g., benzydamine (1) and amiloride (13).

Effect of Oxygen

The photodegradation experiment should be performed under both air-saturated and oxygen-free conditions to determine whether oxygen exerts an influence on the reaction. Oxygen is an effective scavenger of free radicals and some excited state species that have radical character. As a consequence, the product profile and yield can be substantially different, depending upon whether oxygen is present or absent. The involvement of oxygen may occur through direct addition to the photoexcited drug molecule leading to oxidation products of the drug. An example is the formation of chlorpromazine sulfoxide from chlorpromazine (9).

Alternatively, the drug in its triplet state may transfer its energy to the oxygen molecule (also a triplet state and therefore spin matched), with the formation of singlet molecular oxygen. Singlet oxygen is highly reactive toward oxidizable molecules, and this leads to photosensitized oxidation. Many drugs exist in a reduced state, so it is possible that the oxidizable acceptor of singlet oxygen may be the drug itself. To determine whether oxidation of a drug occurs via singlet oxygen as an intermediate, the reaction can be studied in the presence of sodium azide, a quencher of singlet oxygen (14). A decrease in the yield of oxidation product caused by the presence 1 mM sodium azide in the irradiated reaction mixture can be ascribed to involvement of singlet oxygen. Another method is to use a singlet oxygen generating dye such as methylene blue as photosensitizer, with a visible photon source to ascertain whether the same product profile is obtained. By this means, the conversion of 6-mercaptopurine to its sulfinate was found to occur via singlet oxygen addition, while the subsequent conversion of sulfinate to sulfonate does not (15).

Singlet Oxygen Detection

If further information about the generation of singlet oxygen is required, there are several methods for the detection of $^1O_2$ generated in an irradiated solution. A characteristic luminescence at 1270 nm, corresponding to the return of $^1O_2$ to the ground state $(^3O_2)$, can be detected with the appropriate time-resolved equipment (16). While this method is definitive, the equipment is highly specialised. The alternative is to measure the rates of reaction in the presence of molecules that react readily with or quench singlet oxygen. Here the choice depends on the solvent being used, with sodium azide, 2,5-dimethylfuran, and the amino acid histidine being suitably soluble for aqueous systems, while β-carotene, DABCO, and diphenylisobenzofuran (DPBF) being more readily used in organic solvents. Analysis of the reaction rates is achieved in terms of oxygen uptake measured with an oxygen electrode, or by product separation and quantification. DPBF absorbs intensely at 415 nm and reacts rapidly with singlet oxygen to form a colorless intermediate endoperoxide. The DPBF reaction can be used as a benchmark against which the effect of an added quencher is compared. Another method frequently used for singlet oxygen detection involves the bleaching of *p*-nitroso-dimethylaniline in the presence of histidine. A compilation of singlet oxygen yields from biologically relevant molecules has recently been published (17). A number of drugs are included in the listing, which includes results from all the methods mentioned above.

A note of caution must be applied. The use of inhibitors and quenchers alone is not unambiguous in its outcome, and should strictly be supplemented with flash photolysis experiments. Thus, if a photosensitized reaction is quenched by millimolar concentrations of azide ion, it should also be established that azide does not quench the triplet state of the sensitizer directly, because that would also affect the reaction rate. It has also been reported that the furans and histidine can be oxidised to the same products by free-radical processes. Nevertheless, these compounds have such a high reactivity with singlet oxygen that they are very rarely wrong as indicators of its generation by a photosensitizer. Cholesterol is regarded as an unambiguous trapping compound, because singlet oxygen reacts with it to form a single product, the 5-α-hydroperoxide, whereas reaction with radicals gives a mixture of other products (14). This analytical procedure is more technically demanding than that employed with azide, histidine, or the furans, but has less chance of being equivocal. Another useful kinetic technique is to compare the rates determined in heavy water (D_2O) with those found when using normal water, as the lifetime of singlet oxygen is about 10 times greater in D_2O. This technique, however, will only achieve a meaningful result when singlet oxygen deactivation by the solvent is the rate determining process. Frequently, other species in the solution are capable of reacting with singlet oxygen, and the effect of the longer lifetime is not manifest in terms of greater rates or yields of products.

Typical photosensitizers, which generate singlet oxygen, include dyes such as methylene blue, rose bengal, and rhodamine. Many drug molecules such as phenothiazines, quinine and other antimalarials, thiazides, naproxen and other anti-inflammatories, and psoralens have been shown to produce singlet oxygen under the influence UVR of radiation. Environmental contaminants such as the polycyclic aromatic hydrocarbons also are very efficient $^1O_2$ generators. The acridines are of particular interest because the ring nitrogen will protonate in acidic solution, leading to a greatly increased solubility and photosensitizing ability in aqueous environments.

Strictly, it is difficult to compare the photosensitizing ability of a series of compounds because they will all have different absorption spectra (Table 1). In a strict photochemical sense, the quantum yield for the photosensitized oxidation reaction should be calculated. On the other hand, it can be argued from a practical viewpoint that sunlight is the primary source of radiation to which drugs are exposed. Thus, experiments can be performed with the same sunlight-simulating source to irradiate drug solutions prepared to an absorbance of 0.5 at their absorption maximum. The results can be usefully compared to express their relative response to sunlight exposure. A representative group of drugs, which generate singlet oxygen to varying extents (expressed by the rate of oxygen uptake in solutions containing 2,5-dimethylfuran), is given in Table 1.

Free-Radical Formation

The triplet state drug molecule ^3D may interact directly with a suitable acceptor molecule AH, transferring its excess energy with the formation of a free radical $A^{\cdot}$ or radical anion $A^{\cdot -}$ by one of the following reactions 2, 3, or 4:

$$^3D + AH \rightarrow D_0 + A^{\cdot} + H^{\cdot} \tag{R2}$$

$$\text{or } DH^{\cdot} + A^{\cdot} \tag{R3}$$

$$\text{or } D_0 + A^{\cdot -} + H^+ \tag{R4}$$

Table 1 Spectroscopic Characteristics in Aqueous Solution for Representative Photoactive Drugs and Reaction Rates for Photosensitized Oxidation and Polymerisation

Drug	pH	Absorption[a]		Fluorescence		Phosphorescence	Photooxidation rate (dimethylfuran)	Photopolymerisation rate (acrylamide)	References
		λ_{max} nm	ε_{max} M^{-1} cm^{-1}	λ_r nm	ϕ_f	λ_p nm	µM min^{-1}	µM min^{-1}	
Chlorpromazine	7	322	4300	455	0.08	nd	32.8	3.42	18
Furosemide	4	355	5840	415	0.07	nd	2.8	0.59	18
Hydrochlorthiazide	7	318	3200	370	0.007	nd	3.2	1.14	19
Quinine	1	348	5350	450	0.55	nd	48.0	0.10	20
Quinine	6	330	4880	386	0.40	nd	15.0	0.14	20
Chloroquine	6	343	18650	388	0.011	nd	0.3	0.50	20
Chloroquine	10	330	11250	382	0.173	nd	2.7	3.92	20
Naproxen	7	330	2200	352	0.40	510	25.1	0.24	11
Diclofenac	7	285	8220	NF	–	460	3.9	1.79	21
Indomethacin	7	320	7300	NF	–	NP	0.1	0	11
Nalidixic acid	4	335	11600	355	0.022	nd	215	1.35	22
Sulfamethoxazole	5	269	17000	342	0.002	410	8.2	0.72	23
7-Methylbenz[c]acridine	3	372	8200	498	0.82	595	100	0.52	24

[a]Longest wavelength absorption.
Abbreviations: nd, not determined; NF, no detectable fluorescence; NP, no detectable phosphorescence.

The free radical(s) so formed are then scavenged by molecular oxygen forming the peroxy radical or anion (Reaction 5) thus leading to the oxidation of the acceptor AH. This is called the Type I mechanism of photooxidation.

$$A^\bullet (or\ A^{\bullet -}) + O_2 \rightarrow AO_2^\bullet\ (or\ AO_2^{\bullet -}) \tag{R5}$$

According to this reaction scheme, the sensitizer D is returned to the ground state unchanged, although it is possible that the ground state sensitizer may act as its own acceptor and be oxidised by this mechanism.

The electron or proton transfer between an excited state species and a ground state molecule (Reactions 2–4) is frequently observed in the photochemistry of systems containing an electron donor and acceptor combination. Consequently, a pair of radical ions is formed, both of which react with oxygen but with different rates. Electron transfer may occur from a radical anion to ground state oxygen thereby yielding superoxide anion. The superoxide then may add to the radical cation forming DO_2 (Reaction 6). When D is an olefin, the DO_2 formed is a dioxetan that is liable to cleave to yield ketones as products. This contrasts the reaction of singlet oxygen with olefins where a hydroperoxide (DOOH) is the initial product.

$$D^{+\bullet} + O_2^{\bullet -} \rightarrow DO_2 \tag{R6}$$

Photoionization

In the excited singlet state the ionization potential of the molecule is reduced, and the excited electron is more easily removed compared to the ground state. This process of photoionization is also more likely to occur if higher energy UV radiation is used (i.e., wavelengths less than 300 nm) and if the drug molecule is in its anionic state. An example is naproxen in aqueous solution at pH 7, where the process of decarboxylation occurs and a neutral radical is formed (11), Reaction 7:

$$Ar - COO^- + h\nu \rightarrow Ar^\bullet + CO_2 + e^- \tag{R7}$$

Laser Flash Photolysis

The nature of the excited state decay processes is studied by the technique of laser flash photolysis, a full description of which has been given elsewhere (25). Briefly, flash photolysis involves irradiating a sample with a short (nanosecond) intense pulse from a laser, then observing by rapid response spectrophotometry the spectral changes that occur on the time scale nanoseconds to milliseconds.

If a transient species has an absorption spectrum at longer wavelengths than the drug absorber, then the rate of decay of that transient can be studied. The most likely absorptions arising from the irradiated drug are due to the triplet state and/or a cation radical with its associated solvated electron. Several standard tests have been established to aid in the identification of the transient species. Solvated electrons, generated by photoionization in a nitrogen-gassed solution, have a characteristic broad structureless absorption peak at about 700 nm depending on the solvent (720 nm in aqueous solution). Oxygen quenches this absorption and also quenches the triplet state, while nitrous oxide gassing can be used to quench the solvated electron only thereby gaining an indication of any transient absorption, which arises from the triplet state. When a neutral molecule undergoes photoionization, the resulting cation radical usually has a longer lifetime, and is not rapidly quenched by oxygen.

Fluorescence and Phosphorescence

The luminescence phenomena, fluorescence and phosphorescence, are also useful indicators of the potential of a drug to be involved in photochemical reactions. The fact that a molecule dissipates its excitation energy by one of these processes indicates a rigidity of the structure and a longer lifetime of the excited state, providing greater opportunity for interaction with surrounding molecules. The quantum yield of fluorescence is readily determined by reference to quinine as a standard (3). Quantum yields of other excited state processes can only be obtained by difference. Phosphorescence is usually too weak to be observed in solution at room temperature, but it can be measured if the drug is held in a glassy matrix at low temperature. The usual procedure is to dissolve the drug in ethanol and immerse the solution in liquid nitrogen. The phosphorescence accessory of the fluorimeter incorporates a mechanical chopper enabling the phosphorescence to be observed free of interference from any fluorescence. Where available, fluorescence and phosphorescence data is included for the drugs listed in Table 1.

Detection of Free Radicals

The above description is a simplified view of some of the processes that may occur involving free radicals generated from the excited state. A number of techniques have been developed to enable the detection of free radical intermediates in photochemical reactions, including EPR spectroscopy. EPR is useful for detecting and monitoring radicals, which are formed in relatively high concentration and persist for relatively long times. Unfortunately, that is not the case for the great majority of photochemical reactions, and special procedures are necessary such as rigid solution matrix isolation. Addition of free radical trapping compounds to the system (spin traps) is an alternative (26,27). The superoxide anion is also readily trapped and identified by this technique.

An extremely sensitive technique able to detect the nature of radical pairs in a photochemical reaction is called chemically induced dynamic nuclear polarization (CIDNP), which depends on the observation of an enhanced absorption in a nuclear magnetic resonance (NMR) spectrum of the sample, irradiated in situ, in the cavity of a NMR spectrometer. The background to and interpretation of CIDNP are discussed by Gilbert and Baggott (28).

Probably the main technique that has been used to detect free radical intermediates in photochemical reactions is the competitive reaction rate study, in which various free radical scavengers are added to the sample during irradiation. The rate of disappearance of the drug and appearance of particular photodecomposition products is compared to that occurring in the absence of the scavenger. Typical scavengers include ascorbic acid and glutathione for aqueous systems, and 2,6-di-t-butyl-hydroxy-toluene and α-tocopherol for lipophilic systems. However, there is some difficulty in interpreting the results of such a study, because the relative reactivity of both radicals and scavengers are determining the outcome, so that the product profile will invariably change. If the radical intermediates are extremely reactive, they may react with the solvent before they encounter a scavenger molecule, and no change will be observed.

Polymerization as a Detector of Free Radicals

The chain reaction process can be used as a diagnostic aid to determine whether free radicals are generated from a drug when irradiated. Acrylamide is an acrylic monomer, which is widely used in gel electrophoresis, as a polymer formed in situ by peroxide or UV-initiated polymerization. This monomer is a water soluble solid, more easily handled than most other vinyl monomers, and the progress of its polymerization can be readily followed by measuring its contraction in volume utilizing dilatometry, or its increase in viscosity in a viscometer. Details of this experimental technique can be found in Moore and Burt (18).

While the polymerization technique does not give any information as to the identity of the free radical generated by irradiation of the drug it is a chemical amplification process allowing very small concentrations of free radicals to be detected. The rate of polymerization caused by free radicals generated by the UV irradiation of a drug solution containing acrylamide is a reflection not only of the rate of radical generation, but also of their lifetime. Note that oxygen is an efficient scavenger of the polymer radicals and must be excluded from the system. Some representative examples of polymerization rates obtained on irradiation of photoactive drugs are given in Table 1.

Photodehalogenation Reactions

The loss of an aromatic chlorine substituent is a photoreaction type that is frequently observed. Examples of drugs with photolabile chlorine substituents are chlorpromazine (29), hydrochlorthiazide (30), chloroquine (20), furosemide (12), diclofenac (21), and amiloride (13). In each case, when the drug (Aryl-Cl) is photolysed in aqueous or alcoholic (ROH) solution, HCl is liberated (detected by titration with silver nitrate) and a mixture of reduction (Aryl-H) and substitution (Aryl-OR) products is obtained. The photodechlorination reaction occurs for these compounds more strongly in deoxygenated solution. When oxygen is present, it promotes intersystem crossing to the triplet state and the production of singlet oxygen.

The mechanism is by no means completely clear, but the photodehalogenation reaction is postulated to occur through the formation of a pair of radical ions from an exciplex resulting in the excited state (31). The precursor of the reduction product (Aryl-H) is suggested to be a radical anion (Aryl-Cl-) while a radical cation (Aryl-Cl+.) is postulated as the precursor of the substitution product (Aryl-OR). In a less polar solvent, e.g., iso-propanol, direct homolysis of the C–Cl bond occurring from the triplet state has been suggested, based on flash photolysis experiments with chlorpromazine (29). 3,3′,4′,5-Tetrachlorosalicylanilide represents a class of antibacterial agents formerly used in cosmetics and soaps. These compounds were found to undergo sequential photodehalogenation, which was presumed to be related to their capacity to induce skin rashes upon sunlight exposure (32).

Not all chloroaromatic drugs appear to follow this type of reaction. For example, free chloride ion is not formed on irradiation of chlordiazepoxide for which an oxaziridine is the major photoproduct (33). There is variability among reports on other drugs with chlorine substituents. This can arise due to differences in the irradiation conditions. If an unfiltered xenon or mercury arc source is used, the sample will receive 250 to 300 nm irradiation and the C–Cl bond will certainly break, while under longer wavelength irradiation (300 nm) the bond may be stable.

REFERENCES

1. Wang J, Moore DE. A study of the photodegradation of benzydamine in pharmaceutical formulations using HPLC with diode array detection. J Pharm Biomed Anal 1992; 10: 535–540.
2. Gu L, Chiang H-S, Johnson D. Light degradation of ketrolac tromethamine. Int J Pharm 1988; 41:105–113.
3. Calvert JG, Pitts JN. Experimental methods in photochemistry. In: Photochemistry. New York: John Wiley & Sons, 1966:783–804.
4. Moore DE. Principles and practice of drug photodegradation studies. J Pharm Biomed Anal 1987; 5:441–453.
5. Kuhn HJ, Braslavsky SE, Schmidt R. Chemical actinometry. Pure Appl Chem 1989; 61:187–210.
6. Moore DE. Kinetic treatment of photochemical reactions. Int J Pharm 1990; 63:R5–R7.
7. Andersin R, Tammilehto S. Photochemical decomposition of midazolam. II. kinetics in ethanol. Int J Pharm 1989; 56:175–179.
8. Leifer A. The Kinetics of Environmental Aquatic Chemistry – Theory and Practice. New York: American Chemical Society, 1988:255–264.
9. Moore DE, Zhou W. Photodegradation of sulfamethoxazole: a chemical system capable of monitoring seasonal changes in UVB intensity. Photochem Photobiol 1994; 59: 497–502.
10. Greenhill JV, McLelland MA. Photodecomposition of drugs. Progr Med Chem 1990; 27:51–121.
11. Moore DE, Chappuis PP. A comparative study of the photochemistry of the non-steroidal anti-inflammatory drugs, naproxen, benoxaprofen and indomethacin. Photochem Photobiol 1988; 47:73–181.
12. Moore DE, Sithipitaks V. Photolytic degradation of frusemide. J Pharm Pharmacol 1983; 35:489–493.
13. Li YNB, Moore DE, Tattam BN. Photodegradation of amiloride in aqueous solution. Int J Pharm 1999; 183:109–116.
14. Spikes JD. Photosensitization. In: Smith KC, ed. The Science of Photobiology. 2nd ed. New York: Plenum Press, 1989:79–110.
15. Hemmens VJ, Moore DE. Photo-oxidation of 6-mercaptopurine in aqueous solution. J Chem Soc Perkin Trans 1984; I:209–211.
16. Hall RD, Buettner GR, Motten AG, Chignell CF. Near-infrared detection of singlet molecular oxygen produced by photosensitization with promazine and chlorpromazine. Photochem Photobiol 1987; 46:295–300.
17. Redmond RW, Gamlin JN. A compilation of singlet oxygen yields from biologically relevant molecules. Photochem Photobiol 1999; 70:391–475.
18. Moore DE, Burt CD. Photosensitization by drugs in surfactant solutions. Photochem Photobiol 1981; 34:431–439.
19. Moore DE, Tamat SR. Photosensitization by drugs: photolysis of some chlorine-containing drugs. J Pharm Pharmacol 1980; 32:172–177.
20. Moore DE, Hemmens VJ. Photosensitization by antimalarial drugs. Photochem Photobiol 1982; 36:71–77.
21. Moore DE, Roberts-Thomson S, Dong Z, Duke CC. Photochemical studies on the anti-inflammatory drug diclofenac. Photochem Photobiol 1990; 52:685–690.
22. Moore DE, Hemmens VJ, Yip H. Photosensitization by drugs: nalidixic and oxolinic acids. Photochem Photobiol 1984; 39:57–61.
23. Zhou W, Moore DE. Photosensitizing activity of the anti-bacterial drugs sulfamethoxazole and trimethoprim. J Pharm Pharmacol B Biol 1997; 39:63–72.
24. Burt CD, Moore DE. Photochemical sensitization by 7-methylbenz[c]acridine and related compounds. Photochem Photobiol 1987; 45:729–739.
25. Bensasson RV, Land EJ, Truscott TG. Flash Photolysis and Pulse Radiolysis. Oxford: Pergamon Press, 1983:1–19.

26. Mason RP, Chignell CF. Free radicals in pharmacology and toxicology – selected topics. Pharmacol Rev 1982; 33:189–211.
27. Chignell CF, Motten AG, Buettner GR. Photoinduced free radicals from chlorpromazine and related phenothiazines: relationship to phenothiazine-induced photosensitization. Env Health Perspect 1985; 64:103–110.
28. Gilbert A, Baggott J. Essentials of Molecular Photochemistry. Oxford: Blackwell, 1991:145–228.
29. Davies AK, Navaratnam S, Phillips GO. Photochemistry of chlorpromazine (2-chloro-N-(3-dimethyl-aminopropyl) phenothiazine) in propan-2-ol solution. J Chem Soc Perkin Trans 1976; 2:25–29.
30. Tamat SR, Moore DE. Photolytic decomposition of hydrochlorothiazide. J Pharm Sci 1983; 72:180–183.
31. Grimshaw J, de Silva AP. Photochemistry and photocyclization of aryl halides. Chem Soc Rev 1981; 10:181–203.
32. Davies AK, Hilal NS, McKellar JF, Phillips GO. Photodegradation of salicylanilides. Brit J Derm 1975; 92:143–147.
33. Cornelissen PJG, Beijersbergen Van Henegouwen GMJ, Gerritsma KW. Photochemical decomposition of 1,4-benzodiazepines. Chlordiazepoxide. Int J Pharm 1979; 3:205–220.

12 Unexpected Photochemistry in Pharmaceutical Products: A Review on the Role of Diluents, Excipients, and Product Components in Promoting Pharmaceutical Photochemistry

Allen C. Templeton
Pharmaceutical Research and Development, Merck Research Laboratories, Merck and Co., Inc., West Point, Pennsylvania, U.S.A.

William E. Bowen
Global Pharmaceutical Commercialization, Merck Manufacturing Division, Merck and Co., Inc., West Point, Pennsylvania, U.S.A.

Lee J. Klein and Paul A. Harmon
Pharmaceutical Research and Development, Merck Research Laboratories, Merck and Co., Inc., West Point, Pennsylvania, U.S.A.

Yu Lu
Vertex Pharmaceuticals, Cambridge, Massachusetts, U.S.A.

Robert A. Reed
Xenoport, Inc., Santa Clara, California, U.S.A.

INTRODUCTION

The impact of light exposure on drug substance and drug product quality has been known for almost as much time as the modern pharmaceutical industry has been in existence. Although history shows that a wide variety of approaches have been employed to mitigate photostability issues in order to ensure product safety and efficacy, only recently have standardized approaches been in place to assess and manage the impact of drug photosensitivity. The first reported application of a standard photostability chamber was to the study of the fading of 10 water soluble, certified dyes by Kuramoto et al. (1). They found that sugars such as dextrose, lactose, and sucrose increased the rate of fading of these dyes whereas sugar alcohols such as mannitol and sorbitol did not appreciably affect the rate. They also discovered that if trace amounts of strong reducing catalysts remained in the alcohols used they could significantly affect color stability. In the early 1990s, a body of literature began to develop around the most effective ways to test pharmaceutical photostability. Some of the literature considerations discussed in this formative work in the field included discussions on proper lamp selection, relevant light exposure levels, standardized approaches to measuring light levels, and how to deal with the impact of photodriven changes to product quality. The evolution of this thinking led to the development of the International Conference on Harmonization (ICH) Q1B Guidance entitled "Photostability Testing of New Drug Substances and Products," hereafter referred to in this report as Q1B (2).

The Q1B guidance laid a framework for consistency within the industry around photostability testing. Importantly, the document specifies light sources, study types to be conducted (forced stress and confirmatory), the presentation of samples, and the importance of having proper control over the light-dose delivered. Arguably the most important section of the document is the decision flow chart for the photostability testing of drug products. The flow chart provides a stepwise approach to evaluating and mitigating the effects of photostability on drug products and, in doing so, clearly indicates the burden of protection required by pharmaceutical applicants in regulatory submissions. Some of the shortcomings that have been noted in the document are around definition of what constitutes an "acceptable change" during photostability testing, disagreements in the community over light sources recommended, and that the guidance leaves out any mention of the implications of photostability on product administration. Baertschi produced an excellent practical interpretation of the guideline shortly after its publication that provides many useful suggestions for satisfying Q1B requirements (3).

The overall impact of photostability is evident from an examination of the United States Pharmacopoeia (USP) 27 (2004) Reference Table "Containers for Dispensing Capsules and Tablets" (4). Of the 743 pharmaceutical products listed in the table, 248 (33%) require light-resistant packaging. Clearly, developing an improved understanding of photostability would improve the ability of pharmaceutical applicants to effectively control and respond to the specific requirements of each product. Over the last several years, our laboratories have experienced developmental issues regarding photostability of drug products on several fronts. One of these fronts has been in developing approaches to the experimental design and data interpretation of photostability studies as they pertain to protecting photosensitive products during manufacturing, packaging, shelf storage, testing, and administration. Despite the significant lack of Q1B direction, the appropriate conduct of such supporting photostability studies and implementation of necessary protective measures remain an important responsibility of pharmaceutical applicants. In a recent report, we summarized what we felt were important practical considerations in acquiring and using photostability data to make decisions to address the impact of product photosensitivity in each critical area of product development (5). Not only is this an area wherein the Q1B guidance falls short, but it is also extremely important in developing the appropriate information to successfully develop, produce, and market quality pharmaceutical products on a commercial scale.

A second front along which we have launched substantial investigations has been in understanding product systems where the drug itself does not seem to be absorbing incident light, yet the product is found to be photosensitive (6). Even though there is a large and diverse body of literature noting direct degradation of pharmaceutical substances upon exposure to light, including detailed studies of specific degradation pathways, there has been only scant attention paid to the photostability of drug products where the drug is involved in the photochemistry through nonobvious mechanisms. By our estimates, approximately 15% (or 37 products) of the compounds listed as requiring light-resistant packaging in the USP do not have appreciable absorption beyond 300 nm and thus beg the question: What is promoting the photoinstability of the product? The 300 nm wavelength represents a key differentiation determined by a combination of considerations of emission profiles for typical light sources and transmission properties of glasses (e.g., windows and glass envelopes for fluorescent and incandescent bulbs) and typical primary packaging materials. In the following sections of the chapter, we review both our work and the extant literature on unexpected photochemistry in pharmaceutical products.

DILUENT MEDIATED PHOTOCHEMISTRY

Many pharmaceutical products rely on the use of a commercial diluent to either reconstitute a lyophilized formulation or dilute a concentrated formulation prior to administration. Compositions of common diluents include glucose (dextrose), saline, Ringer acetate, and water for injection. The diluents are provided to market as sterile products and can have varied degradation products and extractables/leachables based on the raw materials used, sterilization process/cycle (especially for heat sterilization), and marketed product primary package. Additionally, the degradation products can also introduce buffering capacity to the diluent and thus impact final reconstituted product pH and subsequent photosensitivity. Products that are reconstituted in these diluents may unexpectedly become photosensitive, even when the drug itself does not absorb light of wavelengths greater than 300 nm.

Experience with parenterals has shown that a given drug molecule exposed to a fixed dose of visible or ultraviolet (UV) radiation can exhibit profound photosensitivity in one diluent and yet be completely stable in another. In a recent review of the photostability of parenteral products, Kristensen summarized the effects that a variety of diluent components (including cosolvents, buffers, oxygen, antioxidants, metal ions and chelators, tonicity adjusters, preservatives, bulking agents, protectants, dispersants, and drug carriers such as cyclodextrins) can all have on photostability (7). Despite this background and a relative wealth of experience with the photochemistry of drug products in general (8), there are relatively few examples in the literature of photochemistry mediated by a component of a diluent.

To our knowledge, the first well-defined example of diluent-mediated photodegradation appeared in 2000, when Brustugun et al. documented the sensitivity of epinephrine to bisulfite in infusion media (9). Building on this observation, Brustugun et al. published two papers in 2004 dealing with the photodegradation of other sympathomimetic agents (including epinephrine) in infusion media containing bisulfite (10,11). In the case of epinephrine itself, oxidation in the presence of bisulfite leads to the formation of small quantities of the pigment, adrenochrome sulfonate, which has an absorption maximum at 350 nm and appreciable absorption above 400 nm. Because adrenochrome sulfonate produces singlet oxygen upon irradiation and singlet oxygen quickly oxidizes epinephrine, a catalytic photochemical degradation cycle ensues (Fig. 1).

This cycle demonstrates how one low-level degradate or impurity that exhibits extended absorption can act as a "light antenna," funneling radiant energy into a degradation pathway for an otherwise nonabsorbing starting molecule. It is reasonable to assume that this phenomenon could occur with other catecholamines in sulfite-containing diluents. In general, any weakly-absorbing drug that is easily oxidized to form a stronger chromophore holds the potential to form its own photosensitizer, either with or without the involvement of an excipient.

In the course of the same investigations that implicated bisulfite as shown above, glucose was also shown to have a photodestabilizing effect on epinephrine, even in the absence of sulfites. Further exploring this phenomenon, Brustugun et al. recently demonstrated that 5-hydroxymethyl-2-furaldehyde (5-HMF), generated in situ by the heat sterilization of glucose-containing infusions, was acting as an efficient photosensitizer (12). Thermal degradation of carbohydrates has long been established as a source of 5-HMF and the formation of this aldehyde in heat-sterilized glucose-containing infusions has been known since at least 1955 (13). Levels of 5-HMF are also known to change (both increase and decrease) upon long-term storage of dextrose-containing

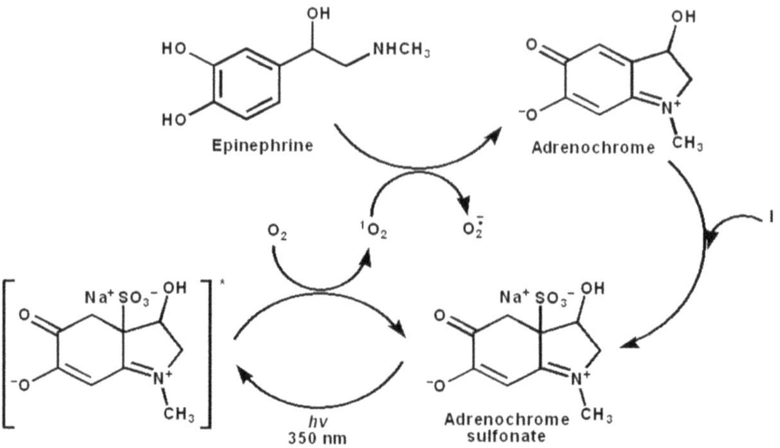

Figure 1   Mechanism for the autocatalytic photodegradation of epinephrine via adrenochrome sulfonate.

injectables (14,15). There has even been a review dedicated entirely to the chemical and physiological properties of 5-HMF in parenteral solutions (16). Given this long history, it is not surprising that several authors have developed increasingly selective and sensitive methods for determining 5-HMF in parenterals and other materials (17–29). Even though there had been some early work showing that levels of 5-HMF in glucose solutions depended at least partly on exposure to UV light (30), it appears that the link between photoexposure, 5-HMF, and drug photostability was not established until 2005 (12).

In this most recent investigation, a series of standard ICH-type photoexposures along with experiments involving steady-state monochromatic photoexposures and laser flash photolysis have shown that 5-HMF is an efficient producer of singlet oxygen ($^1O_2$). Although the highest wavelength absorption maximum for 5-HMF is at 285 nm, significant absorption in condensed phases is apparent up to 340 nm. Thus, the authors found that steady-state exposure of solutions of 5-HMF in D_2O to monochromatic radiation at 333 nm leads to the production of $^1O_2$ as evidenced by luminescence at 1270 nm in proportion to the concentration of 5-HMF, as illustrated in Figure 2.

Involvement of the excited triplet state of 5-HMF in the mechanism above is supported by three additional observations. First, formation of a known excited triplet sensitizer (xanthone) by flash photolysis in the presence of 5-HMF results in a transient absorption spectrum with maxima at 320 and 430 nm. Because the new spectral features are distinct from both xanthone and 5-HMF and are rapidly quenched by oxygen the new spectrum is attributable to the triplet of HMF. Second, laser flash photolysis of 5-HMF at 266 nm produces the same species (having the same absorption spectrum) with virtually unity quantum yield as measured against a chemical actinometer. Subsequent formation of singlet oxygen from the excited triplet was found to have a quantum yield of 0.6. Lastly, irradiation of 5-HMF in the presence of histidine (a singlet oxygen scavenger) at 340 nm in a closed system resulted in a net consumption of oxygen. Though these experiments make it clear that 5-HMF is a producer of singlet oxygen, it is worth noting that

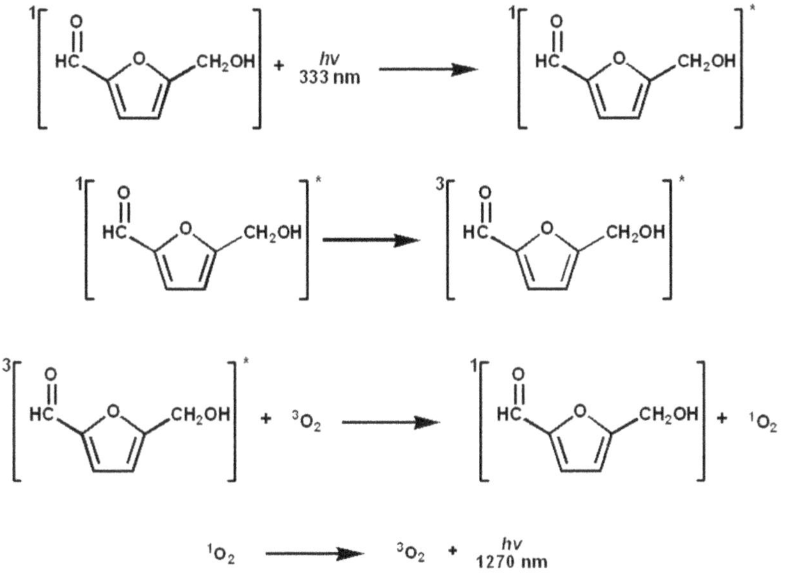

Figure 2 Photochemical generation of singlet oxygen by 5-hydroxymethyl-2-furaldehyde.

it is also rapidly photodegraded in the presence of another known singlet oxygen sensitizer (methylene blue). Although the degradation occurred at a wavelength absorbed only by methylene blue and the rate was not significantly affected by the addition of a superoxide scavenger, it is equally clear that 5-HMF is also a consumer of singlet oxygen.

Finally, the photostability of solutions of isoprenaline is markedly decreased in the presence of 5-HMF at levels typically found in dextrose infusions (3–5 µg/mL). A comparatively higher limit of 22.6 µg/mL can be found in the British Pharmacopeia, whereas the current USP applies only a nonquantitative limit test for 5-HMF in dextrose injections (31). Interestingly, the photostability of a mixture of isoprenaline with 5-HMF is not affected by the addition of a large molar excess of *t*-butanol, which indicates that hydroxyl radicals do not play a significant role in this interaction. On the contrary, the addition of histidine improves the photostability of the mixture, again confirming the role of singlet oxygen in the mechanism.

Owing to the ubiquitous nature of glucose and other carbohydrates in parenteral formulations and the common application of heat sterilization, 5-HMF is certain to be present at varying and significant levels in many diluents and parental formulations. Though 5-HMF is known to generate singlet oxygen (a highly energetic and reactive species capable of interacting with most drug molecules), it seems likely that many more photoinduced oxidations may ultimately be attributed to this molecule in the near future.

EXCIPIENT MEDIATED PHOTOCHEMISTRY

Our first introduction to this topic (see Ref. 6 for a full description of this work) was initiated by pH 6 citrate buffer formulations of drug A (a phenyl ether, shown

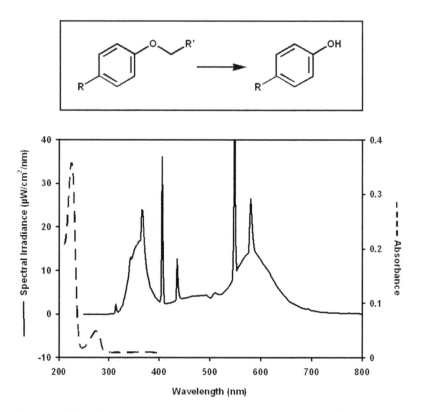

Figure 3 (*Top*) Structures of drug A and the known thermal degradation product, referred to as phenol in this report; (*bottom*) ultraviolet/visible absorption spectrum of 0.064 mM drug A in water (*dashed line*) compared to the combined spectral outputs of the International Conference on Harmonization ultraviolet and visible lamps (*solid line*).

in Fig. 3, top) that were found to be sensitive to light exposure when examined according to ICH-defined light conditions.

The photosensitivity was unexpected as there is negligible overlap of the drug A absorption spectrum and either the ICH visible or UV lamp outputs (Fig. 3, bottom). Furthermore, the molecular components of the matrix (Table 1) are also nonlight absorbing in the 300 to 700 nm exposure regions.

We were then left to explore the nature of the photosensitivity of this citrate-based formulation and examine the generality of the experimental observations to a larger class of buffer systems and, in general, to liquid dosage forms. Furthermore, the role of package components was also investigated and will be discussed as well.

Initial ICH photostability studies on a drug A (Formulation 1, Table 1) solution showed the unexpected formation of several new degradation peaks, the most prominent being a phenol degradate. The formulation solution was exposed in a clear glass vial and exposed to the ICH recommended UV (200 W hr/m²) and visible (1.2 × 10⁶ Lux hour) light doses. The phenol degradation product was clearly evident and present at about 0.7%. Other less prominent species are evident that result from the photochemistry, as well. An identical sample first placed in a protective

Table 1 Summary of the Product Formulation Compositions for Drug A Examined in This Study

Component	Formulation 1		Formulation 2	
	mM	Relative molar amount[a]	mM	Relative molar amount[a]
Drug A	0.57	1.0	0.11	1.0
Citric acid	0.8	1.5	0.17	1.5
Sodium citrate	9.2	16.2	1.83	16.2
Sodium chloride	136	241	153	1357

[a]Molar ratio of specified component relative to drug A free base.

carton and then exposed to the same ICH light conditions does not show any of the proposed photodegradation products. Specifically, the phenol peak is absent, as are the earlier eluting species. Clearly, light exposure impacts the stability of this drug solution and gives rise to elevated levels of the phenol degradation product.

The effect of incremental light exposure levels was examined for formulation solutions for both visible and UV light by titrating the amount of light through full ICH exposure levels. Visible light was increased through 1.2×10^6 Lux hour in 0.3×10^6 Lux hour increments. UV light was increased through $200\,W\,hr/m^2$ in $50\,W\,hr/m^2$ increments. The plot in Figure 4 shows the loss of drug A levels as well as the increase in the phenol degradation product.

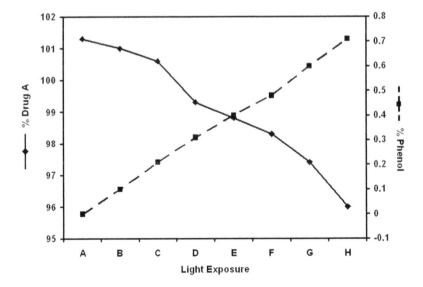

Figure 4 Photodegradation (decreasing drug A levels versus increasing phenol levels) induced by various International Conference on Harmonization visible and ultraviolet light exposures for a solution containing 0.57 mM drug A, 10 mM citrate (pH 6), and 136 mM sodium chloride. A = 0.3×10^6 Lux hour; B = 0.6×10^6 Lux hour; C = 0.9×10^6 Lux hour; D = 1.2×10^6 Lux hour; E = 1.2×10^6 Lux hour followed by 50 W hr/m²; F = 1.2×10^6 Lux hour followed by 100 W hr/m²; G = 1.2×10^6 Lux hour followed by 150 W hr/m²; H = 1.2×10^6 Lux hour followed by 200 W hr/m².

Drug A decreases linearly with visible and UV light exposure from about 101% to 96% levels through full ICH UV and visible light exposure. The plot in Figure 4 also shows that the phenol level increases linearly with UV and visible light exposure to 0.7% levels at full ICH exposure (total degradation products increase to near 3%, not shown). The quantitative results of this study as well as the light exposure conditions given are summarized in Table 2.

Throughout the course of these studies, multiple batches of the same formulation were exposed to either full ICH exposure conditions or only the visible component of ICH light. Substantial variation was observed between different preparations of the formulation. One major trend appears to be that older formulation solutions are more photosensitive than recently prepared solutions (vide supra). After ICH visible exposure, a three-month-old solution has about 0.1% levels of phenol and a 13-month-old solution forms 0.4% phenol, whereas a 23-month-old solution forms 1.2% phenol. The large magnitude of the variation in photosensitivity observed for different manufactured batches as well as for bench top preparations (Table 3) argue that the photodegradation is not a function of a component that is intentionally controlled in the formulation.

Additional experiments were performed to determine the role excipients play in the photodegradation process. Formulations were made in the laboratory that consisted of 0.57 mM concentrations of drug A in (i) 10 mM citrate (pH 6) and 136 mM sodium chloride, (ii) 136 mM sodium chloride, (iii) deionized water, and (iv) 10 mM phosphate (pH 6) buffer. The four solutions were then placed in clear 50 mL USP Type I glass vials, stoppered and placed in the ICH photochamber, and given full ICH light exposure. Photodegradation (0.2% levels of phenol) with full ICH light exposure was observed only for the sample containing citrate. The other three sample preparations did not show any evidence of the photodegradation. These experiments showed that citrate plays a key role in the photosensitivity of drug A.

The appearance of many degradation products along with the phenol provided an opportunity to investigate the photodegradation mechanism. The structures of the early eluting species were therefore investigated by liquid chromatography/mass spectrometry (LC/MS) and UV spectroscopy. The UV

Table 2 Summary of Photodegradation Results Through International Conference on Harmonization Recommended Light Exposures

Exposure condition		% Component formed		
Visible (10^6 Lux hr)	Ultraviolet (W hr/m²)	Drug A	Phenol	Total degradation products
0.3	0	101.3	<LOQ	<LOQ
0.6	0	101.0	0.10	0.10
0.9	0	100.6	0.21	0.33
1.2	0	99.3	0.31	0.81
1.2	50	98.8	0.40	1.13
1.2	100	98.3	0.48	1.35
1.2	150	97.4	0.60	1.58
1.2	200	96.0	0.71	1.86

Abbreviation: LOQ, Limit-of-Quantitation.

Table 3 Batch-to-Batch Variability in Photodegradation of 0.57 mM Drug A, 10 mM Citrate (pH 6), 136 mM Sodium Chloride Solutions with International Conference on Harmonization Recommended Light Exposure

| | | % Phenol observed | | | | | |
| | | Full ICH exposure[a] | | | Visible exposure only[b] | | |
Batch	Solution preparation	Run 1	Run 2	Run 3	Run 1	Run 2	Run 3
1	Manufacturing	2.0%			1.2%		
2	Manufacturing	0.4%[c]	0.7%	0.7%	0.3%	0.2%	0.4%
3	Manufacturing	0.3%			0.1%		
4	Laboratory				<0.1%[d]		
5	Laboratory	0.2%					
6	Laboratory	1.0%					
7	Laboratory	0.3%					

[a]200 W hr/m^2 of ultraviolet light and 1.2 × 10^6 Lux hr of visible light exposure.
[b]1.2 × 10^6 Lux hr of visible light exposure.
[c]Exposure was done using a photochamber at another facility. The remainder of the exposures were performed at Merck Research Laboratories.
[d]1.4 × 10^6 Lux hr of visible light exposure.
Abbreviation: ICH, International Conference on Harmonization.

spectra of all the photodegradates were similar to the drug A parent compound except one, which suggested an additional conjugation of the phenyl ring. Eight species (in addition to the phenol) could be assigned a molecular weight from the LC/MS analysis. Four species have molecular weight gains of 16 mass units over the drug A parent compound. This mass gain is typical of a C–H bond being converted to C–OH bond. Three additional species showed mass gains of 14 mass units, which are interpreted to be conversion of a methylene group (–CH$_2$–) to a ketone (C=O) group. One of the 14 mass gain species showed the conjugated phenyl ring absorbance. Finally, a species showing a +32 mass unit gain over the drug A parent was observed. This species could result from conversion of a C–H group to a peroxy group C–OOH or from two C–H bonds being converted to C–OH bonds. The types of new bonds being formed argue that hydrogen atoms of C–H bonds of drug A are being abstracted by an oxygen-centered radical. The large number of sites of hydrogen atom abstraction further argues that the oxygen-centered radical is not likely hydroperoxy or peroxy radicals (HOO$^{\bullet}$ or ROO$^{\bullet}$), because these radicals are quite selective in their reaction with C–H bonds. Indeed, drug A is stable when exposed to peroxy radicals. In contrast, the rich product distribution could be explained by encounter with hydroxyl radicals (HO$^{\bullet}$) that are much less selective. Abstraction of C–H hydrogen atoms from drug A by hydroxyl radical, followed by oxygen addition, would generate transient peroxy radicals, which would then largely disproportionate as drug A is stable toward peroxy radicals. Disproportionation of the peroxy radicals should then yield equivalent amounts of alcohol C–OH and carbonyl C=O products at each C–H bond site. The dominant 16 and 14 mass unit gains characteristic of the product distribution observed here are consistent with this general pathway and can also account for the formation of the phenol species as shown in Figure 5.

One potential source of hydroxyl radicals is depicted in Figure 6. The reduced form of some transition metals, the most notable and abundant being iron, can react

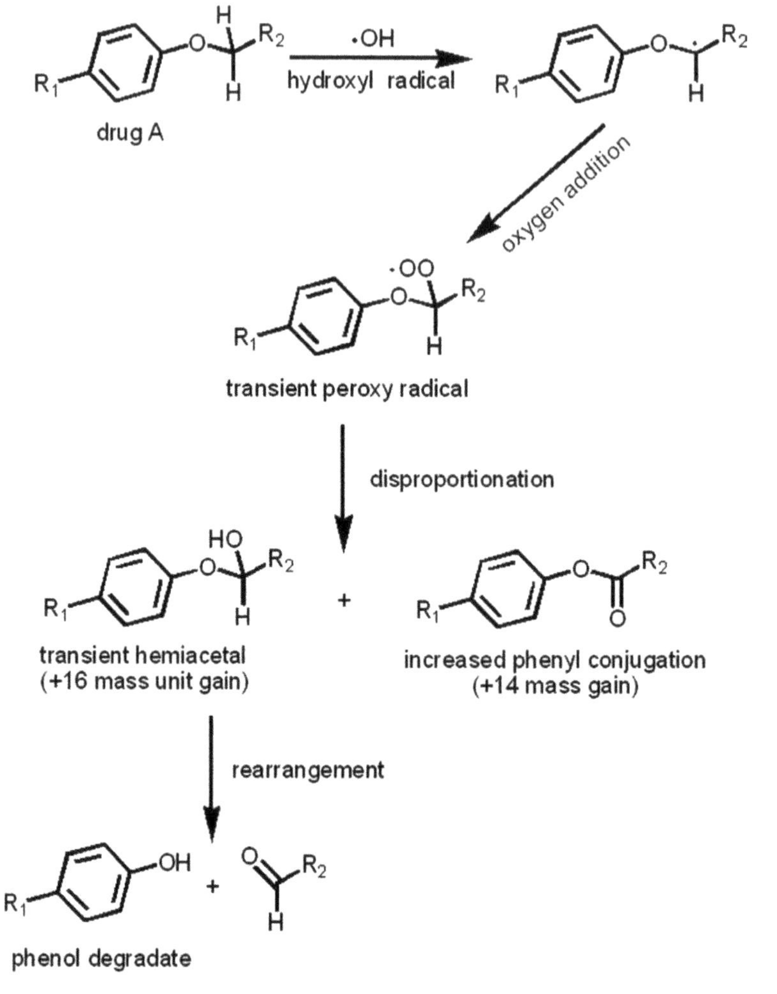

Figure 5 Reaction of hydroxyl radical with drug A to produce phenol degradate and a +14 mass unit gain species with extended phenyl conjugation. Hydroxyl radical abstracts a hydrogen atom leaving a carbon-centered radical to which dissolved oxygen quickly adds to give a peroxy radical. Disproportionation gives an alcohol and a carbonyl species at the original site of hydrogen atom abstraction. In the specific case of the carbon atom shown here, the alcohol is unstable and rearranges to the phenol degradate.

with dissolved oxygen in solution to give superoxide radical, O_2 (32,33). Superoxide radical can then protonate in water to give hydroperoxyl radical. A number of recombinations can then occur to give hydrogen peroxide, H_2O_2. Once hydrogen peroxide is formed, reaction with reduced iron to give hydroxyl radicals is well known and serves as the basis of Fenton's reagent (33–35). That the pathways depicted in Scheme 6 lead to the observed complex drug A degradation profile was easily demonstrated by simply adding a few hundred parts per billion (ppb) of reduced iron (Fe^{2+}) to drug A formulation solutions. Even in the dark, the "photodegradation" profile was faithfully reproduced. Further, it was determined that spiking of Fe^{3+} into drug

Superoxide Radical Formation

$$Fe^{2+} + O_2 \Longrightarrow Fe^{2+}{-}O_2 \Longrightarrow Fe^{3+}{-}O_2^{\overline{\cdot}} \Longrightarrow Fe^{3+} + O_2^{\overline{\cdot}}$$

Superoxide Radical Protonation to Hydroperoxyl Radical

$$O_2^{\overline{\cdot}} + H^+ \Longrightarrow {\cdot}OOH$$

Recombinations to Give Hydrogen Peroxide

$$O_2^{\overline{\cdot}} + H^+ + {\cdot}OOH \Longrightarrow H_2O_2 + O_2$$

$${\cdot}OOH + {\cdot}OOH \Longrightarrow H_2O_2 + O_2$$

Fenton Reaction Liberating Hydroxyl Radicals

$$Fe^{2+} + H_2O_2 \Longrightarrow Fe^{3+} + {\cdot}OH + OH^-$$

Figure 6 Scheme for reduced iron reacting with oxygen to form superoxide and ultimately hydrogen peroxide. Reduced iron then reacts with hydrogen peroxide to liberate hydroxyl radicals.

A product only gave rise to the degradation profile upon exposure to light. One hundred ppb Fe^{3+} was spiked into the 23-month-old product lot and subjected to ICH visible stressing. The same early eluting peak pattern is reproduced, along with increased phenol levels. The photodegradation caused by spiking Fe^{3+} also shows the same citrate and light dependence described above, that is, in drug A solutions in phosphate buffer, added Fe^{3+} does not promote degradation upon ICH visible exposure, and Fe^{3+} spiked into drug A citrate solutions does not promote degradation if the product is kept in the dark. These results indicate that the combination of citrate and light exposure must promote the reduction of Fe^{3+} to Fe^{2+} allowing the reactions in Figure 6 to proceed.

The reactions depicted in Figure 6 were implicated further when it was realized that the photoreduction of Fe^{3+} to Fe^{2+} in citrate complexes is known in the literature and that hydroxyl radical formation has been reported from such systems (33–36). Citrate has long been known to chelate Fe^{3+}, and quantum yields for the photoreduction of these types of iron–carboxylate complexes are known to be pH dependent (34,36). For the case of Fe^{3+} complexes specifically, quantum yields appear higher near pH 4 and decrease near pH 2 to 3; values of 0.1 to 0.3 have been reported from pH 2 to 5 and in the 400 to 450 nm spectral range (34,36). In contrast, it is interesting to note that the recombination reactions leading to hydrogen peroxide in Figure 6 will proceed more readily at lower pH. Thus the overall pH dependence of the hydroxyl radical generation cannot be readily predicted. To this end, the pH dependence of phenol formation was examined in drug A formulation solutions over the pH 2 to 6 range after exposure to 1.2×10^6 Lux-hour of ICH visible light. Figure 7 shows the overall pH dependence of the photodegradation for a fixed iron concentration and light exposure.

This mechanism for the reduction of Fe^{3+} to Fe^{2+} explains the citrate dependence, as well as the seemingly odd wavelength dependence of the observed drug A photoinstability. The spectral output of both ICH UV and ICH visible lamps have intensity in the 400 to 450 nm spectral region in which the Fe^{3+} complex quantum

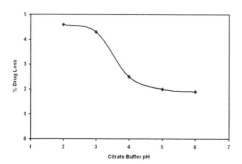

Figure 7 Percentage drug loss observed for 0.57 mM drug A in 10 mM citrate buffer (pH adjusted from 2 to 6) after exposing to International Conference on Harmonization visible lamp (1.2×10^6 Lux hour) in the presence of 10 ppb iron.

yields have been reported. The relative spectral emissions from these lamps are shown in Figure 8.

Also shown in Figure 8 is the absorption spectrum of the iron–citrate complex obtained by preparing 0.1 mM Fe^{3+} in the presence of 10 mM citrate. Figure 8 thus provides a rationale for why the UV and visible ICH stressing conditions will both lead to the observed photodegradation as Fe^{3+} can be photoreduced to Fe^{2+} in the presence of citrate by either lamp output.

Further experiments were carried out to determine if the degradation caused by adding Fe^{3+} to formulation followed by light exposure was dependent on oxygen as predicted by the reaction schemes in Figure 6. Relatively large amounts of Fe^{3+} had to be added to clearly see the effect of oxygen removal by sparging the sample with helium for about 30 minutes. For example, 1000 ppb levels of Fe^{3+} were spiked into two formulation samples. One vial was sparged with helium to reduce the oxygen content, whereas the other vial was unsparged. After ICH visible stressing, the sparged vial showed about eightfold less degradation than the unsparged case, in which about 50% of drug A degraded. If no Fe^{3+} was added, similar sparging failed to substantially reduce the photodegradation. This is readily explained by the relative molar amounts of oxygen and drug A molecules in solution as shown in Table 4.

When 1000 ppb Fe^{3+} is added, the molar amount of drug A that is converted to other species is about 25% of the total oxygen in solution. Getting rid of approximately 90% of the dissolved oxygen in this case has an effect. In the unspiked lot, only a few percentage of drug A is degraded corresponding to about 1% of the dissolved oxygen on a molar basis. Thus, sparging will be much less effective in prohibiting the photodegradation.

Drug A formulation lots were then examined for evidence of iron as well as other transition metal ions. Inductively coupled plasma–mass spectroscopy (ICP-MS) was used initially in a semiquantitative scanning mode. This mode of detection allows for determination of the elements sodium through mercury with detection limits of about 1 ppb with absolute errors typically about ±30%. The three manufactured batches shown in Table 3 were examined. With the exception of iron, no first-, second- or third-row transition metal ion was found in any lot at greater than 2 ppb. Most transition metals were undetectable. Iron, in contrast, was detected at between 10 and 30 ppb. Iron levels were quantitated more accurately by using the method of standard

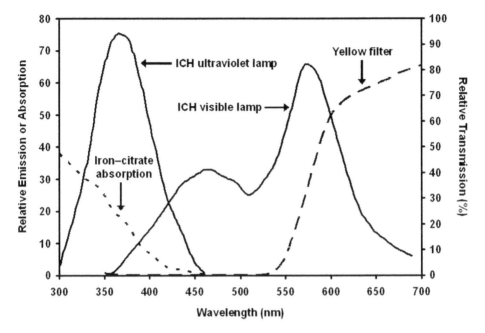

Figure 8 (*Solid lines*) Spectral outputs for International Conference on Harmonization (ICH) visible and ultraviolet lamps with Hg emission lines removed; (*dotted line*) absorption spectrum for the iron–citrate complex; (*dashed line*) transmission profile for the yellow light filters used in this study and common to many manufacturing areas. Note the spectral overlap of both ICH lamps with iron–citrate complex absorption.

addition and by subtracting out $m/z=57$ interferences due to argon oxides in the plasma. A standard addition curve was generated by spiking in 50 and 100 ppb Fe^{3+} into batch 1 along with measurement of the unspiked lot. The iron isotope monitored was the $m/z=57$ species. The interference at the same m/z ratio due to an argon oxide isotope was corrected for by subtracting the $m/z=57$ counts obtained for a 136 mM sodium chloride blank from all samples. Extrapolation of a straight line fit to the data back to the x-axis gives a concentration of about 40 to 45 ppb iron in the unspiked,

Table 4 Molar Ratios of 0.57 mM Drug A, 10 mM Citrate (pH 6), 136 mM Sodium Chloride Formulation Components Normalized to 50 ppb Iron

	Formulation 1		Formulation 2	
Component	mM	Relative molar amount[a]	mM	Relative molar amount[a]
Drug A	0.57	641	0.11	126
Citric acid	0.8	925	0.17	185
Sodium citrate	9.2	10200	1.83	2040
Sodium chloride	136	152222	153	171086
Oxygen	1.3	1444	1.3	1444
Iron	50 ppb	1.0	50 ppb	1.0

[a]Molar ratio of specified component relative to 50 ppb iron.

Table 5 Batch, Age, Levels of Phenol, and ppb Iron by Inductively Coupled Plasma/Optical Emission Spectroscopy

Batch	Age of lot at testing (mo)	Phenol levels[a] (%)	ppb iron
1	23	1.2	45
2	13	0.4	15
3	3	0.1	5

[a]Levels of phenol after 1.2×10^6 Luxhr of visible light exposure.

23-month-old batch 1, 10 to 15 ppb iron in the 13-month-old batch 2, and 0 to 5 ppb in the 3-month-old batch 3. The 5 ppb range in each measurement reflects estimates of possible error due to the background correction as well as the possibility of up to 3 ppb iron levels in the sodium chloride blank. Table 5 shows the batch ID, lot age, and ppb iron level using the 5 ppb iron range determined from the ICP-MS measurements.

The increase in iron levels in the three lots examined by ICP-MS correlates quite well with the observed photosensitivity of each lot, as demonstrated by the two plots in Figure 9.

The square data points show the ppb iron determined by ICP-MS, whereas the circular data points show the % phenol formed in each lot from visible ICH exposure (Table 5). The agreement between the two is excellent. Additionally, the % of drug A loss was examined for solutions of formulation 1 as a function of added Fe^{3+} up through 1000 ppb levels after full ICH visible light exposure (Fig. 10).

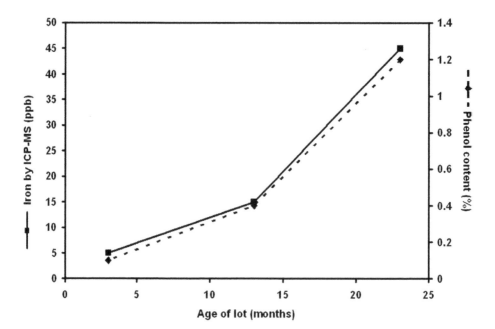

Figure 9 Dual plot of phenol resultant of 1.2×10^6 Lux hour of visible light exposure and ppb iron levels determined for the same three lots by inductively coupled plasma-mass spectroscopy (data from Table 5).

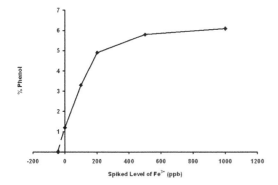

Figure 10 Percentage of phenol formed for 0.57 mM drug A in pH 6 10 mM citrate buffer after exposing to International Conference on Harmonization visible lamp (1.2×10^6 Lux hour) in the presence of varying concentrations of added iron. The level of iron present in the unspiked sample was determined by a standard addition method as 45 ppb (dashed line, Table 5).

The observed photodegradation demonstrates a linear correlation up through about 200 ppb levels of added Fe^{3+}, then becomes increasingly nonlinear. Further evidence of the correlation between iron levels and the extent of drug A photodegradation is that phenol levels, for example, accurately predict iron content. From Figure 10, the % phenol measured in batch 1 with 0, 100, and 200 ppb Fe^{3+} added upon ICH visible exposure if treated as a standard addition curve followed by extrapolation back to the x-axis using a simple polynomial fitting gives an estimate of about 50 ppb iron in the unspiked sample. The 50 ppb iron value based on phenol levels is in excellent agreement with the 40 to 45 ppb iron levels determined by the ICP-MS methods for the same batch. The agreement supports the unique role of iron as the transition metal responsible for the observed photosensitivity.

The rationale for increasing levels of iron with the age of drug A, 10 mM citrate (pH 6), and 136 mM sodium chloride formulations is shown by the data in Table 6.

Table 6 shows the major metal oxides and the iron oxide impurity levels of typical borosilicate Type I glass. Up to 0.05% by weight (500 ppm) iron oxide as Fe_2O_3 may exist in the borosilicate Type I glass. Thus, the increase in iron levels with time likely reflects a slow leaching of iron from the glass vial. Consistent with this explanation is that similar increases in silicon, aluminum, calcium, and barium levels are also observed in older product lots as shown in Table 6. Note that these nontransition metal ions are not known to participate in the type of reactions depicted in Figure 6. Furthermore, it is not clear if the expected increase in iron leaching from amber vials (Table 6) will be readily compensated for by the reduced light transmission at the causative wavelengths offered by utilizing the amber vial as the primary package.

The data described above clearly show that with time, drug A formulation solutions will become more photosensitive due to increasing levels of iron in the product. The increase in photosensitivity over time can be adequately addressed through the appropriate application of a secondary package to provide protection from light exposure during long-term storage.

The primary questions remaining are (i) how photosensitive are formulations at the point of manufacture?, (ii) how much light exposure might occur for a typical

Table 6 Impurity Profiles of Borosilicate Glass and Corresponding ppb Levels of Elements Found in 3-mo-old and 23-mo-old 0.57 mM Drug A, 10 mM Citrate (pH 6), 136 mM Sodium Chloride Formulation Solutions Measured by Inductively Coupled Plasma-Mass Spectroscopy

	Impurity profiles of Type I borosilicate glass, water, and two different product batches				
	Clear glass (ppm)	Amber glass (ppm)	USP water (ppb)	3-mo-old (ppb)	23-mo-old (ppb)
Silicon (as SiO_2)	685,000	664,000	100	300	3600
Aluminum (as Al_2O_3)	60,000	57,000	<1	28	470
Calcium (as CaO)	12,000	17,000	140	150	310
Barium (as BaO)	25,000	12,000	<2	10	120
Iron (as Fe_2O_3)	500	13,000	<2	5	45
Manganese (as MnO)	—	58,000	<2	<2	<2

vial containing drug A, formulation during manufacturing prior to placing each vial into a light protective secondary package? and (iii) how can initial iron levels be controlled in successive formulations? These questions are framed within the context of a target level for phenol in the product of under 0.15%. Phenol levels measured after two years in dark stability chambers have been under 0.1%, reflecting any contributions from the bulk drug substance (as an impurity) as well as thermal degradation that may have occurred over the two-year period. Therefore, about 0.1% levels of phenol can be tolerated due to photodegradation prior to individual packaging.

The data described above clearly correlate the increase in photosensitivity with time to increased iron levels. However, at the time of initial manufacture, iron levels are lower. The best data available currently on initial iron levels is batch 3, which contains 5 to 10 ppb iron approximately three months after manufacture. A substantial portion of this iron is likely due to iron levels in the excipients as described further below. All other transition metal ions are present at less than 1 ppb in the same batch. Iron is perhaps unique in its ability to complex with citrate and be photoreduced, react with oxygen in the reduced form to give superoxide radical, and react in the reduced form with hydrogen peroxide to give hydroxyl radicals. Any transition metal ion must accomplish all three of these tasks in order to give rise to the observed photodegradation. The uniqueness of iron is demonstrated in Figure 11, where salts of some of the more abundant transition metal ions (zinc, copper, cobalt, manganese, and nickel) were spiked into formulation one at 100 ppb levels and then exposed to ICH visible light stressing.

None of the transition metals examined except iron showed increased formation of phenol compared to the unspiked sample of formulation one. Thus, it appears likely that iron levels will remain highly correlated with the photosensitivity of initially prepared drug A formulation solutions.

The photochemical generation of phenol levels in drug A formulation solutions shortly after manufacture is therefore a function of both iron content and light exposure. The data in Figure 4 shows that phenol increases linearly with light exposure. The curve in Figure 10 also shows that phenol increases linearly over the 0 to 50 ppb iron levels at a fixed exposure of 1.2×10^6 Lux hour of visible light. Therefore, the formation of phenol can be predicted in the formulation as a function of both light and iron as illustrated in Table 7.

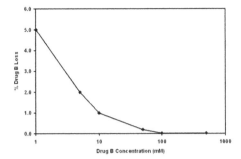

Figure 11 Percentage of drug loss observed for varying concentrations of drug B in 10 mM pH 6 citrate buffer after exposure to 1.2×10^6 Lux hour International Conference on Harmonization visible light in the presence of 50 ppb iron.

If one allows for a maximum of 0.15% of phenol from photodegradation that may occur prior to placing each vial into a light protective secondary package, then the maximum light exposure for a given concentration of iron can be defined as is shown in Table 7 by the solid line. In general, control of iron to lower levels allows more flexibility with regard to the amount of light exposure that the product may receive without concern for appreciable photodegradation prior to packaging each vial in a light protective secondary package.

Light mapping studies at two manufacturing sites have been performed to estimate the worst case scenario for the amount of light exposure. Measurements at one site gave 0.12×10^6 Lux hour (one-tenth of the visible ICH exposure), whereas the other site gave estimates of 0.21×10^6 Lux hour. The value from site one represents the entire manufacturing and packaging exposure, whereas the other site value reflects only the manufacturing process and not the packaging time. If batch 3 is representative of manufacturing capability at 5 to 10 ppb iron, Table 7 shows that much higher light levels than those measured can actually be accommodated without problematic levels of phenol formation.

Table 7 Predicted Phenol Levels as a Function of Iron Content and Unfiltered Visible White Light Exposure

Iron (ppb)	Predicted levels of phenol (%) at increasing visible light exposures (10^6 Lux hr)									
	0.12	0.24	0.36	0.48	0.60	0.72	0.84	0.96	1.08	1.20
5	0.012	0.024	0.036	0.048	0.060	0.072	0.084	0.096	0.108	0.12
10	0.024	0.048	0.072	0.096	0.120	0.144	0.168	0.192	0.216	0.24
15	0.036	0.072	0.108	0.144	0.180	0.216	0.252	0.288	0.324	0.36
20	0.048	0.096	0.144	0.192	0.240	0.288	0.336	0.384	0.432	0.48
25	0.060	0.120	0.180	0.240	0.300	0.360	0.420	0.480	0.540	0.60
30	0.072	0.144	0.216	0.288	0.360	0.432	0.504	0.576	0.648	0.72
35	0.084	0.168	0.252	0.336	0.420	0.504	0.588	0.672	0.756	0.84
40	0.096	0.192	0.288	0.384	0.480	0.576	0.672	0.768	0.864	0.96
45	0.108	0.216	0.324	0.432	0.540	0.648	0.756	0.864	0.972	1.08
50	0.12	0.24	0.36	0.48	0.60	0.72	0.84	0.96	1.08	1.20

Note: The bold line represents the tolerance for light at a given iron level to allow 0.15% formation of phenol.

Alternatively, the benefits of incorporating yellow light filters (or replacing white light with yellow light sources) can be considered for manufacturing and packaging areas. In this analysis, the ICH photochamber was modified to filter visible light through a representative yellow film common in manufacturing/packaging facilities. The light-transmission properties of the specific film used in this study (UV cutoff at 540 nm) is shown in Figure 8 (dashed line). As shown in Table 8, the overall benefit resultant of reducing light exposure specific to the ligand to metal charge transfer transition for the citrate complex with Fe^{3+} is a sixfold reduction in the amount of photodegradation for a given light and iron combination.

The yellow light conditions significantly increase the tolerance for light exposure and iron content for this formulation.

The most logical way to ensure adequate light stability during the manufacturing and packaging process is to control the levels of iron present in the excipients in the formulation. Currently, iron content is controlled only in the sodium chloride that is used in the formulation (as ≤2 ppm). Iron is not controlled in citric acid, sodium citrate, and the bulk drug substance, nor in water. The addition of iron specification controls for the solid components in the formulation are analyzed in Table 9.

Across-the-board specifications of 2 ppm iron for each excipient, for example, would give rise to a maximum of 22 ppb iron in the product assuming each solid component contained iron at the 2 ppm specification limit and that negligible amounts of iron are present in the water. Sodium chloride will dominate the iron contribution if all solid components have the same iron specification as shown in Table 9. A quick survey of typical specifications for maximum iron levels allowed in commercially available materials are noted in Table 9 are 3, 5, and 1 ppm for citric acid, sodium citrate, and sodium chloride, respectively. The maximum possible iron in product due to use of these excipients would be 22 ppb iron as shown in the right hand column. The fact that batch three contains only 5 to 10 ppb iron levels clearly indicates the actual iron levels in our excipients, including water, are significantly below the label specified limits.

The water used in the formulation could potentially contribute significantly and will require iron levels below the maximum allowed ppb iron limit desired in the product. Tables 7 and 8 provide the basis for determining iron control limits for

Table 8 Predicted Phenol Levels as a Function of Iron Content and Yellow-Filtered White Light Exposure

Iron (ppb)	Predicted levels of phenol (%) at increasing visible light exposures (10^6 Lux hr)									
	0.12	0.24	0.36	0.48	0.60	0.72	0.84	0.96	1.08	1.20
5	0.002	0.004	0.006	0.008	0.010	0.012	0.014	0.016	0.018	0.020
10	0.004	0.008	0.012	0.016	0.020	0.024	0.028	0.032	0.036	0.040
15	0.006	0.012	0.018	0.024	0.030	0.036	0.042	0.048	0.054	0.060
20	0.008	0.016	0.024	0.032	0.040	0.048	0.056	0.064	0.072	0.080
25	0.010	0.020	0.030	0.040	0.050	0.060	0.070	0.080	0.090	0.100
30	0.012	0.024	0.036	0.048	0.060	0.072	0.084	0.096	0.108	0.120
35	0.014	0.028	0.042	0.056	0.070	0.084	0.098	0.112	0.126	0.140
40	0.016	0.032	0.048	0.064	0.080	0.096	0.112	0.128	0.144	0.160
45	0.018	0.036	0.054	0.072	0.090	0.108	0.126	0.144	0.162	0.180
50	0.020	0.040	0.060	0.080	0.100	0.120	0.140	0.160	0.180	0.200

Note: The bold line represents the tolerance for light at a given iron level to allow 0.15% formation of phenol.

Table 9 Possible Contribution to Iron Levels in the Formulation from Drug and Excipients at Product Manufacture

Components	Contribution to concentrate at specified iron level for component				
	1 ppm	2 ppm	3 ppm	5 ppm	"Best"
Drug A (ppb)	0.3	0.6	0.9[a]	1.5	0.9
Citric acid (ppb)	0.1	0.3	0.4[a]	0.6	0.4
Sodium citrate (ppb)	2.7	5.4	8.1	13.5[a]	13.5
Sodium chloride (ppb)	8[a]	16	24	40	8
Water	—	—	—	—	—
Total (ppb)	11	22	33	56	23

[a]Indicates level specified in best quality commercially available material.

excipients and water in addition to allowed light exposure limits. Greater tolerance in the specifications for iron in the excipients requires tighter control of the light exposure allowed and vice versa. Together, appropriate iron and light exposure controls can assure continued successful manufacturing batches with negligible phenol levels at the time of final packaging.

The pH dependence of phenol formation was also examined for a series of solutions containing 0.57 mM drug A, 10 mM citrate, and 136 mM sodium chloride over a pH range from 2 to 6 and following exposure to 1.2×10^6 Lux hour of ICH visible light. Figure 7 shows that the photoreaction for a fixed iron and light exposure is pH dependent, with a significant increase in the amount of photodegradation occurring as the pH is decreased from 5 to 3. At pH values above 4 degradation remains low, whereas below a pH of 3 there appears to be a plateau. This transition occurs near two pK_a values for citrate (Table 10) and suggests that the protonation state of citrate plays a significant role in the photocatalytic reaction.

Table 10 Quantum Yields for Photoreduction of Iron in Carboxylate Complexes at pH 2.7 and 4.0

Buffer	pK_a's[b]	Quantum yield[a]		Complexation constant[b]		
		pH 2.7	pH 4.0	No. of ligands	Fe^{3+}	Fe^{2+}
Oxalate	1.3, 4.3	0.65	0.30	2, 3	16.2, 20.2	4.5, 5.2
Succinate	4.2, 5.6			7.5		
Maleate	1.9, 6.3	0.21	0.29	2, 3	15.5	4.4
Tartrate	3.0, 4.4	0.35	0.5		07.5	
Citrate	3.1, 4.8, 6.4	0.28	0.45	1	12.5	3.1
Isocitrate	3.1, 4.8	0.14	0.37			

[a]Quantum yield with ligand-to-iron ration of five to one. Measurements were made with 366 nm irradiation in the absence of oxygen, using phenanthroline as a complexing agent for iron.
[b]Taken from Ref. 37.
Source: From Ref. 36 unless otherwise noted.

One can envision that pH can impact several aspects of this reaction, including the affinity of citrate for both Fe^{3+} and Fe^{2+} (Table 10). Abrahamson et al. studied the quantum yield for citrate complexes of Fe^{3+} and Fe^{2+} as a function of pH and reported that the formation of Fe^{2+} is more efficient at pH 4.0 than at pH 2.7, which would predict the opposite trend observed in Figure 7 (36).

The photodegradation of drug A formulation solutions at a fixed iron content was also examined for a series of 10 mM poly- and monocarboxylate buffers (oxalate, succinate, maleate, tartrate, and formate). The general photocatalysis was observed for all systems examined and varied from 0.40% to 17.6% loss of drug A for solutions containing 50 ppb iron (Table 11).

The following order of photosensitivity was observed:

oxalate > citrate > tartrate > succinate ~ maleate ~ formate

The above observed photosensitivity order does not appear to correlate with the reported quantum yields of some of these carboxylate complexes (Table 10), signaling the significance of the pH dependence of the recombination reactions.

In all cases examined, the level of photodegradation in the presence of 50 ppb iron (representative of potential iron amount after two years storage in borosilicate glass) could have a significant impact on the photostability of a product and may introduce the need to take steps to protect the product from light. At manufacture, formulations containing 10 ppb iron and either succinate, maleate, or formate would likely be sufficiently photostable so as to minimize the need for light exposure control during manufacturing and packaging operations.

The generality of the ability of the iron–citrate–oxygen system to create hydroxyl radicals capable of oxidizing drug substances in solution was then examined for a series of four drugs: drug A (a phenyl ether with no light absorption above 300 nm), drug B (a phenyl piperidone with significant light absorption above 300 nm), drug C (an alkyl piperidone with no light absorption above 300 nm), and drug D (a phenyl thiazole with significant light absorption above 300 nm). All formulations were made at 0.57 mM drug concentration in 10 mM citrate (at pH 3 for drugs A, B, and C and, for solubility limitations, pH 2.5 for drug D) and placed in clear USP Type I glass vials. All solutions were then exposed to 1.2×10^6 Lux hour of visible light in

Table 11 Visible Photodegradation of Drug A as a Function of Carboxylate Buffer System

	% Drug A loss estimated as a function of iron (ppb)		
Buffer[a]	1	10	50
Oxalate	0.351	3.51	17.6
Succinate	0.014	0.14	0.70
Maleate	0.011	0.11	0.55
Tartrate	0.071	0.71	3.55
Citrate	0.187	1.87	9.35
Formate	0.008	0.08	0.40

[a]Buffers were adjusted to pH 6, and solutions were 10 mM for each specified buffer, drug A was maintained at 0.57 mM and 1.2×10^6 Lux hour of visible light exposure was provided to each solution.
[b]pK_a values were taken from Ref. 37.

the ICH chamber with various levels of iron spiked into each preparation. In each case, the level of iron present was determined to account for contributions from excipients and the drug itself. The amount of drug loss observed with 50 ppb levels of iron present ranged from 6.85% to 21.4% and is considered significant for all formulations (Table 12).

Even at 10 ppb iron, the amount of degradation observed is also significant enough to impact light exposure concerns in manufacturing and packaging areas and would likely require a light protective package. The generality of the photo-degradation is expected and strongly supports hydroxyl radicals as the key reactive species, as many types of C–H bonds are susceptible to hydrogen atom abstraction by hydroxyl radicals.

The dependence of the percentage drug lost on the initial drug concentration was examined for a series of solutions containing 10 mM citrate at pH 6, 50 ppb iron, and 1–500 mM drug B. Again, all formulations were placed in clear USP Type I glass vials and the iron content of each formulation was measured, followed by exposure to 1.2×10^6 Lux hour of visible light. Figure 11 illustrates that the extent of photodegradation smoothly becomes increasingly less significant at increasing drug concentrations, becoming largely insignificant at concentrations of 100 mM and above. The trend in Figure 11 is consistent with the observed limitation iron plays at levels under 200 ppb in Figure 8, where a fixed quantity of drug can be consumed at a given iron level and light exposure.

All of the experimental observations are consistent with the scenario stated by Faust and Zepp (34) and depicted in Figure 12. Trace levels of iron in formulations of drug A containing 10 mM citrate (pH 6) and 136 mM sodium chloride increase with time. Solubilized iron complexes with citrate in the formulation. Iron complexed with citrate can be photoreduced by exposure to either ICH visible or UV lamps (Fig. 8). The reduced iron reacts with dissolved oxygen in solution as depicted in Figure 6, to form hydroperoxyl and ultimately hydroxyl radicals via the Fenton reaction. The hydroxyl radicals abstract hydrogen atoms from drug A leading to degradation products as shown in Figure 5. The Fe^{3+} formed can then be photoreduced repeatedly to Fe^{2+}, allowing for continued production of hydroxyl radicals from very low levels of iron. The catalytic role of the iron can be appreciated by examination of Table 3. In batch 1, for example, the 1.2% levels of phenol formed were accompanied by about a 10% loss in drug A label claim. The batch contains near 50 ppb iron. Thus, the molar amount of drug A lost is more than 60 times the molar amount of iron present. It is the photocatalytic nature of

Table 12 Visible Photodegradation of Drugs A to D in 10 mM Citrate Buffer

		% Drug loss estimated as a function of iron (ppb)		
Formulation[a]	Buffer	1	10	50
Drug A (phenyl ether)	10 mM citrate; pH 3.0	0.429	4.29	21.4
Drug B (phenyl piperidone)	10 mM citrate; pH 3.0	0.394	3.94	19.7
Drug C (alkyl piperidone)	10 mM citrate; pH 3.0	0.288	2.88	14.4
Drug D (phenyl thiazole)	10 mM citrate; pH 2.5	0.137	1.37	6.85

[a]Buffers were adjusted to pH 3, with the exception of drug D for which the pH was adjusted to 2.5 to maintain adequate solubility. Each drug concentration was maintained at 0.57 mM and 1.2×10^6 Lux hour of visible light exposure was provided to each solution.

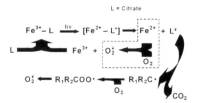

Figure 12 Photocatalytic scheme for iron–citrate mediated generation of Fe^{2+}, which reacts with oxygen to form two equivalents of superoxide radicals for every photon absorbed.

this chemistry that allows ppb levels of iron to oxidize significant quantities of drug A. Furthermore, it is likely that at levels under 200 ppb, iron is the limiting reagent in the drug A formulation and controls the level of photodegradation for a given light exposure. This is born out by the linear region of the slope in Figure 10. However, when the concentration of iron exceeds 200 ppb, it is likely that oxygen becomes the limiting reagent and begins to control the apparent photodegradation reaction rate. The iron-mediated photodegradation of solutions containing polycarboxylate buffers is generally relevant to a wide range of drug substances. Iron has also been shown to be ubiquitous in raw materials, drug substances, and packaging materials. In the present case, the photodegradation observed is significant at ppb levels of iron and thus protection from light is required during manufacturing and packaging operations in addition to shelf storage.

PRODUCT COMPONENT MEDIATED PHOTOCHEMISTRY

Our work in this area originated from observation of ambient laboratory light induced oxidative degradation of a drug A (Fig. 1, *top*) capsule formulation in composite assay sample preparations. As previously noted in Section "Excipient Mediated Photochemistry," this photosensitivity was unexpected as the phenyl ether absorption spectrum of drug A exhibits negligible overlap with either the visible lamp or the UV lamp spectral outputs (Fig. 1, *bottom*). The observed photoinduced oxidation in this instance is attributable to the TiO_2 present in the capsule shell of the formulated product. The following case study describes the nature of the observed TiO_2-induced heterogeneous photocatalytic reaction and comments upon the generality of the experimental observations to other drug formulation sample preparations containing TiO_2.

The rate and extent of oxidative degradation of drug A in capsule formulation composite assay sample preparations is sensitive to ambient laboratory lighting conditions. The drug A capsule formulation consists of neat drug filled into a size #00 hydroxypropylmethyl cellulose (HPMC) capsule. A sample preparation is made by dissolving one drug A capsule in 10 mM phosphate buffer (pH 6.0). A series of chromatograms from high performance liquid chromatography (HPLC) analyses of a drug A capsule formulation sample preparation are provided in Figure 13.

No degradation is observed in a freshly prepared sample solution (Fig. 13, *top*) or in an aliquot of the same sample solution that has sat for five days on a laboratory bench under ambient laboratory light exposure (Fig. 13, *middle*). However, an aliquot of the sample solution that sat on a laboratory windowsill exposed to sunlight for five days contains the phenol degradation product eluting at approximately 13 minutes at 0.7% w/w (Fig. 13, *bottom*). Additional early eluting degradates and one late eluting degradate were also formed and are attributed to other photodegradation products.

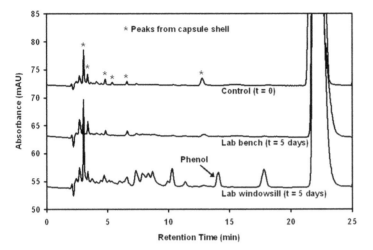

Figure 13 (*Top*) Chromatogram for a solution of drug A in 10 mM phosphate buffer at pH 6 immediately after being prepared; (*middle*) chromatogram for the same solution after being placed on a laboratory bench under ambient laboratory light exposure for five days; (*bottom*) chromatogram for the same solution after being placed on a laboratory windowsill exposed to window glass filtered daylight for five days.

To further explore the photostability, spectral irradiance profiles were collected at bench level for ambient fluorescent laboratory light (Fig. 14, *top*) and on a windowsill for sunlight passing through a window (Fig. 14, *bottom*); absorbance plots for a solution of drug A are included in each case for reference.

Note that there is no spectral overlap between the absorption spectrum of the phenol ether and either irradiance profile. Therefore, the observed photodegradation is not a result of the direct absorption of photons by drug A itself but rather it is derived from some species present in the capsule formulation.

Titanium dioxide is used in the pharmaceutical industry as a white pigment and an opacifying agent in hard and soft capsule shells and in tablet film-coating suspensions (38,39). However, TiO_2 has been extensively studied as a photocatalyst (40–42) and as a means of purifying waste water streams (43–54). Despite the close contact presumably present between the drug substance and TiO_2 in drug formulations, limited studies have been reported in the pharmaceutical industry literature discussing the impact of TiO_2 photocatalytic chemistry on pharmaceutical products. Kakinoki et al. have reported that famotidine is easily discolored by the photocatalytic activity of TiO_2 and that the activity of TiO_2 is strongly affected by the environmental relative humidity (55).

The HPMC capsule shells used in the drug A capsule formulation contain TiO_2, and its confirmation as the source of the photocatalysis was studied by examination of a series of drug A solution photostresses. A stock solution of pure drug in 10 mM phosphate buffer (pH 6.0) was prepared. Three aliquots were spiked with either an HPMC capsule shell, a TiO_2-containing film-coated placebo tablet, or a sample of TiO_2 powder, respectively. An unadulterated aliquot of the stock solution served as a control. All of the sample flasks were subjected to a 20 W hr/m^2 UV light stress in an ICH photochamber. Similar degradation profiles were observed for the capsule shell, the film-coated placebo tablet, and the TiO_2 powder preparations, whereas no

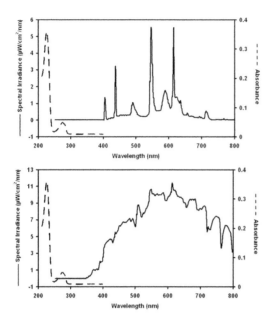

Figure 14 (*Top*) Irradiance profile of ambient light measured on the laboratory bench (Sylvania Octron F043/741 fluorescent bulb, with plastic cover); (*bottom*) irradiance profile of the light measured on the laboratory windowsill (daylight passing through glass window). Ultraviolet/visible absorption spectra for solutions of drug A in water are included for comparison.

degradation was observed in the control sample. Thus, TiO_2 alone is responsible for the observed photochemistry.

The energy band diagram for TiO_2 in pH 7 aqueous solution has been reported (40). Irradiation of TiO_2 generates conductive band electrons (e_{CB}^-) and valance band holes (h_{VB}^+). The redox potentials for conduction band electrons and photogenerated holes are −0.52 and +2.53 V, respectively, versus the standard hydrogen electrode. Upon irradiation with photons (<406 nm), the band gap energy (3.05 eV) is exceeded and an electron–hole pair is generated. The resultant photogenerated electrons can reduce molecular oxygen (O_2) to produce superoxide ($O_2\bullet$) (56), and the photogenerated holes can oxidize water molecules to produce hydroxyl (HO•) radicals (57). This process is described by the series of reactions shown in Figure 15.

The light sensitivity of the observed photo-oxidation of drug A to ambient lighting conditions can be understood through consideration of the TiO_2 surface potential. The output of ambient laboratory light (Fig. 14, top) does not contain components at wavelengths less than 406 nm, hence it is unable to exceed the band gap energy required to generate the electron–hole pair. However, the output of sunlight measured on the laboratory windowsill (Fig. 14, bottom) does contain wavelengths less than 406 nm, thus it is capable of exceeding the band gap energy and initiating TiO_2 photocatalytic activity.

An ICH photostability study was performed to investigate the rate and extent of photodegradation that may be anticipated for extemporaneous preparations of phenyl ether drug A formulations that contain TiO_2. The effect of incremental light exposure levels was examined for both visible and UV light by titrating the amount

$$TiO_2 + hv \longrightarrow e_{CB}^- + h_{VB}^+$$

$$O_2 + e_{CB}^- \longrightarrow O_2^{\cdot -}$$

$$H_2O + h_{VB}^+ \longrightarrow HO\cdot + H^+$$

Figure 15 Photochemical generation of superoxide and hydroxyl radical at TiO_2.

of light through full ICH exposure levels. Visible light was increased through 1.2×10^6 Lux hour in 0.2×10^6 Lux hour increments. UV light was increased through 200 W hr/m^2 in 40 W hr/m^2 increments. The quantitative results of this study as well as the light exposure conditions are summarized in Table 13.

In both UV and visible light stress cases, the percentage labeled claim of drug A decreased from 100% to approximately 40% after full ICH recommended exposure. The occurrence of photodegradation using ICH visible light is attributed to the existence of mercury emission lines in the region less than 406 nm present with the cool white fluorescence bulbs, hence care should be exercised in the interpretation of the visible light stress data. Nevertheless, it may be difficult to meet ICH photostability guidance expectations for drug substance solutions if TiO_2 is present in the sample matrix.

The generality of TiO_2-induced photo-oxidation for drug sample solutions was examined by evaluating drug products containing acetaminophen, naproxen, and pseudoephedrine hydrochloride. All formulations were dissolved in their corresponding assay method diluents (Table 14) and placed in clear volumetric flasks.

Solutions were exposed to 20 W hr/m^2 or 40 W hr/m^2 of UV light in a ICH photochamber and analyzed by HPLC. The amount of drug loss observed for the formulations is significant and ranges from 3.2% to 17.1% label claim (Table 14). The results of this study indicate that media and drug structure may affect the degradation rate but that the underlying photochemistry is likely widely applicable.

Titanium dioxide, found in capsule shells and tablet film coatings, can negatively impact the photostability of pharmaceutical formulation solutions. The observed photodegradation is a heterogeneous photo-oxidation process catalyzed by titanium dioxide. This photo-oxidation process has been shown to occur rapidly during ICH photostability stressing studies and apply to a range of varied drug formulation solutions. As such, protection from light may be

Table 13 Drug Loss as a Function of Incremental Ultraviolet and Visible Light Exposure

UV (W hr/m^2)	% Drug	Visible (10^6 Lux hr)	% Drug
0	100.0	0	100.0
40	80.0	0.27	6.4
80	58.9	0.4	65.6
120	51.9	0.6	57.5
160	46.1	0.8	50.4
200	40.6	1.2	38.8

Abbreviation: UV, ultraviolet.

Table 14 The Impact of TiO$_2$-Induced Photooxidation to Other Drug Products

	Formulation	Media	% Active[a]	
			Control[b]	Sample
Drug A	Film-coated tablet	10 mM phosphate (pH 6)	100.1	84.8
Drug B	Capsule (HPMC)	MeOH/water	101.2	95.6
Drug C	Film-coated tablet	ACN/water/ acetic acid	100.1	82.9
Drug D	Extended release coated tablet	Alcohol	99.7	96.8[c]

[a]All samples were exposed to ultraviolet light for 20 W hr/m^2 in an International Conference on Harmonization (ICH) photochamber.
[b]Control samples were wrapped with aluminum foil.
[c]Sample was exposed to ultraviolet light for 40 W hr/m^2 in an ICH photochamber.
Abbreviation: HPMC, hydroxypropylmethyl cellulose; ACN, acetonitrile.

necessary when generating, storing, or testing TiO$_2$ containing drug product solutions.

CONCLUSION

Photostability testing is required and should not be assumed to be unnecessary, even when the drug molecule itself does not absorb light at wavelengths greater than 300 nm. Diluents, excipients, and product components can all contribute to promote unexpected photochemistry. Careful studies to understand the mechanism of photoinstability can play an important role in guiding appropriate decisions regarding the handling and use of these products to ensure that undesirable photochemistry does not impact either the safety or the efficacy of the product. Finally, many of the photodegradation mechanisms summarized here involve the formation of energetic reactants (e.g., singlet oxygen and hydroxyl radicals) and have been shown to be generally potent enough to react with a variety of drug molecules.

REFERENCES

1. Kuramoto R, Lachman L, Cooper J. The effect of certain pharmaceutical materials on color stability. JAPhA 1958; 47:175–180.
2. Guideline for the Photostability Testing of New Drug Substances and Products. Fed Register 1997; 62:27115–27122.
3. Baertschi SW. Commentary on the quinine actinometry system described in the ICH draft guideline on photostability testing of new drug substances and products. Drug Stab 1997; 1:193–195.
4. Containers for Dispensing Capsules and Tablets. In: USP 27/NF 22. Rockville, MD: United States Pharmacopoeial Convention, Inc., 2004:2741–2746.
5. Templeton AC, Xu H, Placek J, Reed RA. Implications of photostability on the manufacturing, packaging, storage, and testing of formulated pharmaceutical products. Pharm Tech 2005; 29:68–86.
6. Reed RA, Harmon P, Manas D, et al. The role of excipients and package components in the photostability of liquid formulations. PDA J Pharm Sci Tech 2003; 57:351–368.

7. Kristensen S. Photostability of parenteral products. In: Photostability of Drugs and Drug Formulations. 2nd ed. Boca Raton, FL: CRC Press, 2004:303–330.
8. Tonnesen HH, ed. Photostability of Drugs and Drug Formulations. Boca Raton, FL: CRC Press, 2004.
9. Brustugun J, Tonnesen HH, Klem W, Kjonniksen I. Photodestabilization of epinephrine by sodium metabisulfite. PDA J Pharm Sci Tech 2000; 54:136–143.
10. Brustugun J, Kristensen S, Tonnesen HH. Photostability of sympathomimetic agents in commonly used infusion media in the absence and presence of bisulfite. PDA J Pharm Sci Tech 2004; 58:296–308.
11. Brustugun J, Kristensen S, Tonnesen HH. Photostability of epinephrine—the influence of bisulfite and degradation products. Pharmazie 2004; 59:457–463.
12. Brustugun J, Tonnesen HH, Edge R, Navaratnam S. Formation and reactivity of free radicals in 5-hydroxymethyl-2-furaldehyde—the effect on isoprenaline photostability. J Photochem Photobiol B: Biol 2005; 79:109–119.
13. Iwamoto T, Saito M, Taga M. Spectrochemical study of parenteral solutions. I. Dextrose injection. Yakugaku Zasshi 1955; 75:1158–1160.
14. Murty BSR, Kapoor JN, Smith FX. Levels of 5-hydroxymethylfurfural in dextrose injection. Am J Hospital Pharm 1977; 34:205–206.
15. Millar BW, MacLeod TM. A study of 5-hydroxymethylfurfural levels in levulose injections after autoclaving and during storage. J Pharm Pharmacol 1981; 33:100P.
16. Ulbricht RJ, Northup SJ, Thomas JA. A review of 5-hydroxymethylfurfural (HMF) in parenteral solutions. Fund Appl Tox 1984; 4:843–853.
17. Heidemann DR. Rapid, stability-indicating, high-pressure liquid chromatographic determination of theophylline, guaifenesin, and benzoic acid in liquid and solid pharaceutical dosage forms. J Pharm Sci 1979; 68:530–532.
18. Davidson AG, Dawodu TO. Difference spectrophotometric assay of 5-hydroxymethylfurfuraldehyde in hydrolyzed pharmaceutical syrups. II. Isoniazid reagent. J Pharm Biomed Anal 1988; 6:61–66.
19. Davidson AG, Dawodu TO. Difference spectrophotometric assay of 5-hydroxymethylfurfuraldehyde in hydrolyzed pharmaceutical syrups. I. Sodium borohydride reagent. J Pharm Biomed Anal 1987; 5:213–222.
20. Santoro MIRM, Hackmann ERM, Magalhaes JF. Detection of 5-hydroxymethylfurfural by thin-layer chromatography in pharmaceutical preparations containing glucose. Anais de Farmacia e Quimica 1988; Suppl:58–64.
21. Cook AP, MacLeod TM, Appleton JD, Fell AF. HPLC studies on the degradation profiles of glucose 5% solutions subjected to heat sterilization in a microprocessor-controlled autoclave. J Clin Pharm Therap 1989; 14:189–195.
22. Tu D, Xue S, Meng C, Espinosa-Mansilla A, Munoz de la Pena A, Salinas Lopez F. Simultaneous determination of 2-furfuraldehyde and 5-(hydroxymethyl)-2-furfuraldehyde by derivative spectrophotometry. J Agric Food Chem 1992; 40:1022–10255.
23. Espinosa-Mansilla A, Munoz de la Pena A, Salinas F. Semiautomatic determination of furanic aldehydes in food and pharmaceutical samples by a stopped-flow injection analysis method. JAOAC Int 1993; 76:1255–1261.
24. Hewala II, Zoweil AM, Onsi SM. Detection and determination of interfering 5-hydroxymethylfurfural in the analysis of caramel-colored pharmaceutical syrups. J Clin Pharm Therapeut 1993; 18:49–53.
25. Hewala II. Stability-indicating HPLC assay for paracetamol, guaiphenesin, sodium benzoate and oxomemazine in cough syrup. Anal Lett 1994; 27:71–93.
26. Hryncewicz CL, Koberda M, Konkowski MS. Quantitation of 5-hydroxymethylfurfural (5-HMF) and related substances in dextrose injections containing drugs and bisulfite. J Pharm Biomed Anal 1996; 14:429–434.
27. Yuan J, Guo S, Li L. Simultaneous determination of sugars and their degradation product 5-hydroxymethylfurfural in foods and pharmaceuticals by high-performance liquid chromatography. Fenxi Huaxue 1996; 24:57–60.

28. Xu H, Templeton AC, Reed RA. Quantification of 5-HMF and dextrose in commercial aqueous dextrose solutions. J Pharm Biomed Anal 2003; 32:451–459.
29. Zhang H, Shen P, Cheng Y. Identification and determination of the major constituents in traditional Chinese medicine Si-Wu-Tang by HPLC coupled with DAD and ESI-MS. J Pharm Biomed Anal 2004; 34:705–713.
30. Deschreider AR. Photochemical changes of glucose stocks. Rev Ferment Inds Aliment 1957; 12:145–147.
31. Dextrose Injection. In: USP 27/NF 22. Rockville, MD: United States Pharmacopoeial Convention, Inc., 2004:582.
32. Halliwell B, Gutteridge JMC. Free Radicals in Biology and Medicine. Oxford, UK: Oxford University Press, 1989:36–104.
33. Halliwell B, Gutteridge JMC. Free Radicals in Biology and Medicine. Oxford, UK: Oxford University Press, 1991:401–406.
34. Faust BC, Zepp RG. Photochemistry of aqueous iron(III)-polycarboxylate complexes: roles in the chemistry of atmospheric and surface waters. Environ Sci Tech 1993; 27:2517–2522.
35. Eley DD, Pines H, Weisz P, eds. Advances in Catalysis. Vol. 25. New York, NY: Academic Press, 1976:25–26.
36. Abrahamson HB, Rezvani AB, Brushmiller JG. Photochemical and spectroscopic studies of complexes of iron(III) with citric acid and other carboxylic acids. Inorg Chim Acta 1994; 226:117–127.
37. Dean JA. Lange's Handbook of Chemistry. New York, NY: McGraw-Hill, 1999.
38. Rowe RC. Quantitative opacity measurements on tablet film coatings containing titanium dioxide. Int J Pharmaceut 1984; 22:17–23.
39. Béchard SR, Quraishi O, Kwong E. Film coating: effect of titanium dioxide concenration and film thickness on the photostability of nifedipine. Int J Pharmaceut 1992; 87:133–139.
40. Fujishima A, Rao TN, Tryk DA. Titanium dioxide photocatalysis. J Photochem Photobiol C: Photochem Rev 2000; 1:1–21.
41. Pruden ALO, David F. Degradation of chloroform by photoassisted heterogeneous catalysis in dilute aqueous suspensions of titanium dioxide. Environ Sci Tech 1983; 17:628–631.
42. Pruden ALO, David F. Photoassisted heterogeneous catalysis: the degradation of trichloroethylene in water. J Catal 1983; 82:404–417.
43. Konstantinou IK, Albanis TA. Photocatalytic transformation of pesticides in aqueous titanium dioxide suspensions using artificial and solar light: intermediates and degradation pathways. Appl Catal B: Environ 2003; 42:319–335.
44. O'Shea KE. Titanium dioxide-photocatalyzed reactions of organophosphorus compounds in aqueous media. Mol Supramol Photochem 2003; 10(Semiconductor Photochemistry and Photophysics):231–247.
45. Thurnauer MC, Rajh T, Dimitrijevic NM, Nada M. Principles of semiconductor-assisted photocatalysis for waste remediation. Electron Transfer Chem 2001; 5:695–718.
46. Kozhukharov VVP, Stefchev P, Kabasanova E, et al. TiO_2-photocatalyzed oxidative water pollutants degradation: a review of the state-of-art. J Environ Protect Ecol 2001; 2:107–111.
47. Anpo M. Utilization of TiO_2 photocatalysts in green chemistry. Pure Appl Chem 2000; 72:1265–1270.
48. Cunningham JA-SG, Sedlak P, Caffrey J. Aerobic and anaerobic TiO_2-photocatalysed purifications of waters containing organic pollutants. Catal Today 1999; 53:145–158.
49. Penuela GAB, Damia. Photosensitized degradation of organic pollutants in water: processes and analytical applications. Trends Anal Chem 1998; 17:605–612.
50. Mills A, Le Hunte S. J Photochem Photobiol A: Chem 1997; 108:1.
51. Cunningham J, Ghassan, Srijaranal S, Somkiat. Adsorption of model pollutants onto TiO_2 particles in relation to photoremediation of contaminated water. Aquatic Surface Photochem 1994:317–348.

52. Malati MA. The photocatalyzed removal of pollutants from water. Environ Tech; 16:19951093–1099.
53. Pichat PG, Maillard C, Amalric L, D'Oliveira JC. Titanium dioxide photocatalytic destruction of water aromatic pollutants: intermediates; properties-degradability correlation; effects of inorganic ions and titanium dioxide surface area; comparisons with hydrogen peroxide processes. Trace Metals Environ 1993; 3;207–223.
54. Matthews RW. Photocatalytic oxidation of organic contaminants in water: an aid to environmental preservation. Pure Appl Chem 1992; 64:1285–1290.
55. Kakinoki K, Yamane K, Teraoka R, Otsuka M, Matsuda Y. Effect of relative humidity on the photocatalytic activity of Titanium dioxide and photostability of famotidine. J Pharm Sci 2004; 93:582–589.
56. Moro-Oka YC, Pyoung J, Arakawa H, Ikawa T. Chemistry of superoxide ion. Reaction of superoxide ion with substrates having labile hydrogens. Chem Lett 1976; 11:1293–1296.
57. Sheldon RA, Kochi JK. Metal catalyzed oxidations of organic compounds in the liquid phase: a mechanistic approach. Adv Catal 1976; 25:274–413.

13 Chambers and Mapping

Joseph T. Piechocki
Piechocki Associates, Westminster, Maryland, U.S.A.

INTRODUCTION

The ability to have a single compact instrument, for any test, has always been useful in the laboratory. Such a unit helps to limit the number of possible variables that operators can change. This improves reproducibility and allows for the use of less technically qualified personnel, while allowing for the possibility of automation.

As is the case with any new area of interest, many companies seek to adapt their products to meet a new need. This leads to an effort to be first in the market as well as efforts to differentiate one product from another, sometimes without really understanding the problem. Frequently, because of this lack of understanding, misinformation is spread and bad decisions made, to the customers detriment.

Adding to this problem is the lack of unambiguous, scientifically valid guidelines from the International Conference on Harmonization (ICH) guidance (1). This leads to having to choose between what the user thinks the regulators want and the scientific approach.

Baertschi et al. (2), and Sequiera and Volzone (3) have shown problems with the current guideline lamp options. Chapter 3 of this book and Figure 4 of Chapter 6 of this book, from Sequeira and Volzone's paper (3) of this text, illustrate the problems with the current dosage measuring techniques.

The purpose of this chapter is to introduce the reader to the variety of different chambers currently available in the marketplace, worldwide and point out their differentiating characteristics. No comments or recommendations will be given. It is up to each reader to make his/her own decision, based on his/her own circumstances, as to which chamber may be best for them.

The chambers are listed alphabetically and common items such as temperature control, humidification, etc., available as part of the original units or as options have not been listed. Most of this and additional information are available on each vendor's web site.

The mapping of sample chambers will be discussed primarily from the user's viewpoint since few companies publish such detailed information and little spectrophotometric data has been published.

TESTING APPARATUS/CHAMBERS

Most early pharmaceutical photochemists used sunlight or window-glass–filtered sunlight for their testing. Samples were for the most part exposed on a open, sunlit windowsill, or in the same place but with the window closed.

Figure 1 shows one of the first published and patented photostability test apparatus; that of Kallab (4). This apparatus was designed for the testing of materials and because it used sunlight, a lens had to be used with care.

Figure 2 shows the outdoor (A) and indoor (B) cabinets used by Arny (5), Arny, Taub and Steinberg, (6–8) and Arny et al. (9) in their photostability studies.

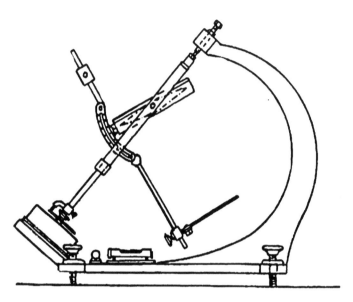

Figure 1 Kallab's original, patented, apparatus for testing the photostability of leather products (4). *Source*: Courtesy Journal of Society Leather Technology and Science.

It is well to note that these studies were all performed behind window-glass and that the diffuse light present in his office cabinet would have been doubly filtered.

   In 1959, after much research and based on the best principles then known, Lachman and Cooper (10), Lachman et al. (11) and Lachman et al. (12) published a series of papers detailing their development of a "Comprehensive Pharmaceutical Stability Testing Laboratory" which included photostability testing capabilities. This

(A)   **(B)**

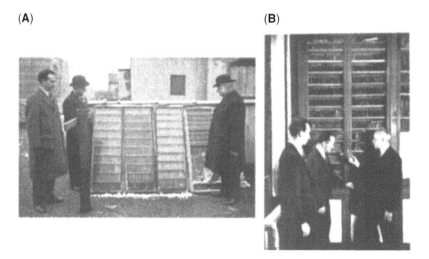

Figure 2 Arny and associates test cabinets: (**A**) rooftop and (**B**) indoor (5).

unit included "cool-white fluorescent lamps," because they were then considered the best replicators of solar radiation, and forced ventilation for temperature control.

Figure 3 is a diagram of the cabinet used the aforecited papers. The importance of this work cannot be under estimated. This cabinet had the effect that the authors had hoped that of standardizing photostability testing. Many units were constructed based on their work and some are still in use today.

There have been pharmaceutical photostability testing apparatus and rooms developed and reported at meetings and in the literature such as those by Forbes (13,14), Cole (15), Turner (16), and Wall et al. (17). Unfortunately, some of these have been proprietary systems, not reported in detail in the literature.

The system developed by Forbes for animal testing utilized a modified Atlas Electric Devices Company's Rm-65 long-arc, water-cooled 6500 kW Xenon system as its radiation source in a temperature-controlled room. R-B Sunburn sensors were

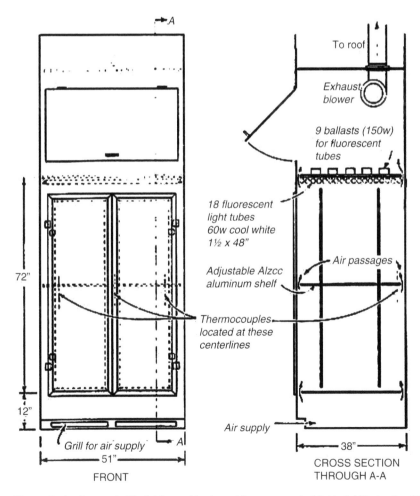

Figure 3 Lachman et al.'s lighting cabinet used for exaggerated light stability testing (12).

mounted in the center of the front each rack of animals for dosage measurement. The radiation system was surrounded by a hexagonal, metal frame, modified to hold large WG 320 Schott filters, 9 in. from the lamp.

Turner (15) reported on an in-house developed fluorescent lamp system. He was interested in evaluating the following (i) how reproducible conditions were from one test to another, (ii) if the ratio of visible/ultraviolet (VIS/UV) was constant at these times, (iii) if the VIS/UV ration varied over the sample surface area, (iv) how often the bulbs need to be changed, (v) if the replacement bulbs were different from the original, and (vi) do the bulbs age at the same rate. His data showed that the intensity pattern of both the UV and VIS lamps did change in both intensity and pattern over a three-month period. He deemed reproducibility not to be an issue but found that the time to reach ICH illuminance and irradiance recommended exposure values differed significantly in the VIS lamps tested. One thing he noted was a "wall effect," caused by absorption of UV radiation by the wall, altering the ratio of VIS/UV. Thus, the reason, along with shading problems is that most chamber manufacturers caution against putting samples too close to the chamber walls. Some even scribe a line on the chamber bottom.

The system of Wall et al. (17) consisted of a large Norlake walk-in temperature and humidity-controlled chamber containing fluorescent lamps and a large number of individual sensors to monitor the chambers 160 shelves. This system was computerized to measure all-important parameters including illumination levels, automatically, continuously.

Boxhammer (18,19) and Boxhammer and Willwoldt (20) reported on their experiences developing photostability chambers for allied industries. This included the use of a number of sources and tabletop as well as large-scale units. Their papers are a good introduction to all of the parameters that should go into the selection of a chamber, of any size.

FLUORESCENT LAMP CHAMBERS

As just presented, Lachman et al. were among the first to promote the use fluorescent lamps for pharmaceutical photostability testing. Throughout the past approximately 50 years, this lamp type has been the most used. Its ready availability, low price, and ready recognition in the normal environment made for easy acceptance. Cabinets utilizing this lamp were easy to fabricate and relatively inexpensive.

Because these lamps were available in many sizes and wattages, environmental chamber manufacturers, responding to the ICH Q1B efforts, were quick to add this feature to their products. This produced a number of different lamps and configuration claims; unfortunately little data is available to support the claims.

A review of many of the chambers currently available and advertised for pharmaceutical photostability follows. Some presentations are lacking because such information was not available at the time of this writing. Readers are advised to check with the individual manufacturers for the latest information. Many of the chambers presented are available with and without certain features such as temperature and humidity control, programmable lighting systems, built in recording devices, etc. For the purpose of this review, only the distinguishing photostability characteristics are noted.

No recommendation is made but the reader is referred to Chapters 2, 3, 4, 6, 7, 10 of this book, and the second half of this chapter on mapping for reference data to aid in their selection.

ATLAS Material Testing Technology LLC
Figure 4 displays the UV2000™ UV condensation Weathering Device of ATLAS Material Testing Technology LLC (21). This unit, originally designed for compliance with numerous international standards such as ASTM G53 and ISO 4892, has been modified for use in pharmaceutical photostability testing. It consists of an A-frame on which are mounted eight-4 ft–40 W fluorescent lamps. The lamps are partially or totally replaced by visible lamps to comply with ICH requirements.

Axyos
Figure 5 presents the four environmental chambers manufactured by Axyos (22) Technologies of Queensland, Australia. This firm serves the East and Near-East markets. They offer both top and rear mounted light options. No other information was available at the time of this publication.

Binder GmbH
Binder (23) offers their KBF Series of test chambers, with ICH compliant lighting, for photostability testing. Their lamps are mounted vertically, in the door, behind a protective panel. All units also have an inner glass door(s). The cabinets have proprietary spherical UV and VIS detectors.

(A) **(B)**

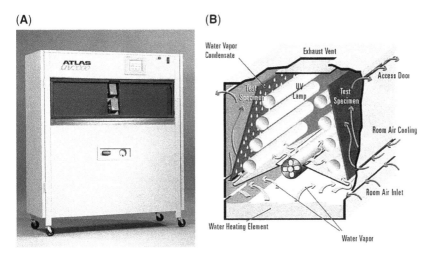

Figure 4 (**A**) UV2000 fluorescent testing apparatus and (**B**) is a cross-sectional diagram of the test chamber before modification for pharmaceutical photostability testing. *Source*: Courtesy of Atlas Material Testing Technology LLC.

Figure 5   Test chamber, without light option. *Source*: From Ref. 21.

Caron

Caron offers four different models, 6510, 6515, 6530, and 6535 as photostability test chambers. (24) The 6510 and 6530 have no humidity control, whereas the 6515 and 6535 do have humidity control. They offer extra UVA and VIS detectors as an option. The interior is constructed using specular reflective aluminum. The lamps are mounted in a "U" shaped opaque cover.

Figure 6   Model KBF climatic chamber with light option, door closed and opened. *Source*: From Ref. 22.

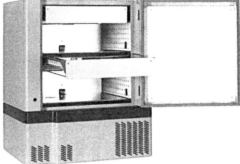

Figure 7   Model 6540A and 6545A photostability chambers. *Source*: Courtesy of Caron Products & Services Inc.

ENVIRONMENTAL GROWTH CHAMBERS (EGC)

Environmental Growth Chambers (25) PST-19 Photostability Chamber, Figure 8, is designed to complete ICH Q1B, Option 2 compliant pharmaceutical testing in as little as 20 hours. It has four shelves having a combination of UV-A and VIS bulbs and each shelf's bulb type is independently controlled. The air-flow to each shelf is independently controllable with dampers.

Environmental Specialties (LUWA)

Figure 9 shows two pharmaceutical photostability chambers from Environmental Specialties (26), their single shelf, ES2000, bench top, chamber (A) and their multishelved

(**A**)   (**B**)

Figure 8   (**A**) PST-19 photostability chamber and (**B**) their custom-built, walk-in, movable, test racks. *Source*: Courtesy of Environmental Growth Chambers (EGC).

(A)   (B)

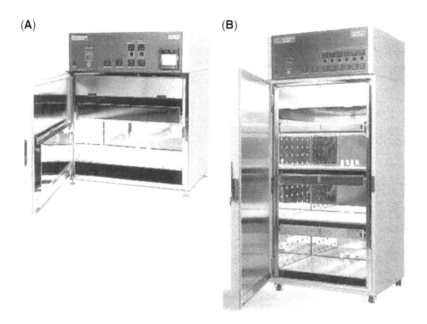

Figure 9 ES2000 reach-in series photostability chambers: (**A**) ES2000 CL-BT, single-
shelf, bench top unit and (**B**) ES2000 CL 4 shelf, upright, floor model. *Source*: Courtesy of
Environmental Specialities (ES) (LUWA).

ES2000 Reach-In chamber (B). The three shelved upright unit is capable of running three
simultaneous studies with each shelf having an independently adjustable light control.

The lamps are covered with a high UV transparent barrier to protect the lamps
and lamp intensities are monitored with a variable height control sensor of custom
design. Lamp intensity is maintained constant by a microprocessor-based controller
and a photodiode sensor to provide a fully automated closed-loop control system.
LUWA also makes custom, large-scale units.

ESI offers a variety of proprietary lamps for use with their instrument. One
lamp is designed to give the correct radiometric and photometric dosage readings
that the company claims allows one to achieve the ICH recommended exposures
in one test.

Hotpack (SP Industries)
Figure 10 shows one of Hotpacks (27) basic environmental stability chambers without
the light option. Hotpack produces five, 19 cu. ft. models of its Illuminated Chambers
line 352632, 352642, 352643, 352622, and 352620. These units provide for independent adjustment of light intensity. All models have a baked enamel finish to provide
maximum light dispersion within the chamber. The model 352642 has four vertically mounted 60 W fluorescent lamps. The model 352643 has four, 40 W fast start
fluorescent and four, 23 W UV (350 nm) lamps. The model 352632 has two vertically
mounted, high output 60 W fluorescent lamps on either side wall. The model 352622
has eight, 20 W fluorescent lamps (two per light bank) on each of four adjustable
shelves, controlled by a separate timer.

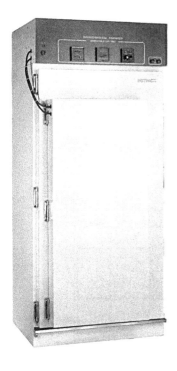

Figure 10 31733, floor model, illuminated chamber.
Source: Courtesy of Hotpack (SP Industries).

Lunaire (Thermal Product Solutions)
Lunaire offers three different model test chambers (28), CEO910W, CEO932W-2, and CEO932W-4, for photostability testing. Figure 11 shows one of these models along with a cross section of the CEO932W-2 model. Light variation over the product area (70% of the available shelf area) will be ±30% of the set point for both the VIS and UV, according to the manufacturer. The product tray is 12 inches below the lamp bank assembly. Up to three lamp packages can be accommodated in the W-2 and W-4 models.

Luzchem Research Inc.
Luzchem manufactures (29) one photoreactor, the ICH-2 (Figure 12) that complies with the ICH Option 2 requirements. This unit can be equipped with 16, T5, 8 W fluorescent lamps, a turntable, and carousel to equalize exposure. It is also equipped with a recessed magnetic stirrer. The interior is brushed with aluminum to provide a quasi-spectrally neutral reflections. A set of four switches allows for alteration of the radiation conditions.

   The unit comes with two complete sets of lamps, one UVA and the other VIS. For VIS illumination, the company recommends Sylvania F8t5/CW lamps and its UV-A lamp is centered around 350 nm. Each unit is individually tested and certified to meet ICH Option 2 guidelines.

Nagano Science
Nagano Science (30) offers three different pharmaceutical photostability testing instruments in their Light-Tron Series, the LT-120 (Figure 13), LTL-200 (Figure 14), and LTL-400 (Figure 15). These instruments can be used to meet ICH Option 1 or 2

(A)   (B)

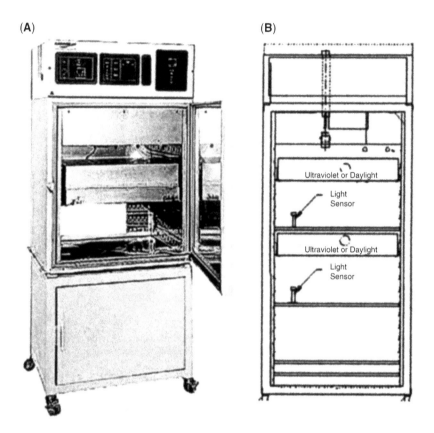

Figure 11 (**A**) CEO910W photostability test chamber and (**B**) a cross sectional drawing of its two shelf configuration. *Source*: Courtesy of Lunaire (Thermal Products Solution).

guideline requirements. The LT-120 holds six, 20 W lamps, the LTL-200, 14 and the LTL-400, 20, 20 W or 10, 40 W. Optional configurations of up to 12, 20 W lamps are available for the LT-120 and LTL-200 models. All units are equipped with carousels, which are discussed in chapter 10 of this text.

Newtronic Equipment Co, PVT. Ltd.
Currently Newtronic offers (31) one model of photstability chamber, which is pictured in Figure 16. This two-chamber unit is designed to meet ICH, Option 2 requirements. It has five, VIS (cool-white) and five, UV-A lamps mounted in an overhead configuration, in their own separate compartments. The inside is all stainless steel.

Percival Scientific, Inc.
Figure 17 shows the Percival Rx3000L (32) photostability chamber. It accommodates two banks of lamps. The unit is designed to comply with ICH Q1B,

(**A**)

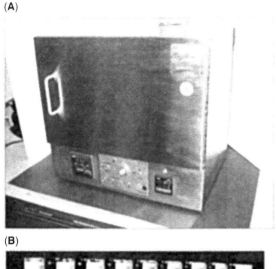

(**B**)

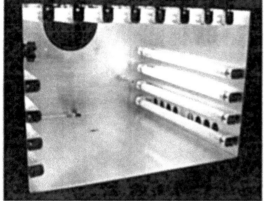

Figure 12 Photoreactor ICH-2, (**A**) door closed and (**B**) open. *Source*: Courtesy of Luzchem Research Inc.

Option 2 requirements. No further information was available at the time of this writing.

Powers Scientific, Inc.

Powers (33) makes two photostability chambers the PST52SD and the PST73SD, which are pictured in Figure 18. These units are designed to meet ICH, Option 2 guidelines. The PST52SD can accommodate four trays, each with a bank of 10, 40 W lamps above them. The PSR73SD holds up to twice that amount. Multilayered hanging screens improve light uniformity across the midpoint of the shelf. Each shelf is configured to have a combination of six cool white and four UV-A lamps and has individual switches to turn off either lamp type.

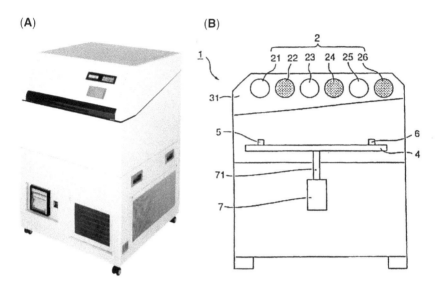

Figure 13 Light-Tron LT-120 series photostability testing device: (**A**) as is and (**B**) in cross sectional diagram. Turntable (*4), UV-A sensor (*5), VIS sensor (*6), turntable motor (*7), VIS (21) and UV-a lamp (22). *Source*: Courtesy of Nagano Science Equipment Mfg. Co. Ltd.

Q-Lab Corporation

Q-Lab (formerly Q-Panel) (34) offers a fluorescent lamp photostability-testing unit, their QUV/cw unit, pictured in Figure 19. This unit, originally designed

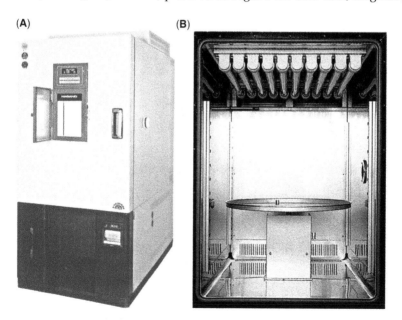

Figure 14 Light-Tron LTL-200 series photostability testing device: (**A**) door closed and (**B**) inside of test chamber. Light-Tron LT-120 series photostability testing device. *Source*: Courtesy of Nagano Science Equipment Mfg. Co. Ltd.

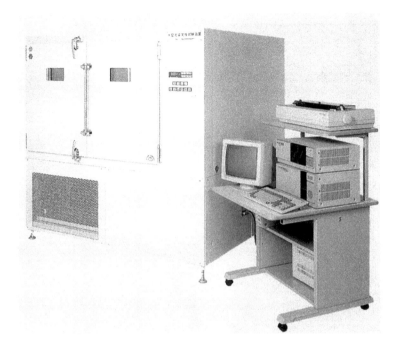

Figure 15 Light-Tron LTL-400 series custom-built, photostability testing device. *Source*: Courtesy of Nagano Science Equipment Mfg. Co. Ltd.

for the electronics industry, has been modified by some for pharmaceutical photostability testing. It can accommodate up to eight 40 W fluorescent lamps in any desired configuration of lamp types. Light intensity is continuously monitored and maintained during testing.

Rumed® (Rubarth Apparate GmbH)

Figure 20 shows several of the photostability chambers manufactured under the Rumed® name (35). The company manufactures nine different models, 1001, 1101, 1201, 1301, 1401, 1501, 1601, 1701, and 1801. The 1001, 1101, 1401, 1501, 1601, 1701, and 1801 models have overhead lighting while the 1201 and 1301 models have side lighting through a protective cover, from both sides. The 1301 model offers a rotational stand option, pictured in Figure 20 (C). A higher-capacity photstability chamber version with up to five shelves accommodating four lamps per shelf is also available. A variety of lamps are available including Daylight, Biolux, and Natural Daylight. Additional details regarding the number of lamps per bank, their wattage, switching etc., were not available at the time of this writing.

Termaks

Termaks (36) produces the Series 6000 and 8000 Environmental Chambers, Figure 22, to meet the needs for control of temperature, humidity, and light. Their interiors are all stainless steel. Light is supplied by 16 (8 per side), vertically mounted lamps, isolated from the interior chamber by a three-layer glass panel. The lamps are regulated by an adjustable electronic control.

Figure 16   Photostability Chamber. *Source*: Courtesy of Newtronic Equipment Co. Pvt. Ltd.

Thermolab Scientific Equipments PVT. Ltd.

Thermolab (37), Figure 23, makes several different chambers for pharmaceutical photostability testing in accordance with ICH guidelines. Their units have mirror polished stainless steel interiors. At the time of writing of this text, no other information was available about these chambers.

Vindon Scientific Ltd.

In Figure 23 is shown Vindon Scientifics Model 1810 (38) photostability cabinet designed to meet the ICH, Option 1 recommendations for photostability testing. The radiation sources used are six 1.2 m-long Artificial Daylight lamps capable of reaching the ICH recommended exposure levels in less than 14 days. The need for a refrigeration system has been eliminated; the cabinet will perform to specifications under normal ambient conditions up to 21°C, according to the manufacturer.

Weiss Gallenkamp

In 2004, Weiss Technik (39) purchased the environmental business of Sanyo Gallenkamp and merged Weiss Technik Ltd. with the Gallenkamp business to

(A) (B)

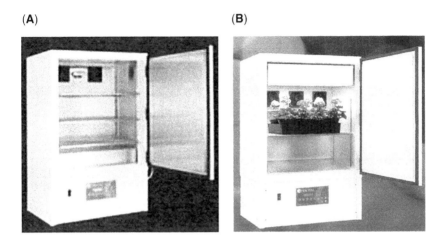

Figure 17 (**A**) Rx3600 series test chamber; (**B**) with light option.
Source: Courtesy of Percival Scientific, Inc.

form Weiss Gallenkamp Ltd. The company markets two lines specifically designed for pharmaceutical photostability testing under their Pharma-Safe line.

In Figure 25 are pictured the table top (A) and three shelf unit (B) of Weiss Gallenkamps Pharma-Safe series of photostability chambers. Both systems are designed to meet ICH Q1B Option 2 requirements and have their unique Exposure Equalization Filter, Figure 25 (C), which the manufacturer claims "brings exceptional light level uniformity across the whole shelf area."

(A) (B)

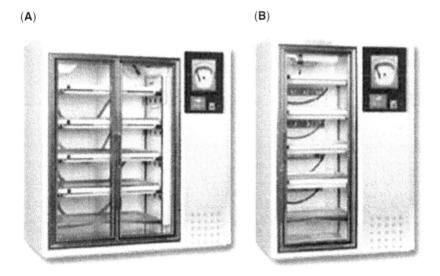

Figure 18 (**A**) PST73D and (**B**) PST52SD photostability chambers.
Source: Courtesy of Power Scientific, Inc.

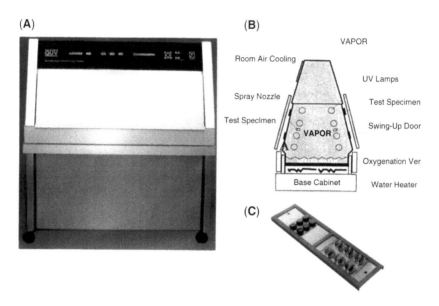

Figure 19   (**A**) QUV/cw, cool white photostability tester: (**B**) a cross sectional view of the test chamber before modification for pharmaceutical testing and (**C**) their tablet sample holder for the unit. *Source*: Courtesy of Q-Lab Corporation.

The three shelves lamps in the large unit can be controlled independently. Each shelf is insulated to reduce heat transfer from the lamps below. The Pharma-Safe configuration is available in larger custom designed, walk-in versions.

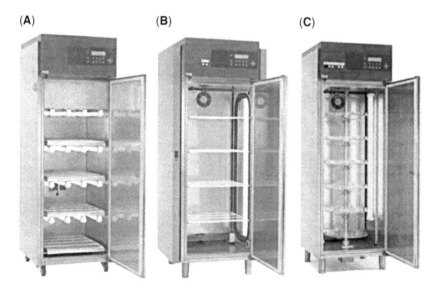

Figure 20   Light thermostats. (**A**) Model 1601 with overhead lighting, (**B**) Model 1301 with lighting from both lateral test room walls and (**C**) a Model 1301 chamber with optional powered, rotational stand. *Source*: Courtesy of Rubarth Apparate GmbH.

(A) (B) (C)

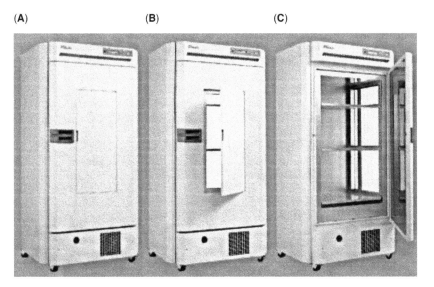

Figure 21 Environmental chambers: (**A**) temperature, (**B**) humidity and (**C**) light chambers. *Source*: Courtesy of Termaks.

XENON CHAMBERS

Atlas Material Testing Technology LLC

In Figure 25 is displayed Atlas's (21) two tabletop models CPS/CPS+ (A) and XLS/XLS+ (B). The "+" models are microprocessor controlled. The source of

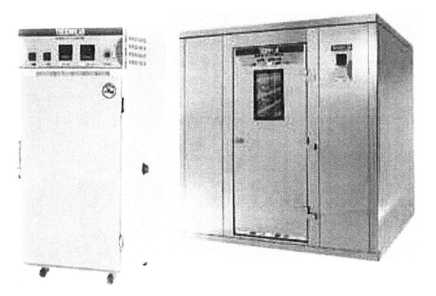

Figure 22 Photostability chamber. *Source*: Coutesy of Thermolab Scientific Equipments Pvt. Ltd.

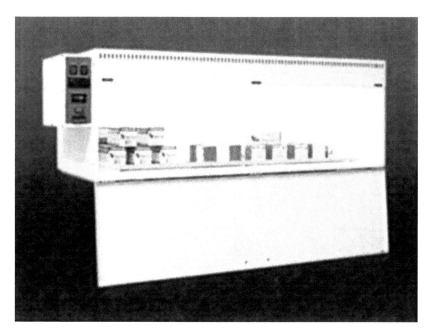

Figure 23   Photostability cabinet. *Source*: Courtesy of Vindon Scientific Ltd.

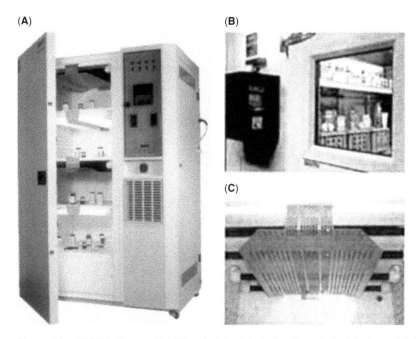

Figure 24   PSC062 Pharma-Safe™. (**A**) Stability test chamber, photostability model; (**B**) custom built walk-in unit; (**C**) exposure equilization filter. *Source*: Courtesy of Weirs Gallenkarmup.

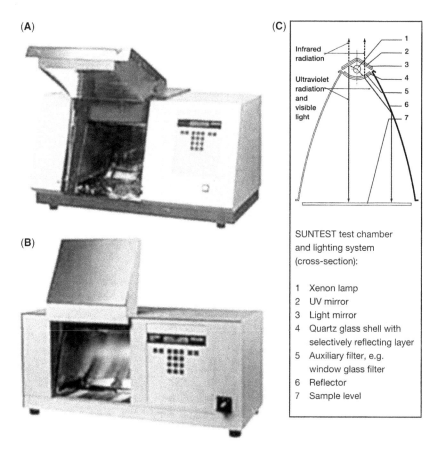

Figure 25 (**A**) Suntest CPS and (**B**) suntest XLS tabletop sunlight exposure systems; (**C**) the optical system used in both systems. *Source*: Courtesy of Atlas Material Testing Technology LLC.

the CPS units is a 1100 W, air-cooled, long-arc, xenon lamp. The larger capacity (144 vs. 86 in.²) XLS/XLS+ has a 2200 W, air-cooled, long-arc xenon lamp. The firm states that it has a unique chamber design for better distribution of the ultraviolet radiation. This unit is available with an optional chiller or water-cooled specimen table.

The optical system of these units, which includes use of a dichroic type filter to transmit IR rather than reflect it is shown as item four on the diagram in Figure 25 (C). Lamp energy is monitored at 340 nm and held constant, automatically, to compensate for any variations during testing.

Atlas produces a number of different size xenon lamp–based units utilizing lamps up to 6500 W, which require water-cooling. These units range in size from small stand-alone units to whole-room systems.

Luzchem Research Inc

In Figure 26 (A) is pictured Luzchem's Model ICH-1 (29) xenon photoreactor which is Option 1 compatible. This unit uses a 175 W PerkinElmer PE 175B series of

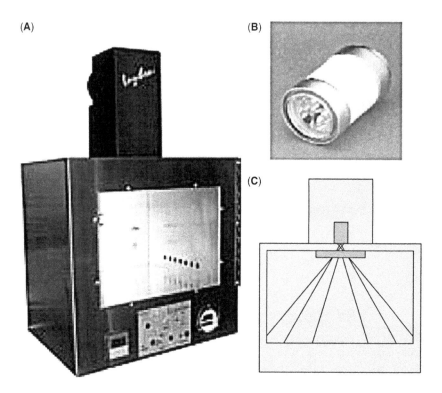

Figure 26 (**A**) LXC-ICH1 Xenon Photoreactor with a picture of (**B**) the Perkin Elmer lamp and (**C**) a diagram of its optical system. *Source*: Courtesy of Luzchem Research Inc.

their CERMAX® short-arc xenon lamp, which is available with either elliptical or parabolic reflector. Since it is low-wattage and the bulb is mounted outside of the test chamber, it requires only air-cooling. A water-cooling option for the lamp as well as filters and diffusers are available options.

Luzchem also makes available the power supply and lamps as well as other accessories needed for other self-designed configurations.

Nagano Science
Nagano (30) markets the Light-Tron Xenon LTX-01 type test chamber, which is pictured in Figure 27 (A), along with a diagram of its optical system (B). The optical system is unique in that it utilizes an isolated, air-cooled, high-powered, short-arc, and filtered xenon lamp as its source. It also utilizes a turntable to compensate for variations in radiation across the sample surface.

Q-Lab
In Figure 28 are pictures of Q-Labs (34) model Xe-1 (A) and Xe-3 (B), Q-Sun Xenon Test Chambers which meet ICH Option 1 requirements. The Xe-1 is a stand-alone tabletop unit, which employs an 1800 W long-arc air-cooled xenon lamp as its source as shown in the schematic diagram in item (C). Item (D) is a diagram of the Xe-3 unit which employs three of the lamp units. Each unit operates independent of the others.

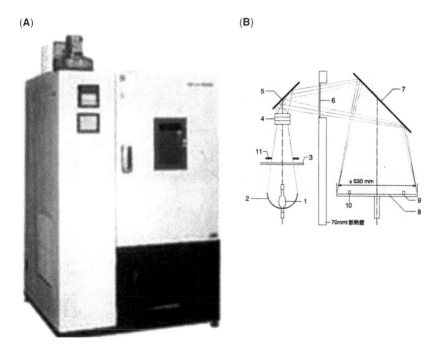

Figure 27 (**A**) Light-tron xenon, LTX-01 type, photostability testing device and (**B**) a diagram of its optical system. *Source*: Courtesy of Nagano Scientific Equipment Mfg. Co. Ltd.

The company offers a choice of sensors, 340 or 420 nm, for monitoring and providing feedback for automatic control of the lamp(s) irradiance.

Suga
Suga (40) makes two light stability xenon testers, which are shown in Figure 29. One (A), the XT2-15 is their tabletop unit called the Table Sun model whereas (B), X25 is a stand-alone, floor unit. The Table Sun has a proprietary 1500 W, long-arc, xenon lamp of a unique physical design, as illustrated in (C). The units come equipped with a turntable to rotate samples and help equalize exposure.

METAL HALIDE LAMPS
Most environmental chamber manufacturers offer metal halide lighting options for their test chambers. The following are some of those companies.

Atlas Material Testing Technology LLC
Atlas (21) manufactures several systems based on metal halide lamps. These include the SC 430, SC 600, SC 1000, and SC 2000 units, which are 340, 600, 1000, and 2000 liters in chamber capacity. Shown in Figure 30 are (A) the SC 2000 and

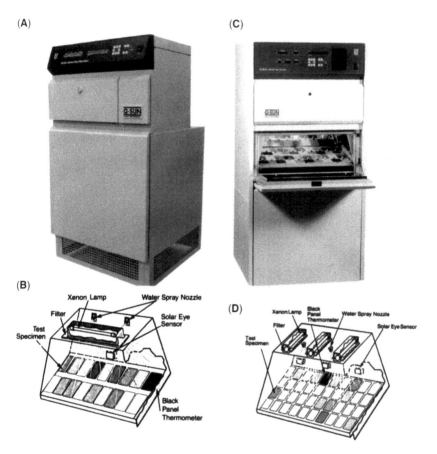

(A)

(C)

(B)

(D)

Figure 28 Q-SUN xenon test chambers: (**A**) Q-Sun/1000, tabletop model shown atop its optional chiller chamber and (**B**) a diagram of its optical system, (**C**) its Q-Sun/3000 full-size chamber and (**D**) the 3000's optical system. *Source*: Courtesy of Q-Lab Corporation.

SC 600 models. Custom-made walk-in units with the metal halide option are also available. The lamp used is a Thorium doped Dysprosium lamp capable of illuminating large areas.

Dr. Hönle

Dr. Hönle (41) produces two basic units which are shown in Figure 31, the Sol 2, (A) Compact Solar Simulator, table-top series and Sol 500, (B), stand-alone modular units. Both units are powered by the company's proprietary, iron-based, air-cooled, metal-halide lamps. The SOL 2 and SOL 500 use 400 W bulbs whereas the SOL 1200 and 2000 units use 1000 and 2000 W air-cooled, bulbs, respectively. All SOL units are ozone free and have a total test area of approximately 1400 cm². The spectral range covered is from 295 to 3000 nm but can be varied with the use of alternate filters, which are available.

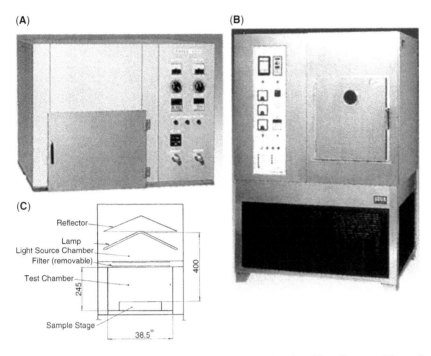

Figure 29 (**A**) TS-1, Table Sun, xenon, tabletop test chamber (**B**) a diagram of the optical system of this units and (**C**) a picture of their stand-alone, FAL-25AX-HC Xenon Long Life Fade Meter. *Source*: Courtesy of Suga Test Instruments Co. Ltd.

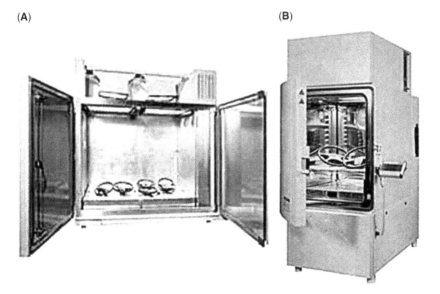

Figure 30 (**A**) Large volume, SC2000, and (**B**) SC600 Solar Simulation Chambers. *Source*: Courtesy of Atlas Material Testing Technology LLC.

(A)   (B)

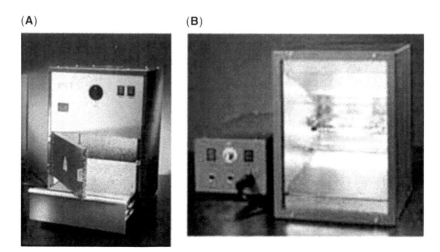

Figure 31   (A) Sol 2, bench top, sunlight test chamber and (B) the Sol 500 universally combinable lamp module and control unit. *Source*: Courtesy of Dr. Hönle AG. UV-Technologie.

Rumed® (Rubarth Apparate GmbH)

Figure 32 shows the Rumed Model 1501 chamber (35). Rumed offers their models 1001, 1101, and 1401 models with either a 3500 K, 4500 K, or 6500 K metal halide lamp option. These lamps are available at two intensity levels, 18 and 40 klux. The lamps appear, by their published spectral power distribution to be of the Thorium doped Dysprosium class.

Figure 32   1501 light thermostat. *Source*: Courtesy of Rubarth Apparate GmbH.

Weiss Gallenkamp

In Figure 33 is pictured Weiss Gallenkamps LSC020.STX.C (A), Light Stress Chamber (39). This unit has a metal halide lamp as its sources for irradiation. This option is also available in their custom-made, walk-in, drug stability rooms.

MAPPING

Mapping of the sample surface of a chamber for the quality and quantity of illumination is the most important qualifying item for photostability testing. Little has been written or presented about it. All manufacturers make claims but only a few present data, as they readily do for temperature and humidity. It is the purpose of this section to present what information does exist on the mapping of chambers and propose what should be done. To our knowledge, few of the calibration companies offer this service.

Currently there are two main methods of mapping in use in the industry, the use of Quinine Actinometry and the use of radiometer/photometer combinations as presented by the ICH (1). Neither of these methods covers the entire range of wavelengths of interest and if modified to do so by employing "quantum detectors" would still not give any wavelength information.

The most reliable method of mapping a chamber is with a spectroradiometer equipped with an integrating sphere. Spectroradiometry has always been regarded as the gold standard for radiometric and photometric measurements because, it gives both wavelength and intensity data in real time. Two items that have plagued its ready acceptance have been size and cost. Both items are no longer a problem.

Figure 33   PSC062,STX.C light stress chamber. *Source*: Courtesy of Weiss Gallenkamp.

Small and relatively inexpensive units designed for this purpose are available from some companies such as Luzchem Research, Inc. (29). There are several other manufacturers of these small, relatively inexpensive, easy to use units on the marketplace. The difference between these manufacturers lies in whether they offer a complete package or components and the amount of support they will provide.

A small, compact spectroradiometer is less accurate than its much larger and more expensive, double-monochromator counterparts. The question to be asked is whether the increased accuracy is really needed or are the smaller units sufficient for our purposes. These smaller units are deemed more accurate because they produce both qualitative and quantitative data at the same instant. These items are important to pharmaceutical photostability testing. As noted in Chapter 3 of this book, current dosage measurements are inaccurate if performed according to the current ICH specified methods of dosage measurement.

Only a spectroradiometer can show the effects, singly or combined, on the various components of a chamber such as reflectors, filters, lamp covers, lamp aging, dimming, mixing lamp types, refraction, reflection, etc. A plot of the spectral power distributions of the various lamps cited by the ICH guideline gives clear evidence of their non-comparability.

Any chamber map will be a function of its resolution. The resolution, in turn, is a function of the detector size and the number of points sampled. It is not unusual, as will be shown, for chamber maps of the same lamp design, to have two different mapping patterns.

Another factor affecting the mapping pattern is the source design. Short-arc (xenon and metal halide) lamps can be considered point sources, which generate bulls-eye patterns whereas long-arc lamps (xenon and fluorescent) will produce oblong patterns. Low-resolution mapping tends to obscure these differences, which are real. How pertinent these results will be to an individual study will depend on a number of factors, one of which will be the samples physical size.

One of the first researchers to report the mapping of a photostability chamber was O'Neill (42). His map (Fig. 34) shows the uneven distribution of UV-A radiation in a tabletop, long-arc, xenon unit and the VIS distribution in a Hotpack Low Temp Light Cabinet (Fig. 35). Both of these figures clearly illustrate the uneven distribution of radiation in these chambers.

Figure 36 shows a diagram of the special exposure unit designed and built by O'Neill, (A), and a photograph of the unit in use (B). The gradations in color intensity are actual color changes that represent the uneven coating of the lamps, a particular problem that increases with the length of the lamp.

Figures 37 A and B were reproduced from a presentation by Turner (16). The tubes used in his studies were 120 cm long. This figure supports the Figure 36B photograph.

At the third International Meeting on the Photostability Drugs and Drug Products, PPS'99, several papers were presented which contained information

Variation in Light Energy Distribution

11 1/2"

887	905	1015	1022	1015	950	908

7 3/4"

875	962	1006	1024	1011	934	894

884	854	1006	1012	988	926	883

Meter = UVX 31 μW/cm²

Figure 34 Radiometric map of a Heraeus Suntest CPS xenon chamber. *Source*: Courtesy of John O'Neill, Sanofi-Winthrop.

Hotpack Low Temp Light Cabinet

Full Lights on, 8" Off Center

S			
O	2020	1020	760
U	2060	1080	1040
R	2060	1080	1040
C	1040	1020	920
E		1261 Ave. Foot Candles	

Figure 35 Photometric map of a hotpack light chamber. *Source*: Courtesy of John O'Neill, Sanofi-Winthrop.

on chamber mapping, by Brumfield (43), Turner (16), Clapham and Sanderson (44) and Baertschi (45), which showed the intensity mapping of several different cabinets. Each of these studies deserves to be discussed separately because they will aid readers in their selection of a chamber and development of their own validation procedure.

Brumfield
Brumfield et al. (43) at Pharmacia Upjohn (now part of Pfizer) undertook a study to evaluate what was the best pharmaceutical photostability chamber for their purposes. The target illumination uniformity level chosen for these tests was ± 10%. One chamber tested in this study had a plastic material on the sidewalls, which reflected VIS radiation well but was not as reflective for the UV radiation.

(A)   **(B)**

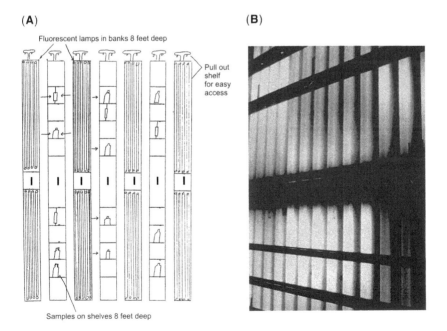

Fluorescent lamps in banks 8 feet deep

Pull out shelf for easy access

Samples on shelves 8 feet deep

Figure 36 **(A)** 8 ft × 8 ft exposure rack diagram and **(B)** picture of part of one bank of lamps, operating. *Source*: Courtesy of John O'Neill, Sanofi-Winthrop.

(A)

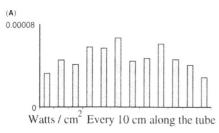

Watts / cm² Every 10 cm along the tube

(B)

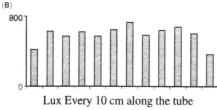

Lux Every 10 cm along the tube

Figure 37 (**A**) Irradiance and (**B**) Illuminance intensity maps of individual 120 cm long fluorescent lamps. *Source*: Courtesy of Nick Turner and GlaxoSmithKline.

Figure 38 shows the maps generated when four Philips 84 HF cool white fluorescent and four, UV-A lamps were used in this cabinet. The UV-A uniformity was not as good as the VIS. Brumfield concluded that sample placement would be critical in the unit.

Figure 39 shows the mapping pattern for four Sylvania, Activa, full-spectrum lamps is the result of using alternating VIS and UV lamp arrangement in this cabinet. The difference between the UV-A and VIS plots is attributed to the reflective coating used in the chamber. These maps were obtained at 50% power. Under these conditions, he found that VIS exposure would require 171 hours, of exposure to meet VIS exposure limits and UV-a, 436 hours. He concluded that he would not employ such an arrangement for his studies.

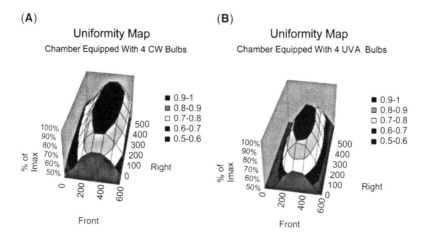

Figure 38 (**A**) Illuminance and (**B**) Irradiance maps of a chamber showing that the two different lamps do not give the same mapping pattern. *Source*: Courtesy of Jay C. Brumfield and Pfizer Ltd.

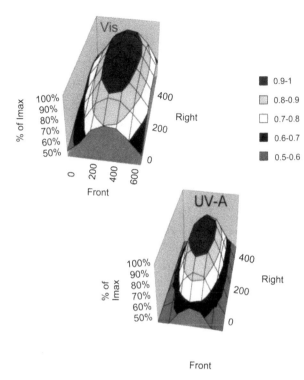

Figure 39 VIS and UVA maps Uniformity of a chamber equipped with four full spectrum lamps. The two maps are not superimpossible, indicating a varying ration of VIV to UVA radiation. *Source*: Courtesy of Jay C. Brumfield and Pfizer Ltd.

Figure 40 is the map of the UV-A radiation obtained when four and seven T-5, full-spectrum lamps are used in two different test chambers. There is a significant difference in the two maps obtained.

Figure 41 is a picture of the chamber equipped with alternating UV-A and VIS lamps.

Figure 42 is a map of a tabletop long-arc, air-cooled approximately 1 kW, xenon test chamber. When the chamber was set at 550 W/m², he calculated that the ICH recommended levels would be reached in 9.7 hours and 6.7 hours for VIS and UV-A, respectively. He noted a rise in temperature and drop in humidity with this basic unit.

Figure 43 is a picture of the insides of the chamber finally selected by Brumfield and his associates for their work. This unit has three shelves, each individually configurable. This chamber uses up to seven 40 W, T-5 fluorescent lamps per shelf. The inside is made of polished aluminum to yield spectral, not diffuse reflectance. The following figures are the maps he obtained for different configurations of this unit.

Figure 44 is the UV-A map he obtained by equipping the chamber with seven full spectrum lamps. The VIS map for this configuration was nearly identical to the UV-A map according to Brumfield.

Figure 45 shows the UV-A maps of the chamber when operated with four, UV-A lamps at 10% and 100%. Little change is noticeable. Brumfield found that this was not the case when using combination UV-A/VIS lamps and full spectrum lamps. He found that the UV-A/VIS ration changed with dimming for one other brand he tested.

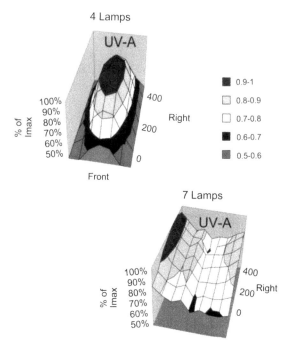

Figure 40 UVA maps of a chamber configured to have alternating UVA and VIS lamps. *Source*: Courtesy of Jay C. Brumfield and Pfizer Ltd.

Figure 46 shows the maps obtained when all seven, HF, full spectrum lamps are used and when only four out of the seven might be used. All of this work shows that while, frequently, the ± 10% specification can be met most of the time,

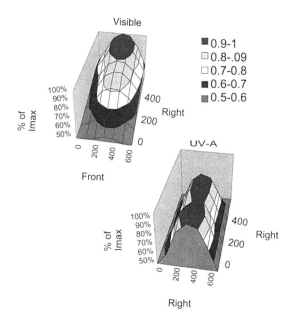

Figure 41 Uniformity map chamber equipped with alternating UVA and CW bulbs: FRONT-UV·VIS-UV-VIS-BACK. *Source*: Courtesy of Jay C. Brumfield and Pfizer Ltd.

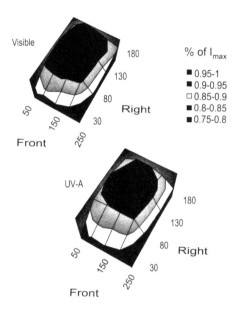

Figure 42 UVA and VIS uniformity maps of a xenon test chamber. *Source*: Courtesy of Jay C. Brumfield and Pfizer Ltd.

there are still other problems to contend with, such are a varying ratio of UV-A to VIS radiation, as previously noted for some full spectrum lamps.

All of these scan were done radiometrically. Spectroradiometric plots or even higher resolution radiometric plots may yield more information that is critical.

Turner
Turner (16), in his presentation at PPS '99, reported his results of studying some of the variables in the guideline which included mapping studies of his, in-house developed photostability chamber. The chamber was 70×140 cm and housed six, 120 cm fluorescent tubes.

Figure 47 (A) shows the cabinet and mapping pattern (B) used in his studies.

Two items of interest were the reproducibility of the lamp illumination patterns and how these might be affected by the aging of the lamps.

Figures 48 and 49 show the results of his studies. Figure 48 shows maps obtained for the UV-A plots of two different lamps at $t = 0$ and $t = 3$ months. Figure 49 is similar data obtained for daylight lamps. Most of the sample surfaces tested appears to be within the $\pm 10\%$ range.

Clapham and Sanderson
Clapham and Sanderson (44) also made a presentation at PPS'99 that included their studies for similar purposes a Brumfield and Henry. Their studies were done at a higher resolution and on three different types, xenon, metal halide, and fluorescent, commercially available units. Radiometric measurements (direct) were taken at equidistant points in the sample surface and spectroradiometric measurements (integrated) at less frequent sampling points.

(A)

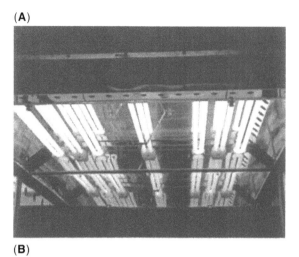

(B)

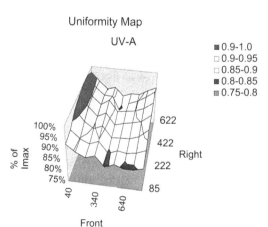

Figure 43 (**A**) Top and (**B**) bottom views of the interior of the chamber selected by Brumfield et al. *Source*: Courtesy of Jay C. Brumfield and Pfizer Ltd.

Figure 44 UV-A map of the selected chamber equipped with seven, full spectrum lamp. *Source*: Courtesy of Jay C. Brumfield and Pfizer Ltd.

Uniformity Map

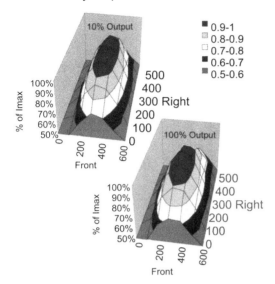

Figure 45 Selected chamber equipped with four, UV-A lamps run at 10 and 100% of full power to determine the effect of dimming on the map obtained. *Source*: Courtesy of Jay C. Brumfield and Pfizer Ltd.

Figure 50 is one of the first to show the oblong irradiation pattern obtained using a Sanyo (now Weiss-Gallenkamp) (39) fluorescent cabinet. This cabinet did not contain the later added "diffuser."

Uniformity Map

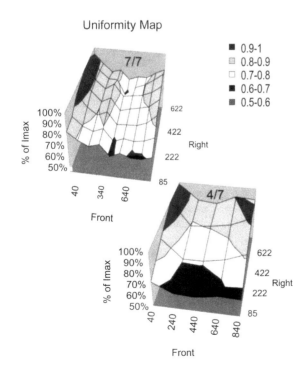

Figure 46 UV-A maps of the selected chamber configured with seven, full spectrum lamps and with only 4 of the seven operating. *Source*: Courtesy of Jay C. Brumfield and Pfizer Ltd.

(A)

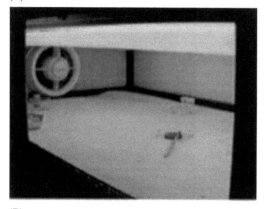

(B)

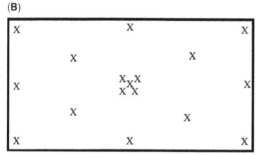

Figure 47 Picture of (**A**) the interior of the chamber used by Turner for his studies and (**B**) the sampling plan used. *Source*: Courtesy of Nick Turner and GalaxoSmithKline.

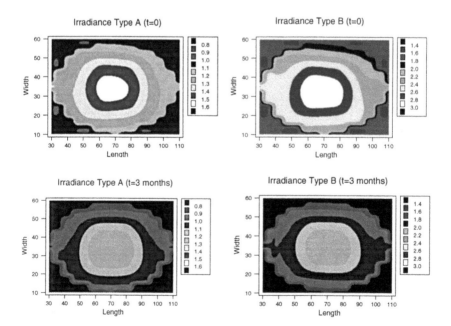

Figure 48 UV-A maps of two different lamps at $t = 0$ and $t = 3$ months. *Source*: Courtesy of Nick Turner and GalaxoSmithKline.

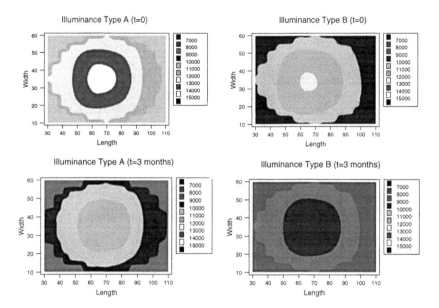

Figure 49 VIS maps of two different lamps at *t* = 0 and *t* = 3 months. *Source*: Courtesy of Nick Turner and Galaxo SmithKline.

The irradiation pattern for all chambers using fluorescent lamps will be similar. Slight variation in the patterns will exist due to lamp shapes, i.e., linear versus bi-axial (U-shaped), but essential they will all be very similar.

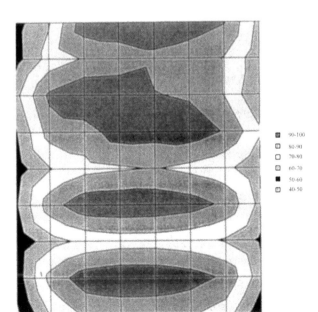

Figure 50 Map of a Sanyo Gallenkamp (now Weiss Gallenkamp) fluorescent chamber with 4 lamps. *Source*: Courtesy of D Clapham and D Sanderson, and GlaxoSmithKline.

(A)   (B)

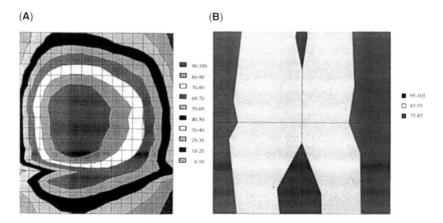

Figure 51 Maps of a Dr. Hönle, Sol 2 unit direct (with a radiometer) and integrated (with a spectroradiometer). *Source*: Courtesy of D Clapham and D Sanderson, and GlaxoSmithKline.

Figure 51 shows the mapping pattern, direct and integrated, obtained using a Dr. Hönle, Sol two metal halide lamp test chamber. A partial explanation for the integrated pattern may be the overload of the detector along with the reduction in the number of sampling points.

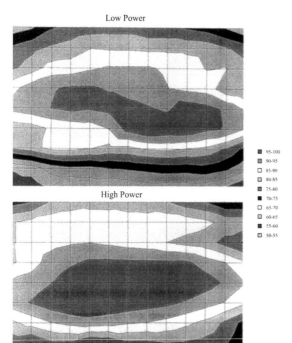

Figure 52 Higher resolution map of a fluorescent lamp showing the pattern obtained with multiple lamps. *Source*: Courtesy of D Clapham and D Sanderson, and GlaxoSmithKline.

Figure 52 is the pattern obtained when a Heraeus DSET, [now part of Atlas (20)], Suntest CPS, long-arc, 1.8 kW, xenon lamp cabinet was mapped at both high and low power settings.

Baertschi

Another participant in PPS '99 was Steve Baertschi (45) who shared his mapping studies. Figure 53 shows his results for a fluorescent unit at his facility. Most striking is the differences to be noted between the VIS and UV-A maps, a phenomena also noted by Brumfield (43).

Figure 54 is Baertschi's UV-A map for an Atlas Suntest chamber. There is a significant difference between this pattern and the one shown in Figure 52. Both units have the same optical arrangement. This is not an unusual finding.

Changes in irradiation patterns can occur with aging, for multi-lamp units, when lamps are changed, in any type unit, or when power levels are adjusted.

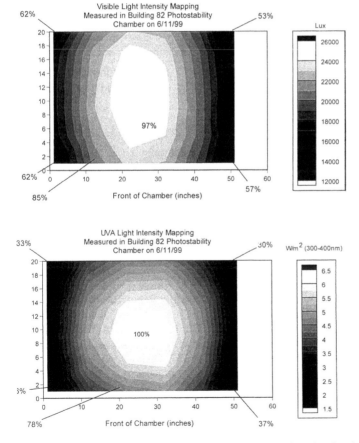

Figure 53 VIS and UV-A maps of a fluorescent test chamber showing that both patterns are not superimposable. *Source*: Courtesy of SW Baertschi and Eli Lilly and Co.

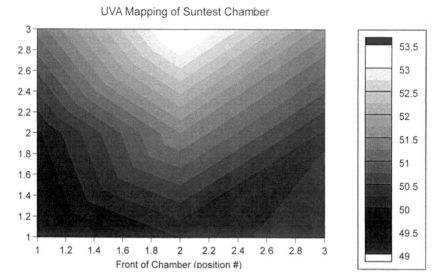

Figure 54   UV-A map of a Suntest test chamber. *Source*: Courtesy of SW Baertschi and Eli Lilly and Co.

One item overlooked and not evident unless studied spectroradiometricaly, is the contribution of the mercury lines to fluorescent and metal halide radiation patterns.

CONCLUSION

It is apparent that the mapping of chambers is necessary to obtain repeatable results. Calibration of the sample area to be used is a good approach; once one can be assured that, the radiation reaching that area is of the right qualitative and quantitative requirements. The use of the "black box" approach, which only dwells on numbers, should be avoided if one wishes to adopt the scientific approach.

The best approach to mapping a chamber is to use a small, computer-controlled spectroradiometer, equipped with a small integrating sphere (1 in.). Mapping plans should be developed for the sample sizes to be tested. More points should be tested for smaller size samples.

Samples should be positioned to give them maximum exposure, not standing on end (or) perpendicular to the source. Mapping should assure that opposite ends of the sample are not in different zones, particularly for semisolid and solid samples.

Following these few rules will assure that the data available from photostability tests will be accurate, of the highest quality and at the lowest cost.

REFERENCES

1. International Conference on Harmonization (ICH). Guideline for the photostability testing of new drug substance and products. Federal Register 1997; 62:27115–27122.
2. Baertschi SW, Kinney H, Snider B. Issues in evaluating the "in-use" photostability of transdermal patches. 3rd International Meeting, Photostability of Drugs and Drug Products, Washington, D.C., July 10–14, 1999.

3. Sequeira F, Vozone C. Photostability studies of drug substances and products. Pharm Tech 2000; 24(8):30–35.
4. Kallab FV. Kallab's belichtungs-apparat, Collegium. J Soc Leather Technol Chem 1912; 287–289.
5. Arny HV. Further studies with light rays. Glass Packer 1933; 12:497–499.
6. Arny HV. Report of Committee on Colored Glass Containers. J Am Pharm Assoc 1928; 20(XVII):1056–1060.
7. Arny HV, Taub A, Steinberg A. Deterioration of certain medicaments under the influence of light. J Am Pharm Assoc 1931; 20(10):1014–1023.
8. Arny HV, Taub A, Steinberg A. Deterioration of certain medicaments under the influence of light. J Am Pharm Assoc 1931; 20(10):1153–1158.
9. Arny HV, Taub A, Blythe RH. Determination of certain medicaments under influence of light. J Am Pharm Assoc 1934; 23(7):672–679.
10. Lachman L, Cooper J. A comprehensive pharmaceutical stability-testing laboratory. I. Physical layout of laboratory and facilities available for stability testing. J Am Pharm Assoc Am Pharm Assoc (Baltim) 1959; 48(4):226–233.
11. Lachman L, Swartz CJ, Urbanyi T, Cooper J. Color stability of tablet formulations II. Influence of light intensity on the fading of several water-soluble dyes. J Am Pharm Assoc Am Pharm Assoc 1960; 49(3):165–169.
12. Lachman L, Swartz CJ, Cooper J. A comprehensive pharmaceutical stability-testing laboratory III. A light stability cabinet for evaluating the photosensitivity of pharmaceuticals. J Am Pharm Assoc Am Pharm Assoc 1960; 49(4):213–218.
13. Forbes PD. Workshop on production and measurement of ultraviolet light: light sources for solar simulation in photocarcinogenesis studies. Natl Cancer Inst Monogr 1978; 50:91–95.
14. Forbes PD. Design limits and qualification issues for room-size solar simulators in a GLP environment. In: Tonnesen HH, ed. Photochemical Stability of Drugs and Drug Formulations. London, U.K.: Taylor & Francis, 1996:288–294.
15. Cole C. Solar simulation with a 6.5 kW xenon arc lamp for skin cancer research. Atlas Sunspots Tech Article Summer. Atlas Electric Devices Co., 1979:1–4.
16. Turner N. Photostability testing-let there be light. 3rd International Meeting, Photostability of Drugs and Drug Products, Washington, D.C., July 10–14, 1999.
17. Wall GM, Morgan J, Wrinkle M, Scott B, Dorsey E, Brammer l. In: The design of a Computerized Pharmaceutical Stability Monitoring System. Scientific Computing and Automation, June 13–16, 1998.
18. Boxhammer J. Design and validation characteristics of environmental chambers for photostability chambers. Photostability: A 1997 Update—Scientific, Regulatory and Practical Issues, AAI 1997 Seminar Series, Arlington, VA, Feb 24–25, 1997.
19. Boxhammer J. Technical requirements and equipment for photostability testing. In: Tonnesen HH, ed. Photochemical Stability of Drugs and Drug Formulations. London, UK: Taylor & Francis, 1996: 39–61.
20. Boxhammer J, Willwoldt C. Design and validation characteristics of environmental chambers for photostability chambers. In: Albini A, Fasani E, eds. Drugs Photochemistry and Photostability. Cambridge, U.K.: Royal Society of Chemistry, 1998:272–287.
21. Atlas Material Testing Technology, LLC., Chicago, IL 60613, U.S.
22. Axyos Technologies, Brebdale, Queensland, Australia 4500.
23. Binder GmbH. D-78532 Tuttlingen, Germany.
24. Caron Products & Services Inc. P.O. Box 715, Marietta, OH 45750, U.S.
25. Environmental Growth Chambers, Chagrin Falls., Ohio 44022, U.S.
26. Environmental Specialties (LUWA), Division of Zellweger Group, 4412 Tryon Road, Raleigh, NC, U.S.
27. Hotpack, SP Industries, Warminster, PA 18974–2811, U.S.
28. Lunaire (Thermal Product Solutions), Williamsport, PA 17701, U.S.
29. Luzchem Research Inc., Unit 12, Ottawa Ontario, Canada, K1J9J9.
30. Nagano Science Equipment Mfg. Co. Ltd., Chidori, Ots-ku, Tokyo, Japan.
31. Newtronic Equipment Co, PVT., Ltd. 117-CD, Government Industrial Estate, Charkop, Kandivali (West), Mumbai-400067, India.

32. Percival Scientific, Inc., Perry, Iowa 50220, U.S.
33. Powers Scientific, Inc., P.O. Box 268, Pipersville, PA 18947, U.S.
34. Q-Lab Corporation, Cleveland, Ohio 44145, U.S.
35. Rumed, Rubarth Apparate GmbH., D-30880 Laatzen, Germany.
36. Termaks. Lien 79, N-5057 Bergen, Norway.
37. Thermolab Scientific Equipments PVT. Ltd. "Thermolab House", Papdi Municipal Ind. Area, Umela Road, Papdi, Vasai (West), Thane, India.
38. Vindon Scientific Ltd., Kiln Green Diggle, OLDHAM, Lancashire, OL3 5JY, U.K.
39. Weiss Gallemkamp, Units 37–38. The Technology Centre, Epinal Way, Loughborough, LE11 3GE, U.K.
40. Suga Test Instruments Co., Ltd., 5-4-14, Shinjuku, Shinjuku-Ku. Tokyo 160, Japan.
41. Dr. Hönle AG. UV-Technologie. Lochhamer Schlag 1, D-82166 Gräfelfing, Germany.
42. O'Neill J. ICH converts light at the end of the tunnel. AAI Seminar Series, Wilmington, NC, 1996.
43. Brumfield JC. The Pharmacia & Upjohn experience in selecting a photostability guideline option. 3rd International Conference on the Photstability of Drugs and Drug Products sponsored by the American Society for Photobiology, Washington, D.C., July 10–15, 1999.
44. Clapham D, Sanderson D. The importance of sample positioning and sample presentation within a light cabinet. 3rd International Conference on the Photstability of Drugs and Drug Products sponsored by the American Society for Photobiology Washington, D.C., July 10–15, 1999.
45. Baertschi S. Practical aspects of pharmaceutical photostability. 3rd International Meeting, Photostability of Drugs and Drug Products, Washington, D.C., July 10–14, 1999.

14 Current Methods for the Analytical Investigation of the Photodegradation of Active Pharmaceutical Ingredients and Products

Karl Thoma[†]
Institut für Pharmazie—Zentrum für Pharmaforschung Lehrstuhl für Pharmazeutische Technologie, Ludwig-Maximilians-Universität, Munich, Germany

Norbert Kübler
Temmler Pharma GmbH and Co. KG, Marburg, Germany

INTRODUCTION

Knowledge concerning the degradation behavior of pharmaceuticals under ultraviolet–visible (UV–VIS) photon exposure is of increasing importance. One reason for this is the steady rise in the number of drug substances with known photo-instability and their potential uses as phototherapeutic agents. On the other hand, there are also legal requirements concerning unknown impurities and the structural elucidation of degradation products (1). Therefore, knowledge of the photolability of a drug substance is not sufficient. There is a strong demand to characterize the photodecomposition products and the photochemical degradation pathways. Furthermore, there are many current publications discussing photon-induced side effects of pharmaceuticals. In this field, there is a need to investigate whether the side effects are due to toxic photodegradation products or due to an invivo reaction, such as a photosensitization.

Although the photostability testing of pharmaceuticals is an essential part of the stability testing of new drug substances and drug products, it was only in 1996 that the International Conference on Harmonization (ICH) established internationally recognized guidelines for the photostability testing procedure (2). An inquiry among pharmaceutical companies in Germany (3) and Great Britain (4) revealed that many different photon sources are used for the photostability of drugs and drug products. Irradiation periods varied by a factor of 100 or more. This complicated the comparison of research results between different working groups considerably. The new ICH guideline facilitates future results and is a necessary step for the pharmaceutical industry.

PROCEDURE FOR PHOTOSTABILITY EXPOSURE

Selection of a Suitable Photostability Testing Cabinet

The photostability testing of drug substances should be performed using a wavelength band that is related to real conditions. A helpful guidance is the definition of the spectral power distribution for solar spectral irradiance defined by the

---

[†]Deceased.

Commission Internationale de l'Éclairage (CIE) (5) and discussed in Chapter 6 of this book.

Daylight exposure as a natural photon source has its restrictions due to within day, daily, seasonal, and terrestrial variation. Artificial photon sources are more reproducible and therefore an indispensable prerequisite for standardized photostability testing.

Many investigators employ xenon lamps because they have a continuous spectrum and cover the UV, VIS, and short infrared (IR) bands similar to solar irradiance. The use of an IR dichroic mirror in combination with a UV filter for the UV band with these lamps produces a very close simulation of the daylight irradiance (Fig. 1).

Figure 2 shows a schematic drawing of a photostability testing cabinet equipped with a xenon lamp.

The effect of the different filters on the resultant spectral power distributions is shown in Figure 3.

For the simulation of indoor, indirect daylight, an appropriate window glass can be used to eliminate UV radiation below 320 nm.

An additional option for accelerated photostability testing is irradiation without filter inserts. In this case, one has to take into account that the different wavelength ranges will influence the types, degradation rates, and the quantitative amounts of products produced. The formation of artifacts, other than those formed under terrestrial short UV irradiation and window glass filtered daylight must always be considered.

Alternative photon sources, such as metal halide and fluorescent lamps, are also in use. The advantage of metal halide lamps is the reduced amount of IR radiation compared to xenon lamps.

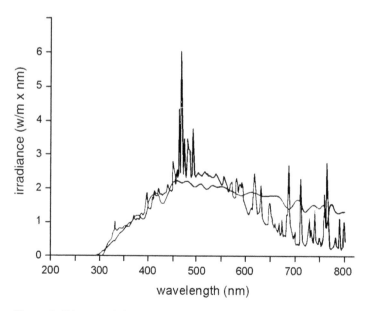

Figure 1 Wavelength distribution of a xenon lamp with a specific ultraviolet filter compared to the solar spectral irradiance according to the Commission Internationale de l'Éclairage.

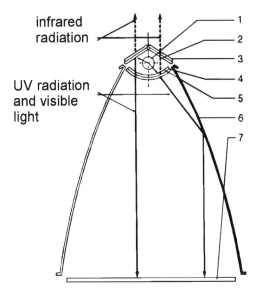

infrared
radiation

UV radiation
and visible
light

Figure 2 Set-up of the photostability testing equipment Suntest CPS. (1) Xenon burner, (2) UV light mirror, (3) visible light mirror, (4) quartz dish, (5) optional light filter for special purposes, (6) parabolic reflector, (7) sample level. *Source*: From Ref. 6.

Fluorescent lamps are available for the near UV range as well as for indoor daylight simulation. Previously cited surveys revealed that this option is used less frequently in Germany (3) than in other countries (4). Monochromatic lamps, e.g., lasers, have their main application in experimental photochemistry and for the calculation of the quantum yields of photochemical reactions (8).

Measurement of the Photon Exposure
The measurement and control of the photon exposure is a principal prerequisite for the performance of reproducible photostability tests. In addition to the decrease of radiation intensity with the lamp aging can cause in addition to the decrease of radiation intensity a change in the spectral power distribution of the lamp (9).

Different units are used by the ICH to measure the amount of UV and VIS photonic energy received by the sample. UV irradiance, measured using a radiometer or spectroradiometer, is expressed in W/m^2. After exposure at a certain irradiance level for a defined time, the total integrated energy (Wh/m^2) can be calculated using the ave rage irradiance level and the time measured.

Another unit, lux, measured using a photometer, is the amount of VIS photonic energy received by stability samples. As used in the ICH document, this measurement does not refer to the entire visible spectrum but rather to that amount measured using a detector with a photonic (eye-like) response. Hence, not all of the radiation present is measured, and that which is measured is not accurate. Chapter 6 of this book discusses this problem in much more detail. A direct conversion of the irradiance into the illuminance is not possible as only the irradiance considers the amount of the energy of a photon corresponding to its wavelength.

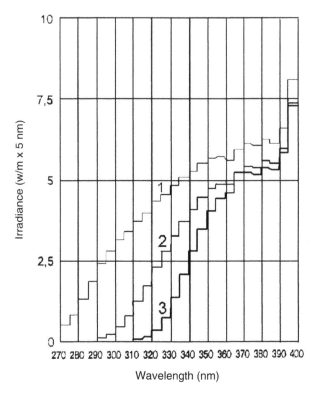

Figure 3 Spectral power distribution with different light filters. (1) Without additional light filters, (2) special UV light filter, (3) window glass filter. *Source*: From Ref. 7.

Besides these physical methods, chemical methods can also be used to monitor the photon exposure of a sample. Chemical actinometry is a very precise method. A publication of IUPAC describes numerous actinometric systems useful for several different wavelength ranges (10). Chemical actinometry is important for the determination of the reaction kinetics and quantum yield of photoreactions in experimental photochemistry as well as for the control and quantification of the overall exposure in photostability testing systems. A more detailed presentation of chemical actinometry is given in Chapter 8 of this book.

Selection of a Suitable Photon Dose
Prior to the publication of the ICH guideline, no valid guideline for the performance of photostability tests was available. Therefore, the total applied amount of photonic energy varied over a wide range. Pharmaceutical companies in Germany (3) used UV–VIS energies ranging from 0.75–250 kWh/m². Another survey (3) reported that the total photon exposure was reported in terms of lux hours. In this case, the figures for different pharmaceutical companies varied between 1.44–108 M lux hours for solid samples and between 1.92 and 108 M lux hours for liquid samples.

The adoption of the internationally recognized ICH guideline (2) in 1996 was a small progressive step in this field. Now, for the first time, it is possible to have a better degree of control between different laboratories and research groups, even though it still maintains a dual system of reporting dosage measurements. The overall illumination is specified at not less than 1.2 M lux hours for the VIS exposure. In the UV band between 320 and 400 nm the integrated energy is to be not less 200 Wh/m². The regulations do not state a total "daylight" exposure value, even though they do allow the use of such lamps.

The guideline also suggests a procedure (not applicable in all cases) for the presentation of the sample using protected samples as dark controls. A detailed decision flow chart for the selection of the testing stages to be performed is also included.

ANALYTICAL METHODS FOR PHOTOSTABILITY TESTING

The requirements for the analytical methods acceptable for photostability testing are increased because the methods have to be more selective, i.e., more indicative of stability. The method must permit the separation, detection, and quantification of all degradation products at very low levels. For unknown impurities, the analytical method must provide as much qualitative information as possible. The following sections give an overview of current analytical techniques used in conjunction with a photostability testing program. Several examples of their application, from our own photostability studies, will also be presented.

Spectrophotometry

Spectrophotometry and high performance liquid chromatography (HPLC) are the most frequently used analytical techniques for monitoring photodegradation processes. The main advantage of spectrophotometric analysis is its speed and simplicity. Sample solutions can often be directly exposed to sources in a quartz cell and subsequently analyzed in this same cell, without further dilution or treatment. Evaluation of the test results can be performed by calculating the change in the absorption maximum of the test substance or all spectral changes in its UV–VIS range (11). Quantitative analysis of the degradation products is only possible for a limited number of cases, i.e., those giving a significant absorbance change and few interfering degradation products. One example is its usage in a study related to the photodegradation of nifedipine. It is possible to quantify nifedipine in parallel to its major photodegradation product, a nitrosophenyl derivative, by calculating the UV absorption ratio at two wavelengths (12).

A major drawback of the spectrophotometric analysis is the fact that many photodegradation reactions provoke no or only negligible spectral changes. This leads to the fact that many photoreactions cannot be registered accordingly.

The irradiation of a solution of ketoconazole will illustrate that situation: The monitoring of the absorption at the spectral maximum of ketoconazole at 242 nm only gives a decrease of 9% after an irradiation period of 100 minutes (Fig. 4A). A comparison of the complete UV absorption spectrum before and after 100 minutes irradiation shows more significant changes in the range around 300 nm (Fig. 4B). But the real extent of decomposition only became apparent with the results of the HPLC stability assay. The HPLC analysis proved that already after 5 minutes 50% of the drug was decomposed by photolysis.

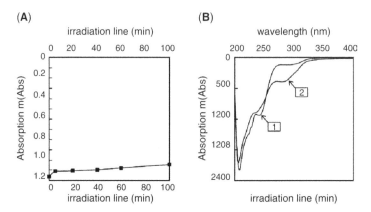

Figure 4 Spectrophotometric investigation of the photolysis of ketoconazole in a methanolic solution (3.6 mg/100 mL; Suntest CPS with UV-filter). (**A**) Absorption change at the spectral maximum of 242 nm, (**B**) UV absorption spectrum of the solution (1) before irradiation; (2) after an irradiation period of 100 minutes.

A comparison of the total UV absorption spectrum before and after irradiation shows that a more significant change occurs around 300 nm (Fig. 4B). However, the real extent of decomposition only becomes apparent after HPLC stability assay. The HPLC analyses reveal that after five minutes 50% of the drug was decomposed.

This example clearly demonstrates that the spectrophotometric analyses are not always suitable for stability measurements. Small changes in the photostability of the sample are not always evident. Analysis using more selective and sensitive chromatographic, stability-indicating techniques give a true picture of the course of events.

High Performance Liquid Chromatography

Even if there is no physicochemical or organoleptic indication of photoinstability, a chromatographic separation method for the detection of a loss in potency or the emergence of photodegradation products is required. HPLC has become the routine technique for analysis for most drug actives and products. The widespread use of HPLC is due to its high separation ability and the wide variety of available sensitive and specific detectors for use with it. This technique has also been shown to have a high degree of precision and accuracy, characteristics necessary for the quantification of substances. In combination with a photodiode-array detector and readily available computer software, analate peak purity can easily be assessed. The absorption spectra obtained using this method of detection can also provide the analyst with useful hints as to chemical structure changes that have occurred.

A good illustration of the benefit of using a photodiode array detector is the investigation of the photodegradation of natamycin. Natamycin is a polyene antibiotic with a tetraene structure (Fig. 5) and a very pronounced photoinstability, mainly, in solutions.

After 20 seconds exposure of a natamycin solution, in a Suntest CPS test chamber, high performance liquid chromatography analysis shows that at least seven degradation products of the drug substance are produced (Fig. 6).

With the help of the photodiode array, the total absorption spectra of all eluting peaks can be taken. From Figure 6, it can be seen that photodegradation

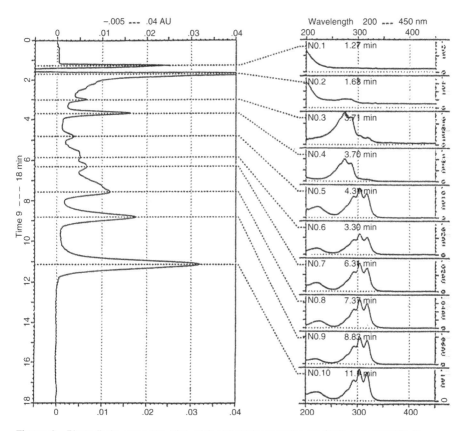

Figure 5 Structural formula of natamycin.

products 3–7 retain the characteristic tetraene absorption band centered at 300 nm. Products 1 and 2, while retaining the overall profile of a tetraene structure, show a distinct bathochromic shift to a band centered at 270 nm. The enone-chromophore,

Figure 6 Photodiode array detection of the photodegradation products of natamycin in methanolic solution after 20 seconds light exposure (Suntest CPS, UV-filter). 1.27 and 1.69 min: solvent, etc. peaks. 3.01 min: Photodegradation product 1, 3.70 min: Photodegradation product 2, 4.81 min: Photodegradation product 3, 5.86 min: Photodegradation product 4, 6.31 min: Photodegradation product 5, 7.57 min: Photodegradation product 6, 8.83 min: Photodegradation product 7, 11.14 min: Natamycin.

with an absorption maximum around 220 nm, is not seen in products 1 and 2 is clearly present in products 3 through 7.

HPLC analyses, after longer irradiation periods, have shown that the tetraene structures are increasingly destroyed upon exposure to photonic energy. The amount of degradation products with triene structure increases steadily. Comparable results have been seen with other polyene antibiotics (13).

Thin Layer Chromatography

Thin layer chromatography (TLC) has continued to be a significant analytical tool despite the widespread use of HPLC. The TLC separation of irradiated samples can cover a wide range of polarities for the degradation products. The so-called on-plate irradiation technique, for studying the photodegradation of drugs, is a procedure where a solution of the drug substance/product solution/extract is placed on the TLC plate, dried, and subsequently irradiated. After irradiation of the plate, the chromatographic separation is performed (14). Using functionally specific spray reagents, functional groups produced as a result of photodegradation can be detected easily.

The use of preparative TLC is, moreover, a way of obtaining sufficiently large amounts of degradation products for the structure elucidation direct probe mass spectrometry or nuclear magnetic resonance spectrometry.

Additional Analytical Methods

Differential Scanning Calorimetry
Differential scanning calorimetry is a technique that has recently been used for the detection of photodegradation products in solid drug substances. Even minor amounts of degradation product within a crystal lattice can provoke a significant decrease in its melting point. This method has successfully been used to monitor the photostability of nifedipine (15).

IR Spectroscopy
The use of IR spectroscopy for the detection of photodegradants in solid drug substances has also been reported in literature (16). The major drawback of this method is its limited sensitivity. Normally changes can only be detected when the degradation products exceed 1%.

Electron/Nuclear Spin Resonance
A more profound knowledge of the photolytic processes of a drug substance under UV–VIS photon exposure can be obtained utilizing this method of experimental photochemistry. The irradiation of many drug substances having photosensitizing properties results in the formation of free radicals as part of their primary photochemical processes (17). The formation of these radicals can be detected and monitored by means of electron spin resonance (ESR) or nuclear resonance spectrometry, especially utilizing the chemically induced dynamic nuclear polarization technique (8).

Liquid Chromatography-Mass Spectrometry
The most frequently used techniques for the structural elucidation of photodegradation products are IR spectroscopy, mass spectrometry, elemental analysis, and nuclear resonance spectroscopy. Degradation products for use with these methods

require isolation/purification by preparative TLC, preparative HPLC, or column partition liquid chromatography. The isolation of sufficiently large enough amounts for analysis can pose serious problems because the samples used for analysis are often complex mixtures containing many different types of functional groups. Solvent impurities can also add impurities to the sample, thereby complicating the identification of a particular analate.

Coupling (hyphenated) techniques, combining both a separation method as well as a detector capable of giving structural information specific to each analate are desirable for use in studying such complex mixtures. Liquid chromatography-mass spectrometry coupling (LC-MS) and liquid chromatography-nuclear magnetic resonance spectroscopy (LC-NMR) are examples of useful developments in this field.

The coupling of LC and MS became possible when the necessary interfaces became available for general application. As LC-MS is a relatively new method, its principles of action are described in the following paragraphs.

The main components of an LC-MS are the HPLC apparatus, an optional UV or photodiode array detector, the interface, the mass spectrometer and a computer system for data management and evaluation. The interface is the key component of the LC-MS system. All other components must be adapted to the particular interface that is used. Most commercially available systems work with thermospray, electrospray, or particle beam interfaces. Each interface has a distinct mode of action and its own operational parameters.

Thermospray

A large number of LC-MS systems are equipped with a thermospray interface, which can accept LC flow rates up to 1 mL/min and is especially useful for the analysis of polar and nonvolatile molecules (18). The major disadvantage of this interface is the mass spectrum produced. The "classical" thermospray system only provides spectra with a single molecular ion and little structural information. Only when used with an additional element like a fragmentor electrode is it possible to obtain detailed information regarding the chemical structure of the molecule (19).

Electrospray

The electrospray interface is a technique where an ionized spray is produced utilizing an electric charge (19). Using this technique it possible to analyze very large molecules, with molecular weights up to 100,000 Daltons. The main application of this interface is for the analysis of biomolecules (20).

Atmospheric Pressure Ionization

Another recent development is the capillary atmospheric pressure ionization (API) interface (20). Most API interfaces are based on the use of a heated capillary from which the mobile phase is sprayed, utilizing a gas into a chamber maintained at atmospheric pressure. The ionization takes place at the corona discharge needle (21). The advantages of this interface are its ability to tolerate high LC flow rates (up to 2 mL/min) and the wide range of chemical substances that it can analyze.

Particle Beam

The particle beam interface is a system in which the solvent of the mobile phase is evaporated from the sample using reduced pressure and an elevated temperature. This requirement places some restrictions on the HPLC-part of the system. The working range, for the flow rate of the HPLC, is specified (19) at 0.1–1.0 mL/min.

In our own investigations we found that the optimum working range for the interface was in the range of 0.4–0.5 mL/min.

A major asset of the particle beam interface is its ability to produce either electron ionization (EI) or chemical ionization (CI) spectra. The EI spectra are very helpful for the structural elucidation of photodegradants. For high molecular weight or thermally labile analates where it is difficult, if not impossible, to obtain a molecular ion using the EI mode, CI is the method of choice. In CI, a low molecular weight gaseous phase proton (M^+) reacts with a gas phase analate molecule to produce a molecular ion with a mass equal to $(M+1)^+$. Two suitable reactant gases are methane and ammonia.

Other Methods
Other commercially available systems which come equipped with the direct liquid injection (DLI) or the continuous flow-fast atom bombardment (CF-FAB) interfaces are used less frequently (21). Because of their limited use, little information has been published about their application.

HPLC SYSTEMS

Chromatographic Columns
Chromatographic columns with an inner diameter of 2 mm are frequently used for LC-MS applications. We have found that columns with a 3 mm inner diameter are more robust and ideally suited for use with a particle beam interface. Flow rates around 0.5 mL/min can be used with conventional HPLC systems as well as LC-MS systems having a particle beam interface. Using this size column and flow rate, no adaptation of the separation method is necessary.

Mobile Phase
The mobile phase used with the interfaces must be completely volatile. A common mobile phase additive is ammonium acetate. For difficult separation problems the addition of trifluoroacetic acid for pH adjustment and heptafluorobutyric acid as an ion pairing agent gave excellent results in our own investigations. The only disadvantage of these additives is their poor UV transparency in the range between 200 and 250 nm.

Monitoring Detectors
During LC-MS analyses, the simultaneous monitoring of the absorption spectrum of the eluate is recommended. The combination of LC-MS coupled with a photo-diode array detector can provide additional information regarding the chemical structure of unknown degradation products. The ability of the photodiode array detector to monitor all wavelengths simultaneously assures that any analate eluting within the selected scan range will be monitored at its maximum sensitivity.

The use of the MS total ion monitor (TIM) output as a detector is recommended to assure the detection of all eluting substance in the sample. It is a "universal" detector, responding to all eluting species.

CONCLUSION

There are several methods used for the analysis of the photodegradation products of drug substances/products. Each method has its own strengths and weaknesses.

It is imperative that analysts be aware of these weaknesses if they are to properly interpret the data they generate.

Hyphenated techniques (HPLC-MS, HPLC-NMR, etc.), which employ a high capacity separation technique in addition to a sensitive and specific detector, give the largest amount of quantitative and qualitative data per analysis.

REFERENCES

1. ICH-Harmonisierung der Anforderungen bei Arzneimittel-Zulassungen Pharm Ind. 1992; 54(3):202–213.
2. International Conference on Harmonisation (ICH) Harmonised Tripartite Guideline Stability Testing: Photostability Testing of New Drug Substances and Products, ICH, Geneva, Switzerland, 1996.
3. Kerker R. Untersuchungen zur Photostabilität von Nifedipin, Glucocorticoiden, Molsidomin und ihren Zubereitungen, Doctoral Thesis, Ludwig Maximillians University, Munich, Germany, 1991.
4. Anderson NH, Johnston D, McLelland MA, Munden P. Photostability testing of drug substances and drug products in UK pharmaceutical laboratories. J Pharm Biomed Anal 1991; 9(6):443–449.
5. Commission Internationale de l'Éclairage (CIE) Technical Report: Solar Spectral Irradiance, Publication CIE No. 85. Vienna, 1989.
6. Suntest CPS+. Technical documentation published by Heraeus Instruments, Hanau, Germany, 1994.
7. Suntest CPS. Technical documentation published by Heraeus Instruments, Hanau, Germany, 1992.
8. Von Bünau G, Wolff T. Photochemie—Grundlagen, Methoden, Anwendungen VCH Verlagsgesellschaft, Weinheim, Germany, 1987.
9. Evaluating chemical photodegradation: instrumental approach, Heraeus Instruments, Heraeus Instruments Symposium, Hanau, Germany, 1992.
10. Kuhn HG, Braslavsky SE, Schmidt R. Chemical actinometry. Pure Appl Chem 1989; 61(2):187–210.
11. Thoma K, Strittmatter T, Steinbach D. Studies on the photostability of antibiotics. Acta Pharm Technol 1980; 26(4):269–272.
12. Al-Turk W, Othman S, Majeed I, Murray W, Newton D. Analytical study of nifedipine and its photooxidized form. Drug Dev Ind Pharm 1989; 15(2):223–233.
13. Thoma K, Kübler N. Untersuchung der Photostabilität von Antimykotika. 2. Mitteilung: Photostabilität von Polyenantibiotika Pharmazie 1997; 52(4):294–302.
14. Reisch J, Topaloglu Y. Ist die Dünnschichtchromatographie zur On-plate-Kontrolle der Lichtstabilität von Arzneistoffen geeignet? Pharm Acta Helv 1986; 61(5,6):142–147.
15. Thoma K, Kerker R. Photoinstabilität von Arzneimitteln 4. Mitteilung: Untersuchungen zu den Zersetzungsprodukten von Nifedipin Pharm Ind 1992; 54(5):465–468.
16. Reisch J, Ulbrich R, Reisch G. IR spectroscopic evaluation of the photostability of crystalline pharmaceuticals. Dt Apoth Ztg 1980; 120(48):2385–2390.
17. Epstein JH, Wintroub BU. Photosensitivity due to drugs. Drugs 1985; 30:42–57.
18. Vander Greef J, Niessen WMA, Tjaden UR. Liquid chromatography-mass spectrometry. The need for a multidimensional approach. J Chromatogr 1989; 477:5–20.
19. Presenting the MS Engine. Technical documentation, Hewlett-Packard Seminar, Weinheim, Germany, 1990.
20. Engelhard UV, Wagner-Redeker W. HPLC-MS-Kopplungen, Technik und Anwendungen. Teil 1: Übersicht über die gebräuchlichsten LC-MS-Kopplungen GIT Spezial Chromatographie 1994; 1:5–9.
21. Tomer KB, Parker CE. Biochemical applications of liquid chromaotography-mass spectrometry. J Chromatogr 1989; 492:189–221.

Photostabilization by Packaging

Karl Thoma[†]
Institut für Pharmazie—Zentrum für Pharmaforschung, Lehrstuhl für Pharmazeutische Technologie, Ludwig-Maximilians-Universität, Munich, Germany

Wolfgang Aman
Research and Development, Gebro Pharma GmbH, Fieberbrunn, Austria

INTRODUCTION

Ultraviolet (UV) and visual (VIS) radiation-resistant packaging is an economical and safe method of protecting photosensitive drugs and preparations. Its photo protective ability is highlighted in the International Conference on Harmonization (ICH) guideline, which states that "... for a photolabile dosage form further testing in the primary and the secondary packaging, respectively, is required to guarantee stability" (1).

One of the primary methods used for protecting liquid and solid photolabile pharmaceuticals and preparations from photodegradation is to package them in colored glass or plastic. Colored/pigmented containers are also suitable for the photostabilization of sterile preparations, such as eye drops.

GLASSES

Colored glass containers are often used to package solutions. However, for e.g., injections/infusions photo protection by colored glass conflicts with the need for VIS control of particles, turbids, and discoloration. Although the VIS control of particles and turbids can be hindered by brown ampoule glass, the French Pharmacopoeia requires the use of colorless ampoule glass (2). For this reason and also economics, secondary packaging, the labeling and carton, and proper handling after opening, is gaining increased significance.

Brown Glass

The European Pharmacopoeia requires testing of the UV and VIS radiation permeability of glasses used for the packaging of pharmaceuticals. Generally, the glass's transmission is measured between 290 and 450 nm.

For colored glass containers, not intended for parenteral preparations, the measured transmission must not exceed a maximum value of 10% at any wavelength between 290 and 450 nm, irrespective of the size or glass type. The limits for colored glass containers, used for parenteral preparations, depend on their intended fill volume. The respective limits for these containers are given in Table 1 (3,4).

As is evident from this table brown glass containers may not provide sufficient photo protection for all drug substances/products. Brown glass absorbs UV and VIS radiation up to about 450 nm. The short wavelength region up to about 350–400 nm is almost quantitatively absorbed, depending on the samples' thickness.

[†]Deceased.

Table 1

	Maximum percentage of transmission at any wavelength between 290 and 450 nm	
Nominal volume (mL)	Flame sealed containers	Containers with closures
Up to 1	50	25
Above 1 and up to 2	45	20
Above 1 and up to 5	40	15
Above 5 and up to 10	35	13
Above 10 and up to 20	30	12
Above 20	15	10

For example, the transmission of 1 mm thick "Fiolax™ braun" glass containers is almost 0% at 350 nm, and only a few percent at 450 nm (5) as illustrated in Figure 1. Therefore, the contents of such a container will be widely protected from UV and VIS radiation up to 450 nm.

Prednisolone phosphate eye drops are very photosensitive when packaged in clear colorless glass ampoules as well as in uncolored, high-density polyethylene containers. When exposed to UV and VIS radiation, the drug substance is completely decomposed within a few hours. However, packaging the prednisolone phosphate solution in brown glass vials, results in satisfactory UV–VIS photo protection (6) as shown in Figure 2.

Similar stabilization has also been achieved with dexamethasone-, physostigmine-, bromhexine-, and nandrolone sulfate eye drops (6), and chlorpromazine- and riboflavine solutions (7). Furthermore, the photo protection of solutions of photosensitive drugs containing the photolabile phenylalkylamine group can be considerably improved by packaging them in brown glass primary packages. In brown glass ampoules, $t_{90\%}$, the time by which 10% of the drug is degraded, is three to five times longer (8).

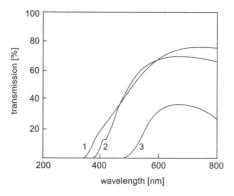

Figure 1 Light permeability of brown glass containers. (1) Ampoule glass (2) Tablet glass containers (3) Infusion bottle. *Source*: From Ref. 5.

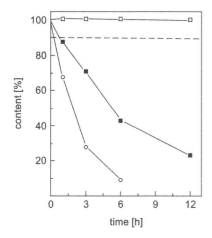

Figure 2 Photostability of prednisolone phosphate-eye drops (0.13% FNA, Formularium der Nederlandse Apothekers) stored in different types of containers in the Novasoltest light testing cabinet. (□) brown glass containers, (■) uncolored high-density polyethylene containers, (○) clear glass ampoules. *Source*: Ref. 6.

Photosensitive herbal medicinal products, such as tincture of colchicum and colchicum pressed juice have been successfully stabilized using this packaging method. Stored in uncolored glass bottles these preparations were strongly photo decomposed after seven days storage in daylight. After six weeks the tincture was completely decomposed. However, when the same solutions were stored in brown glass bottles practically no decrease in the alkaloid content was observed after storage under similar conditions (Fig. 3) (9).

As previously noted, for some very photosensitive drug substances protection by brown glass alone may not be sufficient. This is the reason why, in 1989, molsidomine drops were removed from the German marketplace. After intensive sunlight exposure, detectable amounts of morpholine were found in the product, packaged in its originally approved brown glass bottles (10).

Another drug preparation, for which the storage in brown glass containers is problematic, is chloramphenicol eye drops. After short periods of exposure to UV and VIS radiation, toxic degradation products and significant losses in drug content have been reported (6).

The lack of sufficient stabilizing capacity by brown glass for such drugs is, as previously noted, due to its incomplete filtering of all active wavelengths, i.e., the glass does not completely (100%) filter out all of the incident radiation. For some highly photosensitive drug substances/preparations it may be necessary for them to be labeled, "Store protected from all light."

The residual transmission of brown glass is recognized by both the European and the U.S. Pharmacopoeia. UV–VIS transmission values up to 10% are acceptable. In the case of parenteral preparations transmission values up to 50% are acceptable, depending on the product's fill volume.

Because brown glass shows a strong increase in transmission starting at 400–450 nm (depending on its thickness) it cannot photo protect nitrofurazone-eye drops (6). Such a photosensitive drug product requires protection by other means, e.g., use of a different colored glass or additional packaging.

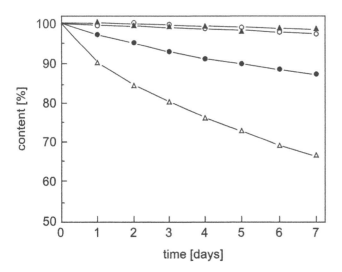

Figure 3 Photodegradation of different colchicine preparations during seven days-storage in diffuse daylight. (△) tincture of colchicum, white glass bottles, (○) tincture of colchicum, brown glass bottles, (●) colchicum pressed juice, white glass bottles, (▲) colchicum pressed juice, brown glass bottles. *Source*: From Ref. 9.

Oral nifedipine drops are packaged in glass bottles, which are strongly pigmented or have a black photo protective coating. Using this procedure, complete photo protection can be guaranteed. Samples of nifedipine drops stored in these containers and exposed for 72 hours to UV–VIS radiation showed no detectable drug loss (11).

For infusion liquids, no completely UV–VIS photon impermeable immediate package can be used. These products require dilution and the inspection for particulates before use; both processes hindered by such packaging. To guarantee safe application/dosing additional photostabilizing packaging is applied for extremely photosensitive drug preparations.

Other Colors

Generally, when people speak of colored, photo protective glass, brown glass is what they mean. This fact is, however, not written into the pharmacopoeias. Different colors of glass can be used for photo protection. The extent of a particular glass's protective effect depends on its spectral characteristics, thickness and spectral properties, and the activation spectra (see Chapter 4 for a discussion of this subject) of the drug substance/product. Therefore, each case must be investigated and judged individually.

The photo protective effect of colored glasses has been known for many years (12–15). Proper selection of the glass color is very important. Green glass, for instance, exhibits little photo protective effect on the yellow photolabile colorant D&C Yellow No. 10. This fact is attributable to the high-degree permeability of green glass in the low wavelength region responsible for the photodegradation of the colorant (16) as shown in Figure 4. Ideally, to provide photo protection, the

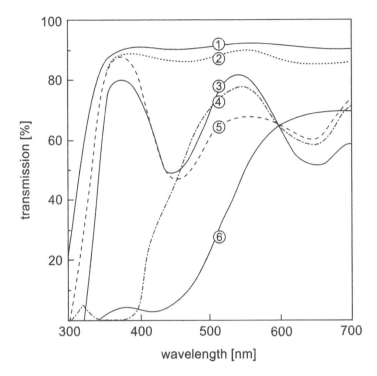

Figure 4 Transmission curves of differently green-colored glass compared to flint and brown glass. (1) Flint, (2) Georgia green, (3) emerald green, (4) UV absorbing green, (5) champagne green, (6) amber. *Source:* From Ref. 16.

colored glass/plastic should completely absorb all of the wavelengths responsible for the photodegradation of the substance/product.

The coloration of glass is achieved by the addition of inorganic metal compounds to the molten glass. The metal ions responsible for the colors of various glasses are presented in Table 2 (17).

The various parameters, which should be taken into consideration when choosing colored glass/plastic for its photo protective properties, are discussed below.

Absorption Profile
The absorption spectrum of the glass depends on the added colorant.

Generally, it is necessary to cover the UV–VIS wavelength region of radiation, as much as possible, for the protection of photolabile substances.

Layer Thickness
According to the Lambert-Beer's law, the absorption increases with the layer thickness. The higher the absorption, the better will be the photo protective properties of the container. Depending on the intended purpose the layer thickness of the glass container is limited to a certain extent. This fact is also considered in the pharmacopoeial tests for the transmission of colored glass containers, where higher transmission values are acceptable (3,4).

Table 2

Colorant		
Metal additive	Ionic charge	Color
Copper	Cu^{2+}	weakly blue
Chromium	Cr^{3+}	green
	Cr^{3+}	yellow
Manganese	Mn^{3+}	violet
	Fe^{3+}	yellow to brown
Iron	Fe^{2+}	blue to green
Cobalt	Co^{2+}	intensely blue, in borax glasses pink
	Co^{3+}	green grey to brown
Nickel	Ni^{2+}	yellow green or blue to violet, depending on the glass matrix
Vanadium	V^{3+}	in silica glass green, in borax glass brown
Titanium	Ti^{3+}	violet (under reducing conditions)
Neodynium	Nd^{3+}	red to violet
Praeseodynium	Pr^{3+}	weakly green

Concentration of the Colorant
By increasing the concentration of the colorant, the UV–VIS photon permeability of the container can be decreased. This parameter can be of importance especially for very photosensitive drug substances and drug preparations, such as injection solutions in ampoules for which the wall thickness may be limited.

PLASTICS

Most neat plastic materials used for packaging of pharmaceutical products absorb UV–VIS radiation below 280 nm. Therefore, photostabilization by plastic packaging is only possible by the addition of colorants, UV absorbers, or pigments to the plastic mass.

Primary Packaging

Immediate Containers for Liquid Dosage Forms
Due to the risk of breakage, glass containers are more and more being replaced by plastic containers. It has been shown that white pigmented, opaque, low-density polyethylene, containers can provide complete photo protection to certain photosensitive products, such as dexamethasone- and predni-solone phosphate-eye drops (6).

Besides opacifying pigments, colorants and UV absorbers may be suitable photo protective additives for plastics, too. This fact was demonstrated using solutions of phenylalkylamines and metoclopramide (8,18). The inclusion of UV absorbers into the plastics, used to manufacture infusion bags and tubes, is one possible means of protecting infusion solutions sensitive to UV–VIS radiation. Using this approach a safe dosing period, over a period of three hours, can be guaranteed, even for the highly photolabile molsidomine-infusion solutions (19).

Immediate Containers for Solid Dosage Forms

Bottles

Plastic bottles are a common form of primary packaging for tablets in many regions because they are lower in weight and less fragile. They may be used as the pure resins or multilayered to reduce their moisture and gas permeability.

Blisters

Polyvinylchloride (PVC) is the favored plastic for the production of blisters. The layer thickness of PVC film used, generally, is about 250 μm before deep drawing. However, due to its higher environmental compatibility, polypropylene (PP) is rapidly becoming the blister material of choice. There are two disadvantages of polypropylene: first it is less tolerant of temperature variations and second it suffers from being more sensitive to molding time variations than PVC films. To control these problems, thicker films of PP—of about 300 μm—are generally used.

This difference in sheet thickness (between these two films) is visible in the blisters produced. Regardless of the thickness of the starting material used, considerable variations in the resulting layer thickness of the deep drawn blisters will be produced. The layer thickness of blister packages from commercial nifedipine tablets was found to vary from 75 to 160 μm for PVC blister films, and from 90–220 μm for PP blister films (20). One should also be aware of the fact that the blister-forming process does not produce blister films of uniform wall thickness. The rim regions of the blister are thinner than the center. The layer thickness in this region can be reduced down to as low as 10 μm, as illustrated in Figure 5.

The influence of variations in the film layer thickness on photo protection was investigated using uncoated nifedipine tablets. Samples of the tablets were completely covered with yellow transparent blister films of layer thicknesses ranging from 30–300 μm. The photostabilizing effect of the yellow blister film was found to increase with film layer thickness. After six hours irradiation the loss of drug content was 24% for the 30 μm films and only 6% for the 300 μm films, (Fig. 6) a substantial improvement in the photo protection of the product (21).

Another investigation of nifedipine tablet commercials demonstrated the photo protective effect of colored blisters (11). The photodegradation of coated nifedipine tablets was decreased from a maximum of slightly more than 5% to a maximum of 2% by use of a colored blister pack. The photo testing used was equivalent to a four- to six-week daylight exposure behind window glass (Fig. 7).

The effect of colored blisters can be even more dramatically seen when packaged, commercial, nifedipine capsules are studied. When stored outside of their protective blister package, the losses of the active content were substantial for both the soft and hard gelatine capsules. More than 20% of the drug substance

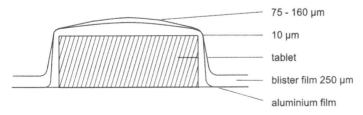

75 - 160 μm

10 μm

tablet

blister film 250 μm

aluminium film

Figure 5 Schematic view of a blister.

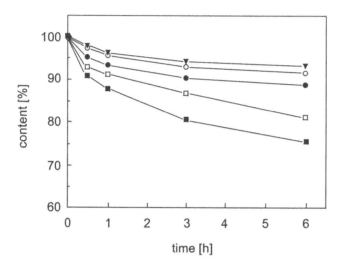

Figure 6 Influence of the film thickness of yellow transparent blister films on the photostability of uncoated nifedipine tablets 20 mg (Suntest CPS+, 720 W/m², UV filter). (▼) 300 μm, (○) 200 μm, (●) 150 μm, (□) 70 μm, (■) 30 μm. *Source*: From Ref. 21.

was photo decomposed. Stored inside their colored blisters losses of drug content were generally less than 5% (11).

Colored blister films are, therefore, used for other highly photosensitive drugs, such as nifedipine, molsidomine, corticosteroids, tetracycline antibiotics, and a number of vitamin preparations, among others, as previously noted. Accelerated studies of uncoated molsidomine tablets produced drug losses ranging from 50% to 60%. Similar studies of the tablets packaged in colored blisters showed a reduction of 30% to 40% in the degree of photodegradation (22). A significant improvement in the photo protection of flordipine tablets, another dihydropyridine derivative, was also found after they were packaged in amber blisters (23).

The color used in the blister plays a major role in the photo protective effect noted. This was demonstrated using uncoated nifedipine tablets. Nifedipine tablet commercials are generally packaged in red blister packs. In addition to this color, photo protective effects were also found for orange-, yellow-, and green-colored blisters. Green blisters showed the highest photo stabilizing effect for nifedipine tablets.

Blue and colorless blisters had little photo protective effect on nifedipine tablets. Therefore, it is evident that the use of a blister of the proper color can be an effective photo protecting measure, but only if its absorption spectrum corresponds to that of the drug substance/product (Fig. 8) (20).

Secondary Packaging

Blisters

Due to their high degree of photosensitivity, molsidomine injection solutions are filled into brown glass ampoules. Additionally, these ampoules are also packaged in green blisters, because photostability studies have revealed a need for this additional secondary packaging. Unprotected molsidomine solutions

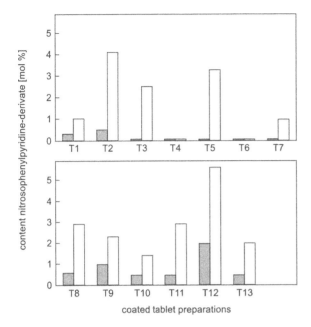

Figure 7 Content of the nifedipine degradation product (nitrosophenylpyridine) in 13 coated nifedipine tablet commercial products after irradiation inside or outside of the blister packaging (Novasoltest™ 72 hours). (■) With blister, (□) without blister. *Source*: From Ref. 11.

were found to still be highly photosensitive when packaged in brown glass ampoules alone.

When the neat solution was exposed to UV–VIS radiation, the stability limit of 90% active was reached in only five minutes. After two hours exposure no intact drug substance was detectable. Packaging the solution into brown glass ampoules resulted in a higher degree of photo protection; however, this package was not totally effective for this product; about 25% of the drug substance was degraded after three hours exposure. When, however, the brown ampoules were also packaged in green blister packs, no photodegradation was observed, even after three hours of irradiation (21) under similar conditions, as shown in Figure 9.

Mantles

Colored

Due to the need to check for particles, fibers, and possible discoloration in drug solutions transparent and colorless packaging materials are desirable. Hence, uncolored glass and plastic, bottles and bags, are commonly used as primary packaging for parenteral. To attain photo protection without losing VIS controllability, transparent containers or covers are necessary.

One major potential problem with the use of additives in primary containers is the possibility of their migration into the drug substance/solution stored therein. To exclude this potential risk, the use of colored plastic sleeves, placed over the primary containers, is recommended. Using this approach, the primary packaging remains unchanged and the photo protective additives do not come in direct

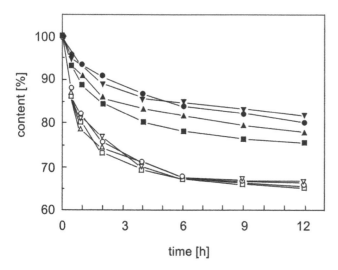

Figure 8 Influence of the blister coloration on the photodegradation of uncoated nifedipine tablets 20 mg (Suntest CPS+, 720 W/m², UV filter). (▼) green blister, (▽) blue blister, (●) yellow blister, (○) colorless blister absorbing UV-light, (▲) red blister, (△) colorless blister, (■) orange blister, (□) without blister. *Source*: From Ref. 20.

contact with the product. Additionally, colored or pigmented sleeves are reusable, once qualified as to their stability. The use of UV absorber containing covers or primary containers (bags), as photo protective devices, have been the subject of several patent applications in Great Britain (24), Japan (25–27), and Europe (28).

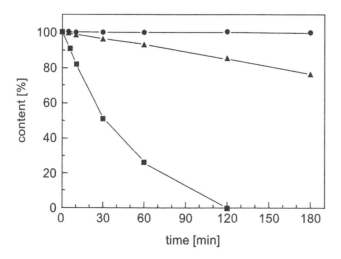

Figure 9 Influence of the packaging on the photostability of molsidomine injection solutions (c=2 mg/mL) (Suntest CPS+, window glass filter, 415 W/m²). (■) neat solution, (▲) packaged in brown glass ampoules, (●) packaged in brown glass ampoules and green blisters. *Source*: From Ref. 21.

The photo protective effectiveness of brown, photon protecting, plastic sleeves, having a high-colorant density, was demonstrated using vitamin K-1 infusion solutions. In this instance, a white-pigmented plastic sleeve did not provide the drug preparation with sufficient photon protection, even though it too had a high-colorant density. Black plastic covers, with a low-colorant density, showed a significant photo protective effect in this study. However, this effect was considerably lower than that provided by the brown plastic covers. Therefore, it can be concluded from the previously cited data that conventional cover sleeves of thin walled polyolefins, of the proper color are very suitable for the photostabilization of infusion liquids (Fig. 10) (29).

Brown glass bottles, covered by yellow plastic mantles, have been used commercially as primary packages for extremely photosensitive solutions, for example, nifedipine infusion solution (Adalata™ pro infusione). Using this photo protective method allows the manufacturer to guarantee the complete efficacy of the nifedipine infusion solution for one hour under normal daylight and six hours under room lighting conditions (30). An investigation of this preparation confirmed the photo protective effect of this packaging. After eight hours, the drug loss amounted to only 5% in a photo-stability testing cabinet and only 3% in daylight (11) as shown in Figure 11.

UV absorber containing covers have also provided photo protection for molsidomine infusions during dosing (31).

Furthermore, the photostabilizing effect of UV absorber containing primary and secondary packaging has been shown to be very photo protective for several UV photosensitive drug infusions (32,33). A UVA or UVB absorber, as well as a mixture of both, was incorporated in polyethylene by melt extrusion, yielding highly transparent plastic films. These films were then used to produce infusion bags or heat-shrunk onto glass bottles. The UV absorber type, concentration, and film thickness all played an important role in the photo protective effect of the covers. No differences in the photo protective properties were observed, whether the film was used as a bag (primary package) or heat-shrunk film on a bottle (secondary package).

It was demonstrated in these studies that a transmission reduction to about 20% was attainable using 45 μm thick films containing 1% UV absorber (Fig. 12). However, depending on the individual activation spectra of the drug substance/

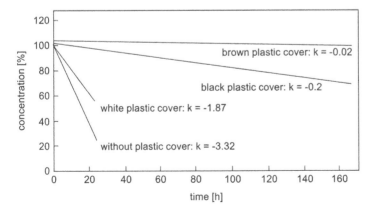

Figure 10 Photodegradation rates of vitamin K-1 infusion solutions with different colored plastic covers. *Source*: From Ref. 29.

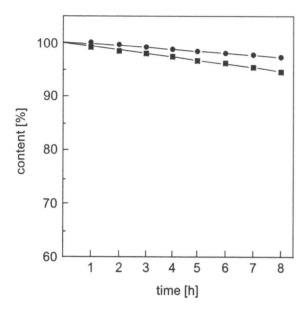

Figure 11 Residual drug amount of nifedipine after irradiation of an infusion solution (0.01%) in its packaging (brown glass with a yellow colored plastic mantle). (●) Daylight, (■) Novasoltest™. *Source*: From Ref. 11.

product and the absorption spectra of the additive, differences in the photo protective properties of the covers were observed.

The 1% UV absorber containing films showed a low residual transmission of less than 4% in the wavelength region 300–375 nm. For the highly photolabile prednisolone infusion, total photostabilization could be achieved using this transparent cover (Fig. 13). Similar results were obtained for cefodizim infusion.

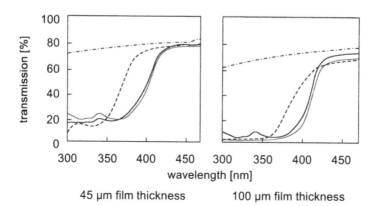

45 µm film thickness   100 µm film thickness

Figure 12 Transmission of transparent films containing 1% UV absorber. 45 µm film thickness: ─·─·─ LDPE film, • • • • 1% UV A absorber; 100 µm film thickness: ▬ ▬ ▬ 1% UV B absorber, ▬▬▬ 1% mixture (1:1). *Source*: From Ref. 33.

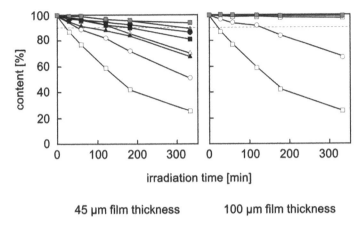

Figure 13 Influence of UV absorber type, concentration, and film thickness on the photo-stabilizing effect of transparent covers on prednisolone phosphate infusions (Suntest CPS+, 415 W/m², window glass filter). 45 μm film thickness: (⊟) without cover, (△) 0,5% UV B absorber, (▲) 0,5% UV A absorber, (★) 0,5% mixture (1:1); 100 μm film thickness (○) 0,25% UV B absorber, (⊟) 1,0% UV B absorber, (⊟) 1,0% UV A absorber, (■) 1,0% mixture (1:1). *Source*: From Ref. 32.

It was necessary to use a film thickness of 200 μm for the total photo protection of multivitamin infusions containing ascorbic acid, thiamine hydrochloride, nicotin-amid, and pyridoxine hydrochloride. For amiodarone hydrochloride and a special quinolone infusion, which showed sensitivity to all wavelengths in the UV region up to 400 nm, the photo protection was considerably lower. However, a reduction of drug losses from 59% and 75% to 5% and 7% was still attainable using this cover thickness. It was concluded, that the limit of application for these particular UV protective covers is photolability of drug solutions with activation spectra above 385 nm.

Opacified
It has been demonstrated, that plastic films containing ultra-fine TiO_2 particles are a good means of photo protection. Due to the use of ultra-fine particles, these films remain transparent to a degree, although pigmented (34). The effectiveness of these films was demonstrated using promethazine HCl, carbidopa, and nifedipine. A variation in this procedure is to include the TiO_2 film as an intermediate layer of a laminate, thus protecting the product from direct contact with the pigment. This procedure is the subject of a Japanese patent (25).

OTHER PACKAGING

Foils
Aluminum is an impermeable UV–VIS photon barrier. Therefore, packaging of this metal, metal tubes, foil or coated on plastic films can provide 100% light protection for the drug preparations stored inside.

The photoprotective effect of the foil was demonstrated in studies using the highly photosenstive riboflavine solution. In spite of intense photo exposure, no pho-todegradation of the aluminum-wrapped riboflavine solution was detectable (35).

Aluminum tubes are often used as the immediate package for dermal preparations. One important point to remember is that after the application of creams, ointments, or gels to the skin, all of the photo protection provided by the packaging is lost and photodegradation can occur. Topically applied drug substances with proven photoinstability, e.g., corticosteroids (36), retinoic acid (37), dithranol (38), and anti-mycotics (natamycine and nystatine) (39) fall into this category.

Tablets can also be photostabilized using aluminum blister packs. This pho-tostabilization method is used by some nifedipine tablet manufacturers. Nifedipine tablets packaged in aluminum foil unit dose blisters do not require any additional photo protection (e.g., coating), providing a significant saving in development and manufacturing costs as well as stability testing (40).

One additional use for aluminum foil is as mantles (sleeves) to cover the pri-mary packaging of photosensitive products. For example, mesalazine rectal solutions have been photostabilized using this approach. Without this protective measure the product, packaged and stored in transparent plastic containers, would develop a strong brown discoloration, when exposed to normal indoor daylight (41).

Cartons

Folding cartons are a frequently used form of secondary packaging. They represent a strong barrier against all radiation, UV and VIS. During storage in a closed carton, no photodegradation is likely to occur.

Photosensitive eye drop products frequently depend on secondary pack-aging (cartons) if the primary containers themselves provide little or no photo protection. Eye drop solutions packaged in virgin, LDPE immediate containers and cardboard carton secondary packaging have been found to be sufficiently photo protected. Chloramphenicol eye drops were exposed in both their immedi-ate and secondary packaging for six weeks to sunlight and diffuse daylight; only those samples stored in cartons were found to be stable. No differences were observed between samples stored in immediate containers of neat or white opaci-fied LDPE bottles, if both of these container types were stored in their secondary packaging (Fig. 14) (6).

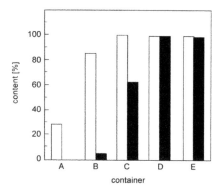

Figure 14 Photodegradation of chloramphenicol eye drops FNA depending on the container. (**A**) White glass ampoule, (**B**) LD-polyethylene container, (**C**) brown glass container, (**D**) LD-polyethylene container, white, opaque, (**E**) LD-polyethylene container in the carton. (□) 6 weeks diffuse daylight, (■) 6 weeks sunlight. *Source*: From Ref. 6.

However good the photo protection achieved by the previously cited means, there is no guarantee of the safe and efficacious dosing by the patient/user, because an integral part of this packaging method is patient/user compliance, i.e., reading and following the labels and storage instructions. In spite of proper storage recommendations and warnings, the patient/user may be seduced into storing the drug preparation outside its protective packaging.

Additionally, older, now generic products may be packaged in both forms, i.e., protected by the immediate package or only by the secondary package. Such duplicity puts an unusual burden on the user/patient to read the labeling every time they refill their prescription. Because random substitution is allowed for all officially listed generics, many dispensers and users/patients may be very unaware of this problem.

CONCLUSION

A number of different methods for the protection of photosensitive products have been presented. No one method is suitable for all products. The method used will depend on many factors.

It is proposed that the most prudent method for preserving the safety and efficacy of drug products is to assure the immediate packaging which provides this protection wherever possible. Even when this goal is achieved, there is no guarantee that the product after application (e.g., creams, ointments, and gels) will remain photostable.

A noninvasive method, applicable to flexible as well as rigid containers is presented. This method, utilizing UV–VIS radiation absorbing plastic sleeves, allows for the inspection of the product before usage or after dilution for particulate matter, while also preserving the efficacy and safety of the product.

REFERENCES

1. ICH Guideline Q1B. Photostability Testing of New Active Substances and Medicinal Products. In: Feiden K, ed. Arzneimittelprüfrichtlinien. Stuttgart: Wissenschaftliche Verlagsgesellschaft, 1998:(2.227)1–9.
2. Ampoules de petite capacité. In: Pharmacopée Française, Edition 8. Paris, 1965:390–1392.
3. Light transmission for colored light-protecting glass containers. In: European Pharmacopoeia, 4th Edition. 2002.
4. Containers: Light Transmission. In: United States Pharmacopoeia (USP), USP 24, 2000.
5. Nürnberg E. Glasbehältnisse zur pharmazeutischen Verwendung. In: Kommentar zur PH.EUR. 1997. Stuttgart / Eschborn: Wissenschaftliche Verlagsgesellschaft / Govi Verlag-Pharmazeutischer Verlag GmbH, 1997:(3.2.1)2.
6. Marschall M. Probleme der Arzneimittelsicherheit bei wässrigen Ophthalmika – Untersuchungen zur stabilitätsspezifischen Analytik, chemischen und physikalischen Stabilität. Ph.D. dissertation, Ludwig-Maximilians-University, Munich, GR, 1988.
7. Surowiecki J, Krówczynski L. Studies on protection of medicinal substances by amber glass against catalytic effects of light. Acta Polon Pharm 1972; 29(4):405–414.
8. Grünert R, Wollmann H. Lichtdurchlässigkeit und Lichtschutzwirkung von Behältnissen für Phenylalkylaminlösungen. Pharmazie 1983; 38(1):30–33.
9. Schönitz S. Stablitätsprobleme bei flüssigen Drogenzubereitungen—Untersuchungen zur stabilitätsspezifischen Analytik, chemischen und physikalischen Stabilität. Ph.D. dissertation, Ludwig-Maximilians-University, Munich, GR, 1990.

10. Schulz M, Dinnendahl V, Braun R. Ruhen der Zulassung von Corvaton™-Tropfen—Hintergrundinformationen. Pharm Ztg 1989; 134(50):6.
11. Thoma K, Kerker R. Photoinstabilität von Arzneimitteln. 3. Mitteilung: Photoinstabilität und Stabilisierung von Nifedipin in Arzneiformen. Pharm Ind 1992; 54(4):359–365.
12. Arny HV, Taub A, Steinberg A. Deterioration of certain medicaments under the influence of light. J Am Pharm Assoc 1931; 20;(10):672–679.
13. Arny HV, Taub A, Steinberg A. Deterioration of certain medicaments under the influence of light. J Am Pharm Assoc 1931; 20(10):1014–1023.
14. Arny HV, Taub A, Steinberg A. Deterioration of certain medicaments under the influence of light. J Am Pharm Assoc 1931; 20(10):1153–1158.
15. Arny HV, Taub A, Blythe RH. Determination of certain medicaments under the influence of light. J Am Pharm Assoc 1934; 23(7):672–679.
16. Swartz CJ, Lachman L, Urbanyi T, Cooper J. Color stability of tablet formulations. IV: protective effect of various colored glasses on the fading of tablets. J Pharm Sci 1961; 50(2):145–148.
17. Schott Glaslexikon, Schott Glas, Hattenbergstr. 10, 55122 Mainz, Germany.
18. Kottke D, Springer M, Pohloudek-Fabini R. Untersuchungen zur Stabilität von Metoclopramidlösungen in Glas- und Plastbehältern. Pharmazie 1978; 33(4):198–201.
19. De Muynck C, Vandenbossche GMR, Remon JP, Colardyn F. Rapid breakdown during exposure to daylight of molsidomine when administered by continuous intravenous infusion. Am J Cardiol 1995; 70:836–837.
20. Thoma K, Aman W. The influence of blister colouration on the photostability of nifedipine tablets. 15th Pharmaceutical Technology Conference, Oxford, UK, March 19–21, 1996.
21. Aman W. Photostabilisierung von Arzneimitteln unter besonderer Berücksichtigung von Tabletten. Ph.D. dissertation, Ludwig-Maximilians-University, Munich, GR 1998.
22. Thoma K, Kerker R. Photoinstabilität von Arzneimitteln. 6. Mitteilung: Untersuchungen zur Photoinstabilität von Molsidomin. Pharm Ind 1992; 54(7):630–638.
23. Narupkar AN, Sheen PC, Bernstein DF, Augustine MA. Studies on the light stability of flordipine tablets in amber blister packaging. Drug Dev Ind Pharm 1986; 12(8–9):1241–1247.
24. Tindale N. Polyesters. Patent Application 1989, GB 2 228 940 A.
25. Kawabata T. Transparent ultraviolet absorbing protective Film. Patent Application 1994; JP 6297630.
26. Seki T, Iwao T, Fujiwara F, Ueda Y, Inoue T. Packaging bag for liquid medicine. Patent Application 1996; JP 8301363.
27. Ikeda N, Hayashi H, Kurihara T. Packaging material insulating ultraviolet ray. Patent Application 1997; JP 9142539.
28. Baker BO. An ultraviolet absorbing and optically transparent packaging material. Patent Application 1995; EP 0 679 506 A1.
29. Martinelli E. Kunststoffumbeutel als Lichtschutz für Infusionen. Krankenhauspharmazie 1995; 16(7):286–289.
30. Directions for use of Adalat™ pro infusione (Bayer AG, Leverkusen, GR).
31. Vandenbossche GMR, De Muynck C, Colardyn F, Remon JP. Light stability of molsidomine in infusion fluids. J Pharm Pharmacol 1993; 45:486–488.
32. Thoma K, Landerer S. Photostabilization of drug solutions by UV protective plastic foils. 3rd International Meeting on the Photostability of Drugs and Drug Products, Washington, DC (USA), July 10–14, 1999.
33. Landerer S. Charakterisierung der Lichtempfindlichkeit und Photostabilisierung von Arzneistoffen und Infusionen Ph.D. dissertation, Ludwig-Maximilians-University, Munich, GR, 2000.
34. Túry G, Szabó GT, Vabrik R, Rusznák I, Nyitrai Z, Víg A. Deceleration of light-induced changes of selected pharmacons by means of light screening films. J Photochem Photobiol A: Chem 1997; 111(1–3):171–179.
35. Waltersson JO, Lundgren P. Studies on light protection of packaging materials—a kinetic approach. Acta Pharm Suec 1984; 21:125–134.

36. Thoma K, Kerker R. Photoinstabilität von arzneimitteln. 5. Mitteilung: untersuchungen zur photostabilisierung von Glukokortikoiden. Pharm Ind 1992; 54(6):551–554.
37. Schaefer H, Zesch A. Penetration of vitamin A acid into human skin. Acta Derm Venerol 1975; 74(suppl):50–55.
38. Thoma K, Holzmann C. Photostability of dithranol. Eur J Pharm Biopharm 1998; 46(2):201–208.
39. Thoma K, Kübler N. Untersuchung der Photostabilität von antimykotika. 2. Mitteilung: photostabilität von polyenantibiotika. Pharmazie 1997; 52(4):294–302.
40. Produktinformation Corinfar™ uno. Asta Medica AWD, Frankfurt.
41. Produktinformation Pentasa™. Ferring Arzneimittel GmbH, Kiel.

16 Photostabilization of Solid and Semisolid Dosage Forms

Karl Thoma[†]
Institut für Pharmazie—Zentrum für Pharmaforschung, Lehrstuhl für Pharmazeutische Technologie, Ludwig-Maximilians-Universität, Munich, Germany

Heiko Spilgies
GlaxoSmithKline, Ware, Herts, U.K.

PHOTOSTABILIZATION AS PART OF THE DRUG FORMULATION

The primary and secondary packaging of drugs can be used as a very effective means of ultraviolet–visible (UV–VIS) radiation protection. However, lacking awareness of a photostability problem during the dosing of drugs may result in the removal of the secondary protective packaging and hence, unintentional destruction of the photostabilizing agent (1). This is particularly important because a survey revealed that even health professionals do not always consider the importance of photoprotection (2). In the case of epinephrin solutions, a loss of up to 55% of the active occurs by the time the drugs were administered (2). To protect against such occurrences, it is therefore advisable to eliminate secondary packaging during photostability testing and consider photoprotection as a necessary part of the drug formulation.

Differences Between the Photostability of Solids and Liquids

The majority of drug photostability studies focus on the photodegradation of drugs in solution rather than the solid state. In both cases, photoreactions are based on the absorption of energy in the form of radiation.

In solutions, absorption can be described by the Lambert–Beer Law and depend on the coefficient of absorption, solution thickness and concentration of the solution

$$A = \varepsilon \times c \times d$$

where, A = sample absorbance, ε = absorption coefficient, c = concentration, and d = sample thickness.

Influence of Concentration

Due to absorption of incident radiation in upper layers, the drug itself may have a photostabilizing effect. Photodegradation often leads to the production of colored products which themselves absorb and further shield the sample. Consequentially, degradation rates decrease with increasing concentrations as shown by the example of nifedipine solutions (Fig. 1) (3).

[†]Deceased.

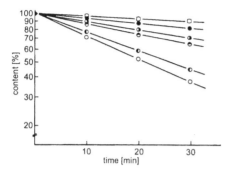

Figure 1 Influence of the initial concentration of active in solution on the photodegradation of nifedipine (Spectrotest, filter 700 to 417 nm, depth of solution 4.8 cm). (□) 300 mg/100 mL, (●) 120 mg/100 mL, (◑) 60 mg/100 mL, (◓) 30 mg/100 mL, (◖) 6 mg/100 mL, (○) 3 mg/100 mL. *Source*: From Ref. 3.

A similar effect was observed for the highly photosensitive nifedipine and molsidomine tablets where the loss of content decreased with concentration (Fig. 2) (Thoma K, Aman W. In preparation) (4).

Influence of Sample Geometry

Compared to solutions, the influence of UV and VIS radiation on solids is much more complicated and depends on a variety of different factors. The irradiation effect on solids is a function of the geometry of the sample, in particular of the particle size. Solid particles form an irregular surface, which produces a large amount of reflection and diffraction of the incident radiation. Furthermore, the effect of incident radiation is restricted as to its penetration depth, which is also a function of wavelength.

This protective effect is negligible in solutions, where the substance being irradiated is exposed with its maximum surface area and the incident radiation penetrates to a greater depth due to the dilution of the drug. Even at high concentrations where it is possible for all of the incident radiation to be absorbed in the top layer, undegraded drug from deeper layers will always diffuse into the top most layers, hence, exposing more undegraded compounds.

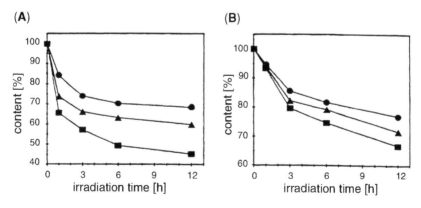

Figure 2 Influence of the total drug content of similarly sized tablets on the photodegradation (Suntest CPS+). (**A**) Nifedipine tablets: (720 W/m², UV special glass filter), (●) 20 mg, (▲) 10 mg, (■) 4 mg. (**B**) Molsidomine tablets: (415 W/m², window glass filter), (●) 20 mg, (▲) 8 mg, (■) 4 mg. [Thoma K, Aman W. In preparation].

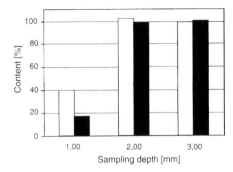

Figure 3 Influence of and sampling depth on the noted percentage photodegradation of nifedipine and molsidomine tablets. (□) Nifedipine tablets (Suntest CPS+, 720 W/m², UV special filter, 12 hours). (■) Molsidomine tablets (Suntest CPS+, 415 W/m², UV special filter, 48 hours). *Source*: From Ref. 4.

In contrast, penetration into solids is normally restricted to a depth much less than 1 mm. Only the top most layers of nifedipine and molsidomine tablets showed photodegradation (Thoma K, Aman W. In preparation) (4). This can easily be seen with nifedipine tablets where photodegradation results in coloration. After 12 hours, in samples stored in a photostability testing cabinet, no further degradation could be observed, as shown in Figure 3. Microscopic examination of these samples showed this coloration was restricted to a depth of less than 400 μm, as shown in the microphotograph of Figure 4.

A deviation from this rule is found if degradation products of the drug are volatile, which leads to a change of the surface during the irradiation process. An example of this phenomenon is found in studies with molsidomine where the volatiles, carbon dioxide, nitrogen, ethanol, and morpholine are formed (5). This change in topology can easily be detected visually because the originally homogeneous surface shows yellow spots after irradiation (Fig. 5).

Influence of the Sample Surface and Volume
Because the penetration depth of the incident radiation into a drug dosage form is constant for a specific compound and formulation, the degree of photodegradation of

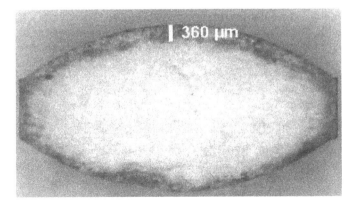

Figure 4 Cross-section of an irradiated 20 mg nifedipine tablet showing the visual depth of photodegradation (bi-convex, 6 mm diameter) (Suntest CPS+, 12 hours, 720 W/m², UV special filter). *Source*: From Ref. 4.

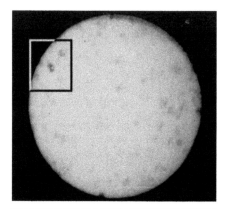

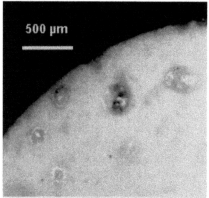

Figure 5 Microphotograph of the photodegraded surface of irradiated 4 mg molsidomine
tablets (bi-plane, d = 8 mm, Suntest CPS+,
12 hours, 415 W/m², window glass filter).
Source: From Ref. 4.

the solids depends on the exposed surface area. Hence, increase in the percentage
of photodegradation will be noted with the use of smaller particle sizes of the active
and smaller tablet sizes.

Sample volume is an important consideration when dealing with solids. Because
photodegradation of solids is a surface phenomenon, any additional layers of the substance simply act to dilute the test results as was illustrated in Figure 3. Conversely, if
one wishes to achieve the highest degree of photodegradation, the thinnest sample of
the smallest particle size should be used. A good graphical illustration of this volume
effect can be found in the article by Matsuda (6) and is reproduced in Figure 6.

The impact of the surface area can also be shown by the fact that once a
formulation is pressed into tablets, little difference between the particle sizes of
the drug compound is evident, as then the surface area of the tablet becomes more
important. Figure 7 shows the comparison of two different sizes of nifedipine
active and tablets produced using these materials. This comparison also reveals a
decrease in the photodegradation rate, which can be attributed to the decrease in
the drug surface exposed to incident radiation. Similar results were obtained with
molsidomine (Thoma K, Aman W. In preparation) (4).

Influence of the Drug Physical Form

Furthermore, the photostability of solids also depends on the physical state of the
drug, amorphous or crystalline. It can be assumed that the thermodynamically less

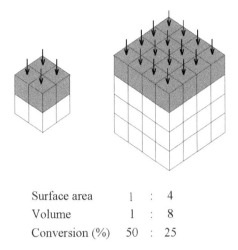

Surface area	1	:	4
Volume	1	:	8
Conversion (%)	50	:	25

Figure 6 Influence of sample volume on the percent on degradation. *Source*: From Ref. 5.

stable amorphous form is more susceptible to photodegradation than a crystalline form as shown by the examples of ergocalciferole (7) and the beta-lactamase inhibitor BRL42715 (Thoma K, Spilgies H. In preparation) (8). In this context, it should be considered that methods used to generate amorphous drug forms, e.g. lyophilization, might result in a considerable increase in surface area, thus enhancing the problem of photodegradation.

Additionally, different chemical modifications e.g., isomer, salt, inclusion compound, etc., vary in their photosensitivity. It is a common practice in modern drug development to prepare several different modifications of a drug and identify that modification with the most suitable properties, mainly relating to bioavailability. However, the photosensitivity of these different forms may vary markedly, even to the degree of forming different photodegradation products (9).

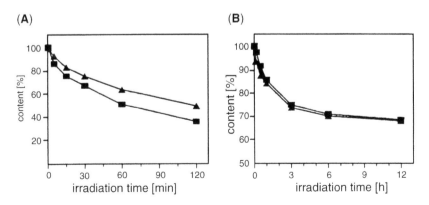

Figure 7 Influence of the particle size on the photostability of nifedipine active and tablets (Suntest CPS+, 720 W/m^2, UV special glass filter). (**A**) Drug compound (▲) 220 μm, (■) 25 μm. (**B**) Tablets 20 mg (bi-plane, thickness 8 mm) (▲) 220 μm, (■) 25 μm. *Source*: From Ref. 4.

In contrast to solutions, solid drugs have a fixed conformation resulting in topochemical reactions. The majority of photoreactions in the solid state, described in the literature, deal with lattice-controlled examples and photodimerizations. A precondition for these reactions is the parallel position of the double bond of two adjacent molecules in the crystal lattice as shown by the example of the trimorphic, *trans*-cinnamic acid. Irradiation of the α- and the β-modifications causes the formation of α-truxillic acid and β-truxinic acid, respectively, whereas the γ-modification is photostable due to the distance of the double bonds fixed by the lattice (Fig. 8) (10).

It should be considered that even pseudo-polymorphs might react differently upon irradiation. The example in Figure 9 compares the HPLC plots of irradiated hydrate and anhydrous form of a steroid. Upon irradiation, the hydrate generates photoproducts whereas its anhydrous form does not (11).

Influence of Humidity
The photostability of solids may also be affected by the presence of moisture. As the photodegradation rates in solution are largely dependent on the concentration, it is very likely that the water sorption effects photodegradation in the solid state. However, this factor has so far been neglected in literature. The influence of moisture is difficult to investigate due to the heat generated by the photon sources used. Rising temperatures in hygrostats result in changes in the relative humidity (Fig. 10).

The effect of even a very small amount of moisture on the photodegradation rate of BRL42715 has to be considered under this aspect (Fig. 11) (Thoma K, Spilgies H. In preparation) (8).

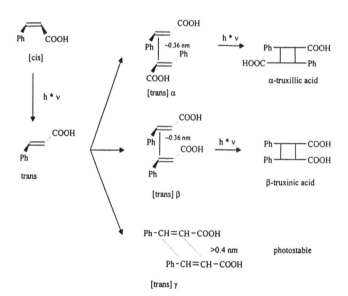

Figure 8 Photochemistry of cinnamic acids. influence of the isomeric form on photoreactivity. *Source*: From Ref. 10.

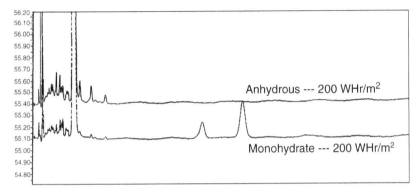

Figure 9 Effect of hydration on the photostability of a steroid. *Source*: From Ref. 11.

Apart from a direct impact of humidity on the photostability of drugs, humidity may alter the physical form of drugs (hydration/dehydration), which may then result in different photostability properties as described above and demonstrated in Figure 9.

PHOTOSTABILIZATION OF TABLETS

The influence of the drug concentration, surface area, and sample volume has already been described. As a result, the photostability of a tablet formulation can positively be enhanced by developing small tablets with a high drug concentration (Thoma K, Aman W. In preparation) (4).

Ultraviolet Absorbers

In order to protect photosensitive drugs, the amount of undesirable radiation penetrating the drug molecules must be reduced. One way of protecting drugs is the photostabilization by spectral overlay, which has been successfully applied to drug solutions (12). According to this principle, excipients, with absorption spectra

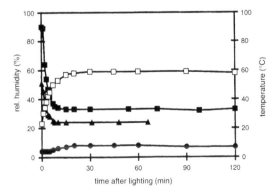

Figure 10 Change of the temperature and the relative humidity as a function of time in a hygrostat during xenon lamp irradiation (Suntest CPS+, 540 W/m², UV special glass, hygrostat covered by quartz glass). (■) hygrostat solution for 90% r.h., (▲) hygrostat solution for 45% r.h., (●) solid calcium chloride anhydrous for 0% r.h., (○) temperature in the hygrostat (right axis). *Source*: From Ref. 8.

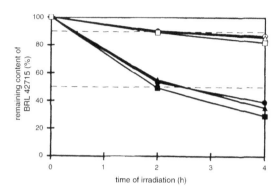

Figure 11 Influence of the form of solid and of moisture on the photostability of a beta-lactamase inhibitor BRL42715A (amorphous) and BRL42715B (sodium monohydrate). **(A)** (■) 90% r.h., **(B)** (□) 90% r.h., (▲) 45% r.h., (△) 45% r.h., (●) 0% r.h., (○) 0% r.h. *Source*: From Ref. 9.

similar to that of the photosensitive drug are added to the formulation, thereby reducing the amount of radiation interacting with the drug molecules (Fig. 12).

An example of where this principle was successfully applied to tablets is molsidomine (Thoma K, Aman W. In preparation) (4). The absorption spectra of the UV-B absorber Eusolex® 6300 [3-(4-methylidene)-camphor], the UV-A absorber Eusolex 9020 [4-(t-butyl-4′-methoxydibenzoyl)-methane], and ferulic acid have spectra which overlap with that of molsidomine resulting in tablets that are less sensitive to incident radiation (Fig. 13). Of course, the permissibility of the excipients used for photostabilization must be guaranteed.

This example clearly illustrates the importance of spectral correlation. Daylight behind window glass only contains radiation of wavelengths above and about 310 nm. Due to its better overlay with the absorption spectrum of molsidomine in this region, Eusolex 9020 is a more effective photostabilizer for molsidomine compared to Eusolex 6300 and ferulic acid.

A combination of the two Eusolex types results in even better photostabilization. The mixture has a higher absorption in the region between 310 nm (lowest wavelength emitted by daylight if filtered by window glass) and 350 nm (absorption limit of molsidomine).

These two UV-absorbers are registered, often, for external use only. In pharmaceutical practice, food colorants can be used as photostabilizing excipients

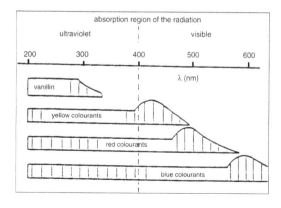

Figure 12 Principle of photostabilization through spectral overlap with absorbing excipients. *Source*: From Ref. 12.

(**A**)

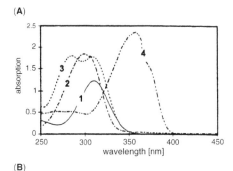

(**B**)

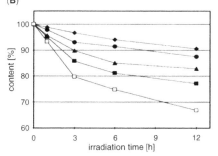

Figure 13   Photostabilization of molsidomine tablets by UV absorbers. (**A**) Comparison of the absorption spectra of UV absorbers and molsidomine ($c = 20\,\mu g/mL$). (*1*) Molsidomine, (*2*) Eusolex® 6300, (*3*) Ferulic acid, (*4*) Eusolex 9020. (**B**) Photostabilization of molsidomine tablets 4 mg (bi-plane, $d = 8\,mm$) by UV absorbers Suntest CPS+, 415 W/m², window glass filter). (◆) Eusolex 9020 + Eusolex 6300, (●) Eusolex 9020, (▲) ferulic acid, (■) Eusolex 6300, (□) without stabilizer. *Source*: From Ref. 4.

for oral dosage forms. Their colors are due to photonic energy absorption in the visible range. However, because they also absorb radiation at lower wavelengths, they may also be suitable for the photoprotection of compounds sensitive to UV radiation.

In the case of molsidomine, curcumin (E 100), riboflavin (E 101), and azorubin (E 122) proved to be as effective as Eusolex 9020. It is remarkable that molsidomine can even be stabilized by riboflavin which itself degrades in solution, acting as a photosensitizer and resulting in accelerated photodegradation (13). In solution, riboflavin transfers photonic energy onto drug molecules and forms radicals, which then react with the drug. These reactions are inhibited in the solid state (Thoma K, Aman W. In preparation) (4).

Food Colorants

As the molsidomine case shows, the photodegradation of drugs in the solid state can be enhanced by the use of colorants. Another example of this approach are ethinyl estradiol tablets which can be stabilized by the addition of erythrosine (FD&C Red No. 3) but degrade faster if they contain 1-p-sulfophenylazo-2-naphthol-6 sulfonate, di-sodium salt (FD&C Yellow No. 6) (14).

Opacifiers (Pigments)

Tablet Core

Besides UV absorbers and food colorants, use of opacifiers is another way of photostabilizing photosensitive tablets. The protecting effect of yellow, red and black iron oxide could be shown for tablets containing the antiviral drug sorivudine, as well as molsidomine and nifedipine tablets (15). Iron oxides were found to be more effective than titanium dioxide, especially if the drug is particularly sensitive to radiation of wavelengths between 400 and 420 nm (e.g., nifedipine), where titanium dioxide shows an "absorption gap" (Thoma K, Aman W. In preparation) (4).

The photoprotective effect of pigments depends on their particle size, which in turn affects their radiation absorption and scattering properties. For example, the optimum particle diameter for titanium dioxide is about 200 nm according to the manufacturer (16). A point that has to be considered in this context is that a loss of the photoprotective properties of the pigments may result from secondary agglomeration. Consequently, pigments are most effective if they are evenly distributed over the surface of the individual photosensitive drug particles in the tablet (Thoma K, Aman W. In preparation) (4).

Film Coatings

Instead of adding radiation-attenuating excipients to the matrix of a drug, photosensitive drug compounds can also be protected by film coatings with reflecting or absorbing properties. The most important characteristic of film coatings related to their protective effect is their absorption, which is affected by both the film thickness and the concentration of the absorbing/reflecting excipients. The photostabilizing effect of oxybenzone films on the coloration of sulfisomidine tablets is shown in Figure 14 (17).

Investigations, related to the photoprotection of nifedipine tablets using films containing titanium dioxide and/or tartrazine, revealed that the photoprotective effect of a film could be evaluated by using its concentration of colorant (C) and thickness (L) value. This value is the product of concentration of the colorant C and the film thickness L. Tablets coated with films having the same CL value had the same photodegradation rates. Degradation rates were found to be proportional to the CL value for every colorant system tested (18).

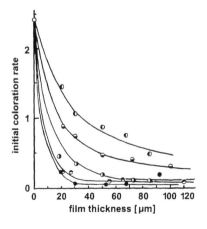

Figure 14 Effect of film thickness and concentration of oxybenzone on coloration rate of sulfisomidine tablets. (○) Control (one point), (◑) 2%, (●) 0.5%, (◐) 5%, (□) 1%, (●) 10%. *Source*: From Ref. 17.

These experiments also revealed that, according to the principle of spectral overlay for maximum effectiveness, the transmission properties of the film must be adjusted to the absorption spectrum of the photosensitive drug. The combination of tartrazine and titanium dioxide, in film coatings for nifedipine tablets, shows lower transmittance in the important wavelength range between 290 and 450 nm than the coatings containing only one of these excipients. Accordingly, film coatings containing this combination gave the best photoprotection for the tablets.

Another example, which demonstrates the importance of the spectral overlay principle, is the fact that film coatings containing titanium dioxide provide sufficient protection for molsidomine (Thoma K, Aman W. In preparation) (4), whereas nifedipine tablets coated with such films still show photodegradation (19).

The photoprotecting effect of opacifiers is based primarily upon their total absorption and/or reflection of incident radiation. Another approach to predict the photostabilizing effect of a film is therefore to determine its opacity.

In the paint industry, the degree of opacity of a film is determined as the ratio of the measured reflectance of the incident when the film is placed on a black substrate, to the reflectance obtained on a white substrate (20). Applying this method to film coatings containing different pigments/fillers revealed that yellow, red, and black iron oxides are the best opacifiers with contrast ratios greater than 98%. Titanium dioxide, because of its transmission properties in the UV, was less effective (90%). This agrees with the findings that films containing yellow iron oxides give better photoprotection for nifedipine tablets than those containing titanium dioxide (21).

Films containing the fillers—calcium carbonate, calcium sulphate, and talc were all relatively transparent. The opacity of films containing colored lake pigments depends on both, the parent dye and the dye concentration, the degree of opacity decreasing as blue > red > orange > yellow (22). The opacity of film coatings can be determined with detached films but also—more practically—without removing the film from the tablet (23).

Film coatings offer a unique form of photoprotection, as they can be made impervious to UV and VIS radiation. They are capable of protecting the most photosensitive products.

Photoinstability of Dyes

Even though dyes can be used for the stabilization of photosensitive drugs, it should be noted that they too could undergo photodegradation. This is important because the photostabilizing effect of the dye is lost and new photodegradants can be formed which then must be not only analyzed for but also qualified as to their toxicity, allergenicity, teratogenicity, carcinogenicity, etc.

Dyes are often used to differentiate between different tablet strengths so their use as photoprotective agents is merely the extension of an old practice. Photofading can result in confusion, leading to incorrect dosing of a drug. The choice of color and its concentration should be done carefully.

Several investigations on the photostability of certified dyes have been conducted. As with drugs, the photostability of the different colorants depends on test conditions employed (pH, temperature, and radiation intensity) (24) and the presence of excipients in the formulation (25–27). Colorants absorb both UV and VIS radiation. The UV radiation contributes significantly to fading. Hence, UV absorbers (28) and sunscreen agents (29–31) can be used to increase their colorfastness. It is interesting to note that lakes are similarly (32) or even less photostable (33) than the corresponding water-soluble dyes. Even inorganic yellow iron oxide was reported to show fading (21).

These results reveal that it is important to test the colorfastness of each drug formulation. The fading of tablets can be determined by three different techniques. The change of absorbance in solution, the change of reflectance of radiation, or the color change as determined by trimstimulus color measurements (34). When conducting such photostability tests, it is important to consider that fluorescent tubes used may cause more damage to the colorants than artificial daylight, due to the spikes in the emission spectra of such lamps in the visible region. An example of this problem is the studies done using indigo carmine (FD&C Blue No. 2), the only universally acceptable blue dye for pharmaceutical use (35).

Cyclodextrin Complexation

Cyclodextrins can form inclusion complexes with some drugs, modifying their physical and chemical properties. Because cyclodextrins are mainly used to increase the solubility of poorly soluble drugs, most investigations in this field focus on photostability of drugs in solution. Here, the positive effect of cyclodextrins on the photostability of colchicine (36), emetine and cephaeline (37) could be shown. However, these results show that the photoprotective effect depends on the particular cyclodextrin used. For some forms of cyclodextrin, an increase of the photodegradation rate can be obtained, as the example of molsidomine shows (13).

A limited amount of published information on the effect of cyclodextrins on the photostability in the solid state is available. Results similar to those reported for solutions, related to the choice of cyclodextrin type can be expected for the solid state. This was confirmed by Tomono et al.who showed that the remaining content of four different drugs increased or decreased after irradiation, depending on the selected cyclodextrin type (38).

PHOTOSTABILIZATION OF CAPSULES

Shell Additives

Despite the radiation-scattering effect of the gelatine shell, photonic energy can penetrate into the capsule core. Hence, capsules of photosensitive drugs need further photoprotection. An investigation of the photostability of commercial ubidecarenone products revealed that soft gelatine capsules were the most unstable when compared to tablets, hard capsules, and granules (39).

Similar to film coatings, the photostabilizing effect of the gelatine shell depends on its thickness and its concentration of photon absorbing excipients. This was demonstrated using indomethacin in gelatine capsules, stabilized by the incorporation of titanium dioxide. The coloration rate of the photosensitive drug was directly dependent on the average transmittance of the capsule shell, which was varied using different thicknesses and titanium dioxide concentrations (40).

For the photoprotection of drugs in soft gelatine capsules, the shell thickness must be sufficient and homogeneous. Lower shell thickness may occur especially in the sealed area. This thickness in homogeneity can lead to varying photodegradation rates depending on the irradiated area.

The photostability testing of commercial nifedipine in soft gelatine capsules revealed remarkable differences if the capsules were irradiated in parallel to the seal. One preparation (W 10) produced a 10-fold greater amount of the nitrosophenyl

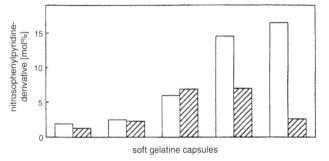

Figure 15 Content of nitrosophenylpyridine derivative in soft gelatine capsule contents after irradiation of different capsule shell areas. (□) In the seal area, (▨) out of the seal area. *Source:* From Ref. 41.

degradation product because the shell thickness dropped from 260 to 85 μm around the seal (Fig. 15) (41).

In general, photoprotecting photon absorbers can be incorporated into the capsule shell as well as into the capsule filling. Stabilization experiments of nifedipine in soft gelatine capsules using the food colorant acid yellow (E 105) showed that similar results were obtained using either procedure. As illustrated in Figure 16, the best results are obtained by stabilization of both the capsule shell and the filling (12).

Irradiation of gelatine capsules may not only result in photodegradation of the drug but may also affect the bioavailability of the drug. Irradiated gelatine films showed a decrease of their water vapor transmission. Even though this effect did not occur with indomethacin capsules, long-term irradiation resulted in increased dissolution rates (42).

Microencapsulation

Microencapsulation of photosensitive drugs is another way of photostabilization. The photostability of ubidecarenone microcapsules could be improved by incorporating

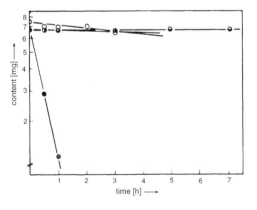

Figure 16 Stabilization of nifedipine in soft gelatine capsules using Acid Yellow (E 105). (○) dye in capsule shell, (□) dye in capsule shell and filling, (◑) dye in capsule filling, (●) without dye. *Source:* From Ref. 12.

the fat-soluble vitamins phytonadione and tocopherol, which themselves undergo photodegradation. Tocopherol acetate did not have a stabilizing effect (43). In dry emulsion particles, the photostability of ubidecarenone could be considerably enhanced through the presence of colorants dissolved in their oily carrier (44).

PHOTOSTABILIZATION OF DERMAL PREPARATIONS

In contrast to solid dosage forms, semisolid dosage forms are mainly applied externally and are therefore susceptible to photodegradation even after application. Despite their exposure during application to the uncovered skin and the resulting high risk of photodegradation, little attention has been spent on investigations of the photostability of dermatics and the means of protecting these products during use. This might be due to the false assumption that creams, gels, and ointments are sufficiently protected by their packs, mainly UV–VIS photon impermeable aluminium tubes. On the other hand, the in-use exposure to UV–VIS radiation removes stabilization provided by the packaging materials. Stabilizing of the drug is only possible by proper product formulation.

Testing of Dermal Preparations

Because semisolid drugs require a different handling than solid drugs, it is important to consider the experimental conditions for in vitro photostability testing. Due to the limited penetration depth of the incident radiation, it is important to provide samples of constant thickness. As noted previously, the "volume effect" can affect the result obtained. Furthermore, the sample area must also be kept constant if repeatable results are to be obtained.

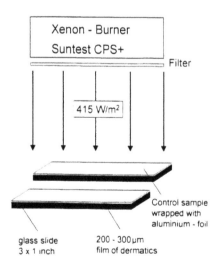

Black Standard Temperature: 30°C
Distance Burner-Sample: 26 cm

Figure 17   In vitro method for the photostability testing of dermatics. *Source*: From Ref. 45.

The experimental set-up used by Thoma et al. (Fig. 17) can be used as a reference. Samples are spread, as a film, 200 to 300 μm in depth, on a glass slide of defined area. A second sample, similarly prepared, is covered with aluminium foil and exposed along side of the test sample, as a control (45).

Under certain circumstances, it may be necessary to control the relative humidity to prevent a change of the formulation during the experiment, particularly if the water content of the drug preparation is high. This problem might be solved by using humidity-controlled stability testing cabinets. Alternatively, a sealed sample cell can be constructed by covering with quartz glass as a fully permeable cover for UV–VIS radiation.

The problem of temperature control needs to be addressed because even below the melting point of semisolid preparations, their structure may be affected by the temperature and thus the photostability of the formulation will change.

Formulation Factors

In the vast majority of creams, the drug is dispersed. However, even if the drug is photostable in the solid state, it cannot be concluded that the incorporated drug does not need specific photoprotection. An example where the photostability of a drug decreases markedly when dispersed in a cream base is retinoic acid. This drug substance is fairly photostable in the solid state. In solution, however, the remaining drug content was only 20% after 20 minutes of irradiation. Figure 18 shows the difference in the remaining contents of retinoic acid for four different cream formulations of the German Pharmacopoeia and Extrapharmacopoeia. The amount of retinoic acid remaining after two hours of irradiation at $415\,W/m^2$ ranged from 78% to 98%. The influence of formulation differences can be explained by the different solubilities of the drug in different bases. The solubility of retinoic acid in cream A is $5.1\,mg/100\,mL$ compared to only $0.08\,mg/100\,mL$ in cream B (46).

Another example for the importance of the solubility of the drug is corticosteroids. The different photosensitivities of three marketed hydrocortisone oil-in-water creams as shown in Figure 19, is due to the drug concentration as well as the composition of the preparation. Similar degradation rates were found for creams with triamcinolone-acetonid and betamethasone esters (47).

It can be assumed that the hydrocortisone's solubility will vary in different formulations. This was shown by the following experiment. The solubility and photostability of triamcinolone acetonid in water were tested using three concentrations of the surfactant sorbitan monostearate. The drug solubility was found to increase with the concentration of the surfactant. Therefore, the extent

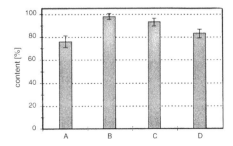

Figure 18 Influence of different cream bases on the photostability of Retinoic Acid. **A,** Basis cream DAC; **B**, hydrophilic ointment with water DAB; **C**, nonionic hydrophilic cream DAB; **D**, macrogolic ointment DAB. *Source*: From Ref. 46.

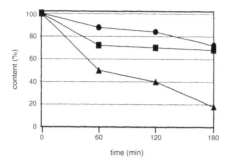

Figure 19   Photodegradation of hydrocortisone in different oil-in-water products. (●) Product A 10 mg/g, (■) product B 10 mg/g, (▲) product C 5 mg/g. *Source*: From Ref. 47.

of photodegradation of triamcinoloneacetonid correlates with the solubility of the drug (Fig. 20) (47).

Photostabilization by Spectral Overlay

Photosensitive drugs in topical formulations can also be stabilized according to the principle of spectral overlay. Because the risk of the absorption of additives through the skin is reduced compared to oral drugs, a larger choice of suitable UV absorbers and colorants is available. Particularly, sunblockers, which are commonly used in sun creams, offer a variety of different UV spectra, which can even be combined to match the spectrum of a drug.

Diclofenac sodium is the active of a variety of different topical, mainly gel, preparations. These preparations are as photosensitive as aqueous solutions, which show a 30% decrease in content after one hour. The absorption of diclofenac up to about 330 nm requires the use of compounds absorbing in the UV region from 300 to 350 nm.

Figure 21 shows the absorption spectra of diclofenac, a UV-B absorber (camphor derivative), a UV-A absorber (dibenzoylmethane derivative), and a broad-spectrum absorber (benzophenone derivative) (45).

As can be expected from the principle of spectral overlay, all sunblockers exhibit a photoprotective effect (Fig. 22). The remaining content after two hours can

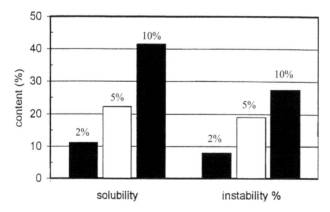

Figure 20   Influence of the concentration of sorbitan monostearate (2%, 5%, 10%) on the solubility and the photodegradation of triamcinoloneacetonide in water. *Source*: From Ref. 47.

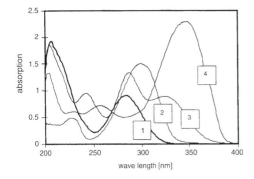

Figure 21 Comparison of the absorption spectra of UV absorbers and diclofenac sodium. (*1*) Diclofenac sodium, (*2*) 3-(4-methylbenzyliden)-camphor 0.002% (Eusolex® 6300), (*3*) 2-hydroxy-4-methoxybenzophenone 0.002% (Eusolex 4360), (*4*) 4-isopropyldibenzoylmethane 0.002% (Eusolex 8020). *Source*: From Ref. 45.

be improved from 53% in the unprotected cream to approximately 80% with the additives. However, total protection cannot be achieved.

Photostabilization by Opacification

In contrast to colorants, which stabilize by photon absorption, pigments are photon impermeable, for the most part, and stabilize by reflection and diffusion. Retinoic acid in Basis creme DAC is an example where the inclusion of titanium dioxide pigments improves the photostability of the product markedly. The drug content in the unprotected cream decreases by 28% with two hours of irradiation. Adding 2% of hydrophilic titanium dioxide (formulation E) reduced degradation to only 9% (Fig. 23) (45). The authors observed corresponding results with 2% zinc oxide and with 0.5% yellow iron oxide.

The reason for the incomplete protection of retinoic acid can be found by examining the absorption spectrum of the pigments used. Zinc oxide and titanium dioxide both block off some but not all UV radiation. They both have absorption cutoffs at about 350 and 380 nm, respectively (Fig. 24) (45).

Due to these characteristics, these pigments are useful for the photostabilization of some white compounds, e.g. corticosteroids, whose photodegradation is only caused by absorption of ultraviolet radiation. As shown in Figure 25, the addition of 0.5% of titanium dioxide or zinc oxide leads to the effective photoprotection of

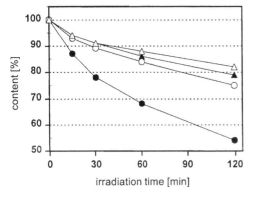

Figure 22 Effect of UV absorbers on the photostability of diclofenac sodium (0.002% in ethanolic solution) (Suntest CPS+, UV special filter, 415 W/m²). (●) Without UV-absorber, (□) with 4-isopropyldibenzoylmethane 0.002% (Eusolex® 8020), (▲) with 2-hydroxy-4- methoxybenzophenone 0.002% (Eusolex 4360), (Δ) with 3-(4-methylbenzyliden)-camphor 0.002% (Eusolex 6300). *Source*: From Ref. 45.

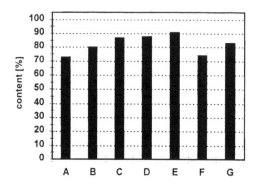

Figure 23 Influence of several types of titanium dioxide pigments and their concentration on the photostability of retinoic acid (0.03%) in Basis creme DAC (Suntest CPS+, window glass filter, 415 W/m², 120 minutes). **A,** Without pigment; **B,** titanium dioxide 0.5%; **C,** titanium dioxide 2%; **D,** micronized hydrophil. mod. TiO₂ 0.5% (Hombitec® H); **E,** micronized hydrophil. mod. TiO₂ 2% (Hombitec H); **F,** micronized hydrophob. mod. TiO₂ 0.5% (Hombitec L5); **G,** micronized hydrophob. mod. TiO₂ 2% (Hombitec L5). *Source*: From Ref. 45.

triamcinoloneacetonide in Basis creme DAC. The drug content after three hours of irradiation was 38% in unprotected creams and 95% in pigmented creams (45).

Photostabilization by Cyclodextrin and Liposome Inclusion

Liposomes are widely used in dermal preparations (dermatics) for various reasons. Cyclodextrins play a role in the solubilization of poorly soluble drugs. Hence, their use in dermatics is rarely restricted.

If a photosensitive drug is sufficiently lipophilic and can form an inclusion complex with either of these class of compounds, it may be protected from photonic energy by sterical shielding. This effect, however, is not predictable because the general mechanism of protection is not yet fully understood.

Diclofenac sodium is an example where cyclodextrin inclusion has a positive effect on the photostability of the gel formulation. Spray dried diclofenac-cyclodextrin complexes in hydroxyethylcellulose gels showed, after three hours of irradiation,

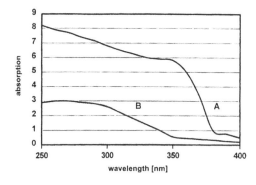

Figure 24 Absorption spectra of micronized pigments. **A,** Micronized zinc oxide; **B,** micronized titanium dioxide. *Source*: From Ref. 45.

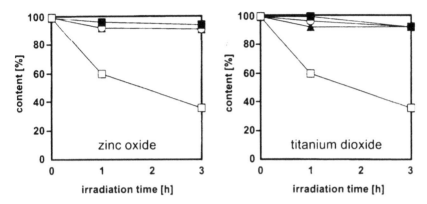

Figure 25 Influence of zinc oxide and titanium dioxide pigments on the photostability of triamcinoloneacetonide in Basis creme DAC (Suntest CPS+, 415 W/m², window glass filter, film 200 μm). (■) 4% pigment, (▲) 0.5% pigment, (○) 2% pigment, (□) without pigment. *Source*: From Ref. 45.

an increase in the remaining content from 78% to 98%, for the neat versus the complexed drug. In contrast, drug-loaded liposomes showed little photostabilizing effect (Fig. 26) (45).

CONCLUSION

The various means of photostabilization of solid and semisolid dosage forms, based on the principle of spectral overlay and opacification, have been reviewed and presented. Selection of one particular technique will depend on the particular drug substance or ingredients causing the photodegradation and its particular spectral characteristics.

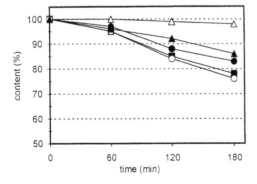

Figure 26 Influence of several different excipients on the photostability of diclofenac sodium (1% in hydroxyethylcellulose gel) (Suntest CPS+, window glass filter, 415 W/m²). (■) Without excipients, (●) unsaturated Sojalecithin (30 mg/mL), (○) saturated Sojalecithin (30 mg/mL), (▲) 2-phenylbenzimidazole-5-sulfonic acid 3% (Eusolex® 232), (△) hydroxypropyl-γ-cyclodextrin 2%. *Source*: From Ref. 45.

REFERENCES

1. Krämer I, Frank P. Lichtschutzbedürftigkeit photoinstabiler arzneimittel zur infusion. Krankenhauspharmazie 1987; 8(5):137–143.
2. Tønnesen HH. The ICH guideline: 6 months after its adoption. 3rd International Meeting on Photostability of Drugs—Photostability 99, Washington D.C., 1999.
3. Thoma K, Klimek R. Untersuchungen zur photoinstabilität von nifedipin 2. mitt.: einfluss von milieubedingungen. Pharm Ind 1985; 47(3):319–327.
4. Aman W. Photostabilisierung von Arzneimitteln unter Besonderer Berücksichtigung von Tabletten. Ph.D. dissertation, Ludwig Maximilians University, München, Germany, 1998.
5. Asahi Y, Shinoyaki K, Nagaoka M. Chemical and kinetic study on stabilities of 3-morpholinosydnonimine and its N-ethoxycarbonyl derivative. Chem Pharm Bull 1971; 19(6):1079–1088.
6. Matsuda Y. Some aspects on the evaluation of photostability of solid-state drugs and pharmaceutical preparations. Pharm Tech Japan 1994; 10(7):7(739)–17(749).
7. Hüttenrauch R, Fricke S, Knop M. Bedeutung des molekularen ordnungszustands für die lichtempfindlichkeit fester wirkstoffe. Pharmazie 1986; 41(9):664–665.
8. Spilgies H. Investigations on the Photostability of Cephalosporins and Beta-Lactamase Inhibitors. Ph.D. dissertation, Ludwig Maximilians University, München, Germany, 1998.
9. Nyqvist H, Wadsten T. Preformulation of solid dosage forms: light stability testing of polymorphs as a part of a preformulation program. Acta Pharm Technol 1986; 32(3):130–132.
10. Reisch J. Topochemische Lichtreaktionen an arzneistoffen und arzneizubereitungen. Dtsch Apoth Ztg 1979; 119(1):1–4.
11. DeAngelis N. Photostability—industry practice and experiences. 3rd International Meeting on Photostability of Drugs—Photostability 99. Washington D.C., 1999.
12. Thoma K, Klimek R. Photostabilization of drugs in dosage forms without protection from packaging materials. Int J Pharm 1991; 67:169–175.
13. Thoma K, Kübler N. Einfluß von hilfsstoffen auf die photozersetzung von arzneistoffen. Pharmazie 1997; 52(2):122–129.
14. Kaminsky EE, Cohn RM, McGuire JL, Carstensen JT. Light stability of norethindrone and ethinyl estradiol formulated with FD&C colorants. J Pharm Sci 1979; 68(3):368–370.
15. Desai DS, Abdelnasser MA, Rubitski BA, Varia SA. Photostabilization of uncoated tablets of sorivudine and nifedipine by incorporation of synthetic iron oxides. Int J Pharm 1994; 103:69–76.
16. KRONOS Leitfaden Grundlagen und Anwendung. KRONOS Titandioxid Kölnische Verlagsdruckerei, Köln, 1967.
17. Matsuda Y, Inouye H, Nakanishi R. Stabilization of sulfisomidine tablets by use of film coating containing UV absorber: protection of coloration and photolytic degradation from exaggerated light. J Pharm Sci 1978; 67(2):196–201.
18. Teraoka R, Matsuda Y, Sugimoto I. Quantitative design for photostabilization of nifedipine by using titanium dioxide and/or tartrazine as colourants in model film coating systems. J Pharm Pharmacol 1988; 41:293–297.
19. Béchard SR, Quraishi O, Kwong E. Film coating: effect of titanium dioxide concentration and film thickness on the photostability of nifedipine. Int J Pharm 1992; 87:133–139.
20. Mitton PB. Opacity, hiding power, and tinting strength In: Patton TC, ed. Pigment Handbook, Vol. III, Characterization and Physical Relationships. New York: Wiley, 1973:287–339.
21. Nyqvist H, Nicklasson M, Lundgren P. Studies on the physical properties of tablets and tablet excipients. V. Film coating for protection of a light-sensitive tablet-formulation. Acta Pharm Suec 1982; 19:223–228.
22. Rowe RC. The opacity of tablet film coatings. J Pharm Pharmacol 1984; 36:569–572.
23. Rowe RC. The measurement of the opacity of tablet film coatings in-situ. Acta Pharm Suec 1984; 21:201–204.
24. Everhard ME, Goodhart FW. Stability of certified dyes in tablets I—fading of FD&C red No. 3 in tablets as a function of concentration, time, and light intensity. J Pharm Sci 1963; 52(3):281–283.

25. Kuramoto R, Lachman L, Cooper J. A study of the effect of certain pharmaceutical materials on color stability. J Am Pharm Assn 1958; 47(3):175–180.
26. Brownley CA, Lachman L. Preliminary report on the comparative stability of certified colorants with lactose in aqueous solution. J Pharm Sci 1963; 52(1):86–93.
27. Scott MW, Goudie AJ, Huetteman AJ. Accelerated color loss of certified dyes in the presence of nonionic surfactants. J Am Pharm Assn 1960; 49(7):467–472.
28. Lachman L, Urbanyi T, Weinstein S, Cooper J, Swartz CJ. Color stability of tablet formulations. V—effect of ultraviolet absorbers on the photostability of colored tablets. J Pharm Sci 1962; 51(4):321–326.
29. Hajratwala BR. Influence of sunscreening agents on color stability of tablets coated with certified dyes I: FD&C red No. 3. J Pharm Sci 1974; 63(1):129–132.
30. Hajratwala BR. Influence of sunscreening agents on color stability of tablets coated with certified dyes II: FD&C blue No. 1. J Pharm Sci 1974; 63(12):1927–1930.
31. Hajratwala BR. Influence of sunscreening agents on color stability of tablets coated with certified dyes III: FD&C yellow No. 6. J Pharm Sci 1977; 66(1):107–110.
32. Goodhart RW, Everhard ME, Dickcius DA. Stability of certified dyes in tablets II—fading of FD&C red No. 3 and FD&C red No. 3 aluminium lake in three tablet formulations. J Pharm Sci 1964; 53(3):338–340.
33. Lachman L, Weinstein S, Swartz CJ, Urbanyi T, Cooper J. Color stability of tablet formulations III—comparative fastness of several water-soluble dyes and their corresponding lakes. J Pharm Sci 1961; 50(2):141–144.
34. Hajratwala BR. Stability of colors. STP Pharma 1989; 1(6):539–544.
35. Clapham D, Gower I, Hicks SJ, Thomson CM. A case study of colour stability of blue and green aqueous film coats containing indigo carmine (FD&C blue No. 2). 3rd International Meeting on Photostability of Drugs—Photostability 1999. Washington D.C., 1990.
36. Ammar HO, El-Nahhas SA. Improvement of some pharmaceutical properties of drugs by cyclodextrin complexation. 2. Colchicine. Pharmazie 1995; 50(4):269–272.
37. Teshima D, Otsubo K, Higuchi S, Hirayama F, Uekama K, Aoyama T. Effects of cyclodextrins on degradations of emetine and cephaeline in aqueous solution. Chem Pharm Bull 1989; 37(6):1591–1594.
38. Tomono K, Gotoh H, Okamura M, Ueda H, Saitoh T, Nagai T. Effect of B-cyclodextrin and its derivatives on the photostability of photosensitive drugs. Yakuzaigaku 1988; 48(4):322–325.
39. Ogawa M, Hirahara M, Shigematsu M, Nagayama K, Miyabe I, Hori K. Quality tests of commercial ubidecarenone products. J Nippon Hosp Pharm Assn Sci 1982; 8(4):282–287.
40. Matsuda Y, Itooka T, Mitsuhashi Y. Photostability of indomethacin in model gelatin capsules: effects of film thickness and concentration of titanium dioxide on the coloration and photolytic degradation. Chem Pharm Bull 1980; 28(9):2665–2671.
41. Thoma K, Kerker R. Photoinstabilität von arzneimitteln 3. Mitteilung: photoinstabilität und stabilisierung von nifedipin in arzneiformen. Pharm Ind 1992; 54:359–365.
42. Matsuda Y, Kouzuki K, Tanaka M, Tanaka Y, Tanigaki J. Photostability of gelatin capsules: effect of ultraviolet irradiation on the water vapor transmission properties and dissolution rates of indomethacin. Yakugaku Zasshi 1979; 99(9):907–913.
43. Matsuda Y, Teraoka R. Improvement of the photostability of ubidecarenone microcapsules by incorporating fat-soluble vitamins. Int J Pharm 1985; 26:289–301.
44. Takeuchi H, Sasaki H, Niwa T, et al. Improvement of photostability of ubidecarenone in the formulation of a novel powdered dosage form termed redispersible dry emulsion. Int J Pharm 1992; 86:25–33.
45. Thoma K. Protection of drugs against light—actual problems and possibilities of stabilization. 3rd Int. Meeting on Photostability of Drugs—Photostability 99. Washington D.C., 1999.
46. Neumayer I. Chemische Stabilität von Wirkstoffen in Dermatika in Abhängigkeit von der Zusammensetzung der Grundlage. Ph.D. dissertation, Ludwig Maximilians University, München, Germany, 1999.
47. Sandmann P. Untersuchungen zur Stabilität und den Abbauprodukten von Corticosteroiden in Dermatika. Ph.D. dissertation, Ludwig Maximilians University, ünchen, Germany, 1999.

17 Photostability Studies of Solutions and Methods of Preventing Photodegradation

Ahmed F. Asker

College of Pharmacy and Pharmaceutical Sciences, Florida A&M University, Tallahassee, Florida, U.S.A.

INTRODUCTION

Many drugs in various dosage forms, particularly in solutions, have been reported to undergo photodecomposition under the influence of various radiation sources. In the past, the investigation of photostability of drugs was more or less ignored since more attention was paid to the effects of temperature, humidity, and oxygen on drug stability. The number of therapeutically active drugs reported to be photolabile has increased considerably over the last three decades. They include antibiotics, antibacterials, analgesics, antidepressants, antihypertensive agents, antihistamines, steroids, nonsteroidal anti-inflammatory agents, vitamins, and antineoplastic agents. Nontherapeutic agents, e.g., excipients, impurities, and dyes used to color various pharmaceutical preparations are also known to undergo photodecomposition.

The study of the degradation of drug substances and products under the influence of ultraviolet–visible (UV–VIS) radiation should be an integral component of the pharmaceutical product development process in order to ensure safety, efficacy, and acceptability of the product to the patient throughout its proposed shelf life. Products that undergo photodegradation may cause a decrease, increase, or complete loss of the desired effect and/or alteration in the intended pharmacological activity, as well as undesirable changes in color and taste.

PHOTODEGRADATION EFFECT

Change of Potency

A significant loss of insulin activity occurred when solutions of amorphous insulin were exposed to ultraviolet radiation (1). The ultraviolet irradiation of phenylephrine HCl results in the formation of a product with demonstrated bronchodilator potency greater than that of the original solution, when tested on perfused guinea pig lungs (2). A solution of epinephrine exposed to radiation from a quartz mercury burner for 35 minutes was reported to produce practically no pressor, but depressor action, when tested on an anesthetized dog. However, the nonirradiated solution produced a marked pressor action with little depressor effect (3).

The photooxidation of angiotensinamide and its effects on the pressor, oxytocic and gut-stimulating activities of the peptide have been investigated (4). In these studies, control samples showed no loss of oxytocic, pressor and gut-stimulating activities, whereas solutions exposed to VIS radiation for 10 and 30 minutes had losses of 45% and 75%, respectively.

The antibacterial activity of curcumin is greatly enhanced by UV–VIS radiation (5). Photodegraded solutions of thiomerosal were found to be more active against *Streptococcus aureus* and *Pseudomonas aerogenosa* (6). The photoproducts of ciprofloxacin infusion solution, formed under the influence of window-glass filtered natural irradiation may be responsible for the loss of the antibacterial activity of the drug, as well as the occurrence of side effects, e.g., phototoxicity, as well as others (7).

Organoleptic Changes

The pharmacodynamically active molecular structure of the calcium channel blocker, nifedipine, is photodecomposed by UV–VIS radiation (8). This effect is manifested by a change in color (from yellow to colorless) or the fading of colored preparations, formation of precipitation and alteration of the taste of the product. Such changes during storage may give patients the sense of uneasiness, confusion, and doubt as to the safety and efficacy of the product.

Solutions of phenylephrine hydrochloride, for example, develop a brown color as a result of photodegradation (2). Many dyes, including FD&C dyes, used for coloring liquid pharmaceutical preparations such as syrups and elixirs, fade on exposure to light (9). Curcumin, a natural food colorant, fades on exposure to UV–VIS radiation (10).

Methylene blue, a colorant used in preparations for external use, undergoes photo-oxidation (11) and photoreduction in the presence of ethylenediamine tetraacetic acid (EDTA) (12). The photodecomposition of vanillin solutions in ethanol is accompanied by the formation of a yellow color and the development of a slightly bitter taste (13).

Photosensitization

Photoproducts can cause adverse in vivo photosensitization. A photodecomposition product of dacarbazine is reported to cause pain at the site of injection (14). Ciproflaxcin causes an increased photosensitivity to sunlight (7).

Phototoxicity

The release of cyanide and ferricyanide ions in photodecomposed solutions of sodium nitroprusside is considered to be a great health hazard (15). One of the photoproducts of benzydamine hydrochloride is reported to be responsible for its reported phototoxic effect (16).

Photodecomposed chloramphenicol used in eye drops is reported to cause ocular damage in experimental animals due to the *p*-nitrobenzaldehyde formed (17). Oxaziridine, a photoproduct of chlordiazepoxide, is more phototoxic than the parent compound (18). An isolated photoproduct of fenofibrate is reported to induce hemolysis (19).

Some of the dyes that are photolabile have been reported to cause phototoxic effects. For example, curcumin has demonstrated phototoxicity to mammalian cells (20). Methylene blue in eye solutions was reported to induce toxic effects (21). It is important, therefore, that drugs and excipients used in product formulations be screened for their photostability as well as photosensitizing effects.

Photocarcinogenicity

Photocarcinogenicity is another serious adverse effect of photoproducts, which has recently received attention (22). It has been shown that the irradiation of sodium naphthionate, a photolysis product of amaranth, gives rise to α-naphthylamine, a known carcinogen (23).

ROUTES OF PHOTOCHEMICAL DEGRADATION

In order to be able to suggest methods of protection against drug photodecomposition, one needs to be aware of the routes of photodegradation involved. Photochemical degradation occurs by two mechanisms: primary photochemical decomposition and/or secondary or photosensitizer photochemical decomposition (24).

Direct Photodegradation

Primary photochemical decomposition occurs when the drug molecule itself absorbs the photonic energy. Such absorption takes place when the absorption band of the molecule overlaps, to some extent, with the incident energy. The absorbed energy may be dissipated as fluorescence, phosphorescence, thermal energy (producing an increase in temperature of the surrounding medium), or chemical energy (initiating chemical decomposition with formation of photoproducts). Chemical routes include intramolecular rearrangements or the formation of free radicals and excited molecules, which may react in secondary photochemical reactions to form new products.

Photosensitized Degradation

Photons, therefore, provide the activation energy needed for the initiation of a reaction. Whether or not the drug itself will absorb energy can be deduced from an examination of its UV–VIS spectrum. Since the incident photonic energy may be converted to heat, a photolytic reaction may be accompanied by a thermal reaction, resulting in a photothermal effect. Once a thermal reaction is initiated, it may proceed even in the dark (25).

Secondary or photosensitizer-initiated decompositions occur when the photonic energy is absorbed by atoms/molecules other than the drugs. These excited molecules/atoms pass their increased energy to the drug molecules, which subsequently undergo chemical degradation. The photon-absorbing molecule/atom in this case is said to play the role of a sensitizer. Small amounts of impurities found in certain compounds or drug substances may act as photosensitizers or catalysts for the photochemical reactions especially if one of the components in the mixture is colored. For example, small amounts of riboflavin can act as a photosensitizer in the decomposition of ascorbic acid or folic acid (24).

A photosensitizer may behave as a catalyst if not consumed or regenerated in the reaction. Ferrous ions are photooxidized to ferric ions in the presence of acid. Therefore, lack of absorption bands in the UV–VIS region of the spectrum neither ensures photostability (24) nor indicates that a product will be photostable. The presence of chlorominoxidil as an impurity in minoxidil has been reported to act as a sensitizer in the photodecomposition of minoxidil (personal communication from the Pharmacia & Upjohn Company, Kalamazoo, Michigan). It is essential, therefore, that the purity of the drug substance should be determined to ensure freedom from photosensitizing impurities.

Dyes used to impart color to various liquid dosage forms may have a photosensitizing effect in the decomposition of certain drugs. Erythrosine, eosine, rose bengal, mercurochome, methylene blue, azure A and azure B, all accelerate the photodecomposition of solutions of pyridoxine HCl and pyridoxamine HCl. The singlet oxygen generated by these photon-excited dyes plays a role as a mediator in the decomposition of pyridoxine HCl (26).

Erythrosine was also found to promote the decomposition of phenylbutazone, when irradiated with unfiltered light from a 300 W projector bulb. In the absence of the dye, the drug is stable (27).

The photolytic degradations of drugs are complex, and include reactions such as oxidations, reductions, *cis–trans* isomerizations, structural rearrangements, hydrolyses, dechlorinations, etc. Oxidation is one of the major pathways of photochemical reactions. Table 1 lists the pathways of photochemical decomposition of some therapeutic and nontherapeutic agents.

The most common change resulting from exposure of a solution to UV–VIS energy is fading or a change in color. Generally, fading or discoloration can serve as a preliminary indication of photodecomposition. In addition to visual observation of color change, spectrophotometric data from the UV–VIS regions of the irradiated and nonirradiated solutions provide a simple, quick method of detecting photodecomposition. After irradiation, prochlorperazine loses its characteristic

Table 1 Photodegradative Pathways of Some Therapeutic and Nontherapeutic Agents

Pathway(s)	Agent	References
Binding	Riboflavin	(28)
Decarboxylation	Naproxen	(29)
Decarboxylation and oxidation	Vietorolac tromethamine	(30)
Dechlorination and hydrolysis	Hydrochlorothiazide	(31)
Dehalogenation	Chlorpromazine, prochlopromazine, frusemide, and hydrochlorothiazide	(32)
Isomerization	Benorylate	(33)
	Flupenthixol, clopenthixol, and chloroprothixene	(34)
Isomerization and photolysis	Cerufoxime axetil	(35)
	Cefotaxime	(36)
Oxidation	Nifedipine	(37)
	Angiotensin II	(4)
	Chlorpromazine	(38)
	Vanillin	(13)
	Methylene blue	(11)
Oxidation, reduction, and condensation	Chloramphenical	(39,40)
Polymerization	Doxepin	(41)
Rearrangement and side chain fission	Diphenhydramine	(42)
Reduction	Nimodipine	(43)
Reduction and hydrolysis	Frusemide	(44)

absorption maxima at 255 and 310 nm, and develops a new absorption maximum at 500 nm (45).

A change in absorbance will only occur if the chemical reaction takes place at a site involving the chromophore or an associated auxochrome. For example, the photolysis of frusemide involves loss of Cl from the parent chromophore, producing a small, but discernible change in its major absorption peak at 271 nm (44).

A number of spectral changes have been reported for the photolysis of several barbituric acid derivatives. These changes are manifested by a decrease in absorption peaks and the appearance of new bands (46).

A series of UV spectra, which show the gradual decrease in the λ_{max} of a drug, has been utilized to monitor the extent of photodecomposition of several drugs including doxorubicin hydrochloride (47) and minoxidil (48). Typical plots from such a study of these two drugs are presented in Figures 1 and 2.

The extent of photodegradation of riboflavin is not evident by either visual or spectrophotometric means (49). These results clearly demonstrate that these techniques, though rapid and easy to perform, are not appropriate in all circumstances and that good, validated, stability-indicating analytical techniques should be used to quantitatively determine the extent of photodecomposition.

PHOTOSTABILITY TESTING

A considerable amount of information has been accumulating in the literature regarding the analytical methods used to determine the extent of photodecomposition of drugs. Examples of these analytical techniques used are high pressure liquid chromatography (HPLC) (7,31,34–36), gas liquid chromatography (GLC) (50,51),

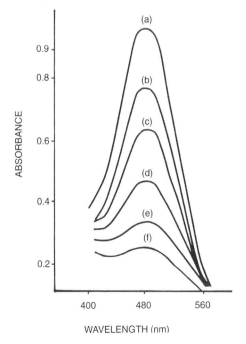

Figure 1 Absorption spectra of adriamycin in phosphate buffer of pH 7 during photolysis under fluorescent light. (a) 0 minutes; (b) 15 minutes; (c) 25 minutes; (d) 35 minutes; (e) 45 minutes; (f) 60 minutes. *Source*: From Ref. 47.

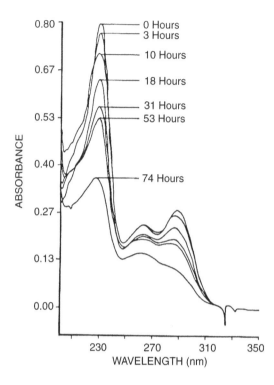

Figure 2 Ultraviolet spectra obtained during photolysis of minoxidil in pH 7.0 phosphate buffer under fluorescent radiation. *Source*: From Ref. 48.

fluorometry (52), thin layer chromatography (TLC) (49,53,54), polarography (55), and UV–VIS spectrophotometry (46–48,56,57). As previously noted, for certain drugs, the latter method is simple to carry out since photodegradation can be monitored at different time intervals of exposure, by recording the change in absorbance, and hence drug content can be calculated, but has its limitations.

A review of current literature reveals a wide variation in the design of photostability cabinets promoted for the testing of pharmaceuticals. Also, the photon source, its intensity, and bandwidth, in addition to the methods used to measure photon intensity, temperature, and time of exposure, vary considerably. This variation is a result of the lack of agreement regarding product exposure and the absence of definitive regulatory guidelines and standards.

The need for a universal standardized approach to photostability testing of pharmaceuticals exists. The International Conference on Harmonization (ICH) has acknowledged such need, and recommended for adoption of the ICH Tripartite Guideline Stability Testing: Photostability Testing New Drug Substances and Products in November 1996. The guideline makes recommendations for photon sources, sample presentation, procedures for testing, and methods of assessment of results. A procedure for standardizing UV source calibration based on chemical actinometry is also included as an annex (58). Some details regarding the development and content of this document are discussed within this book, especially in Chapters 3, 4 and 8.

An extensive review of photostability testing has been published by Sandeep et al. (59). They discussed the properties of various photon sources and cabinets, along with the effects of wavelength, distance from the photon source, exposure time, and type of container on the results obtained. A photostability study protocol is suggested, which would include various parameters such as the appropriate source, calibration of the system, defining the illuminance level and the exposure time, determining the transmittance of the container, preparing control samples, controlling and recording the relevant environmental conditions, controlling of storage variables and selecting lots for challenge.

FACTORS AFFECTING PHOTODEGRADATION IN SOLUTIONS

Formulation

pH

Many solution reactions are catalyzed by hydrogen or hydroxyl ions and consequently may undergo accelerated decomposition upon the addition of acids or bases. The catalysis of a reaction by hydrogen or hydroxyl ions is known as acid–base specific catalysis. In many cases, in addition to the effect of pH on reaction rate, there may be catalysis by one or more species of the buffer system. This type of catalysis is known as the acid–base general catalysis. Solutions of vitamin B_6 were found to be most stable at low pH values when exposed to VIS radiation (60).

The photodegradation of doxorubicin, daunorubicin, and epirubicin is accelerated by increases in the pH of their solutions (61). Furosemide is stable in alkaline solutions, but quickly degraded in acid solutions (62).

The photolysis of aqueous solutions of sulfacetamide and sulfanilamide was studied at pH 3 to 11. The rate of photolysis was found to be highest around pH's 5.0 and 3.0 for sulfacetamide and sulfanilamide, respectively (63). Thoma and Klimek (64) found that photodegradation of nifedipine is not influenced by changes of pH in the range from 2 to 12. Similarly, the rate of photochemical decomposition of sodium nalidixate does not depend on the pH of the buffer within the pH range 8.2 to 11.0 (65). The rate of photodegradation of folic acid was high in the pH range of 2 to 4 (66).

Drugs whose decomposition rates are influenced by the pH include doxorubicin hydrochloride (47), minoxidil (48), menadione sodium bisulfite (67), metronidazole (68), riboflavin (69), colchicine (70), tetracycline hydrochloride (71), phenobarbital (72), dacarbazine (73), furosemide (74), daunorubicin hydrochloride (75), and demeclocycline hydrochloride (76). Figures 3 to 6 illustrate how the pH of the solution can influence the degradation rate of some of the reported photolabile drugs.

Buffer Species and Ionic Strength (μ)

Buffer Species

Drug molecules in solution are also catalyzed by the actual buffer species chosen and its molarity, in addition to the normal hydrogen and hydroxyl ions catalysis. The catalytic effect of various buffer species on the photodegradation rate of some drugs has been reported in the literature.

Citrate and phosphate ions were found to catalyze the photodegradation of sulfacetamide and sulfanilamide (63). Adriamycin is most stable in a citrate buffer, followed by phosphate and then acetate buffer (47). The order of the catalytic effect of buffer species for metronidazole was found to be: citrate > acetate > phosphate buffer (68). Acetate ions were found to be most detrimental to the photostability of

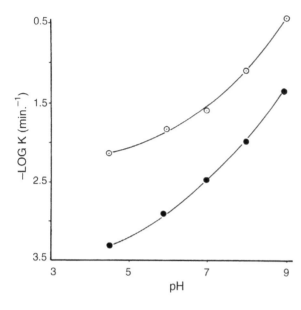

Figure 3 pH-rate profile for the photodegradation of adriamycin in phosphate buffers exposed to fluorescent radiation. ○ Without PABA; ● with PABA. *Abbreviation*: PABA, *p*-Aminobenzoic acid. *Source*: From Ref. 77.

tetracycline hydrochloride when compared to the phosphate and the citrate ions (71). The photodegradation rate of dacarbazine is reported not to be influenced by the buffer species (73).

The photolysis of demeclocycline hydrochloride, in the presence and absence of glutathione as a photoprotective agent, was lowest in citrate buffer when compared to acetate or phosphate buffers (76). Table 2 shows the magnitude of this effect.

The photodegradation of daunorubicin hydrochloride is influenced by both the buffer species and its concentration. Of those buffers tested, divalent phosphate

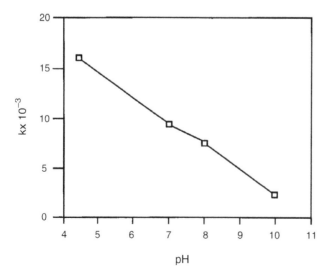

Figure 4 pH-rate profile for the photodecomposition of minoxidil in phosphate buffers exposed to fluorescent radiation. *Source*: From Ref. 48.

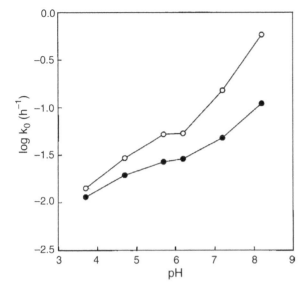

Figure 5 pH-rate profile of daunorubicin hydrochloride solutions in phosphate buffers exposed to fluorescent radiation. ○ Without sodium sulfite; ● with sodium sulfite. *Source*: From Ref. 75.

anion had the greatest catalyzing effect (75). Figure 7 shows the effect of buffer concentration on daunorubicin photostability in the presence and absence of sodium sulfite as a stabilizer.

Ionic Strength (μ)

Ionic strength is a measure of the intensity of the electrical field in a solution and can considerably change the properties of water as a solvent. By varying the ionic strength of the medium through the addition of an inert electrolyte such as NaCl and KCl while keeping other concentrations constant, the rate of reaction will either increase, decrease, or remain the same. The photostability of nifedipine at pH 6.8 is not influenced by changes in ionic strength produced using potassium chloride (8).

The photostability of daunorubicin hydrochloride in pH 5.5 acetate buffer, increases with increasing ionic strength, from 0.1 to 0.5 μ. The presence of sodium

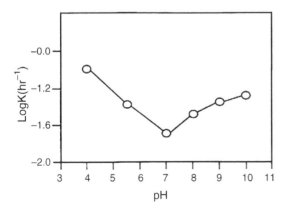

Figure 6 pH-rate profile for the photodegradation of furosemide in phosphate buffers exposed to fluorescent radiation. *Source*: From Ref. 74.

Table 2 Effect of Buffer Species and Glutathione (GTH) on the Photodegradation of Demeclocycline Hydrochloride (DCL) Exposed to Fluorescent Irradiation

| | Rate constants[b] | | | | |
| | Neat | | With GTH | | Percentage increase[c] in photostability |
Buffer[a]	K_c	$t_{1/2}$	K_p	$t_{1/2}$	
Phosphate	9.14	758.2	7.34	943.6	19.7
Acetate	5.60	1237.3	4.96	1398.3	11.4
Citrate	3.68	1885.4	3.38	2052.6	8.2

[a]In pH=4.5 and μ=0.1.
[b]$K \times 10^4$ min^{-1}.
[c]$[(K_c-K_p)/K_c] \times 100$.
Source: From Ref. 78.

sulfite incorporated as a stabilizer had no effect on this process (75). Figure 8 illustrates this effect.

Likewise, increases in the ionic strength of furosemide (pH 8 phosphate buffer) and demeclocycline (phosphate buffer at pH 6) solutions (76), from 0.1 to 0.5 µ, had no effect on their degradation rate constants (74).

Increasing the ionic strength produced a photostabilizing effect on minoxidil solution (pH 8 phosphate buffer) up to a value of 0.17 µ; then the rate constant remained unchanged with further increase of ionic strength (48), as can be seen in Figure 9.

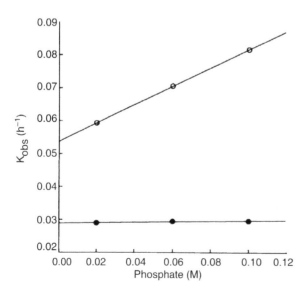

Figure 7 Effect of pH 6.2 phosphate buffer concentration and sodium sulfite on the observed first-order rate constant for the photodegradation of daunorubicin hydrochloride. ○ Without PABA; ● with PABA. *Abbreviation*: PABA, *p*-Aminobenzoic acid. *Source*: From Ref. 75.

Dielectric Constant (ε)

The dielectric constant of a solvent plays a significant role in solubility and stability of the solute. Few studies relating to the dielectric constant of the solvent medium to the photodegradation rate have been undertaken. Thoma and Klimek (8) reported that the improvement in the photostability of nifedipine noted by them is due to an increase in dielectric constant in both ethanol/water and propylene glycol/water mixtures.

A mixture of 50% v/v propylene glycol or 50% v/v polyethylene glycol (PEG) 400 with phosphate buffer of pH 7 increased the photodegradation rate constant of democlocycline hydrochloride by 1.8 and 2.5 times, respectively, compared to the phosphate buffer alone (76).

Additives

Additives incorporated into solutions of photolabile drugs can adversely or favorably affect the photostability of the drug. Heavy metal ions, particularly those possessing two or more valence states with a suitable oxidation–reduction potential between them, e.g., copper, iron, and nickel, can catalyze oxidative decomposition. Their effect is to increase the rate of formation of free radicals.

Trace metal impurities in buffer salts were found to enhance the oxidative degradation of prednisolone (78). Phenylephrine hydrochloride decomposes in the presence of heavy metal ions (79). The autooxidation of procaterol, a sympathomimetic amine, is enhanced in the presence of ferric ions (80). The photodegradation of riboflavin is enhanced by the presence of polysorbate 80 and sodium lauryl sulfate (81).

The effect of several commonly employed macromolecules on the apparent first-order rate constant of photobinding of riboflavin solutions has been reported (28). In this report, it was reported that polyethylene glycol 4000 slightly enhanced, and sodium decylsulfate enhanced the aerobic photofading of riboflavin, whereas

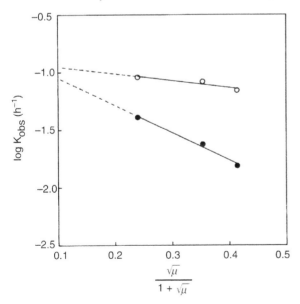

Figure 8 Effect of ionic strength on the photodegradation rate of daunorubicin hydrochloride in pH 5.5 phosphate buffer. ○ Without PABA; ● with PABA. *Abbreviation:* PABA, *p*-Aminobenzoic acid. *Source:* From Ref. 75.

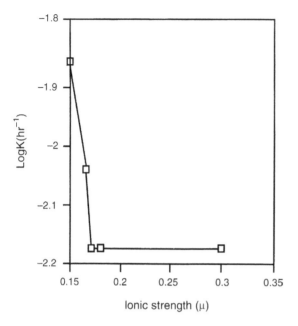

Figure 9 Log *K* of the photodegradation of minoxidil in pH 8 phosphate buffer versus ionic strength of the solution. *Source*: From Ref. 48.

polysorbate 80 and polyvinylpyrrolidone (PVP) greatly accelerated the rate. No significant effect was noted for carbitol or methyl cellulose.

The catalysis of riboflavin by the macromolecules is partially attributed to reversible binding of excited riboflavin molecules to macromolecules, which produce longer-lived excited species. Evidence is presented to indicate involvement of a triplet state in both the photobinding and photodecomposition of riboflavin. The enhanced binding of riboflavin to macromolecules during irradiation was detected by both precipitation and resin desorption techniques.

Asker and Colbert (82) found that Tween 80 enhanced the photodegradation of FD&C Blue No. 2 dye. In another report, Asker and Canady (83) reported that Tween 80 had little effect on the photostability of sodium nitroprusside solutions, while polyethylene glycol 300 enhanced it, possibly due to its chelating potential.

The photodegradation of pyridoxine is noticeable in sugar and saline infusion solutions, but much less in amino acid infusion solutions. The presence of tyrosine and tryptophan inhibits photodecomposition, whereas flavine mononucleotide accelerates it (84).

The photodecomposition of adriamycin was found to be dependent on the composition of the vehicle used. The relative order of a vehicle protecting ability is ethanol>0.9% NaCl>distilled water>Ringer–Krebs bicarbonate (52).

The degradation rate of vitamin A palmitate under UV radiation was significantly lower in the presence of mixtures of amino acids (85). Tween 80 provides a good inhibitory effect on chloramphenicol photolysis due to incorporation of the drug within the micelles (86).

The rate of photolysis of nifedipine was considerably reduced in the presence of PVP. In this report, PEG 20 sorbitan monolaurate was found to enhance the photostability of nifedipine. This was attributed to the concentration of nifedipine in the lipophilic core area of the micelles and absorption properties of PEG 20 sorbitan monolaurate in the wavelength range 400 to 450 nm at which the stability of nifedipine is photolabile.

Self-Absorption
Color additives have been reported to also either enhance drug photodegradation or protect against it. Erythrosin, eosine, role bengal, methylene blue, azure A, and azure B accelerate the photodecomposition of pyridoxine hydrochloride and pyridoxamine HCl (26). Erythrosine was found to enhance the decomposition of phenylbutazone when irradiated with unfiltered light from a 300 W projector bulb, but was stable in the absence of the dye (27). Certain dyes such as methylene blue have been found to enhance the photostability of solutions of daunorubicin (87).

Drug Concentration
Most drugs undergo zero, first, and pseudo first-order reactions in solution. Reactions beyond second-order are uncommon in pharmaceutical solutions. Zero-order reactions occur when the reaction rate is independent of the concentration of the reactant, but affected by other factors such as solubility, the amount of radiation absorbed in a photochemical reaction, or the amount of catalyst in a catalytic reaction.

Zero-order reaction can be expressed by the equation, $dc/dt = -K$, where dc/dt is the rate of change of concentration with time and k is the rate constant. In a zero-order reaction, a plot of concentration against time will produce a straight line whose slope is equal to $-K$ and the half-life of the reactant is equal to $\frac{1}{2} Co/K$ where Co is the initial concentration of the reactant.

A first-order reaction can be expressed by the equation, $dc/dt = -KC$, which states that the rate is dependent on the concentration of the reactant. In a first-order reaction, a plot of log C against time will yield a straight line whose slope is equal to $K/2.303$ and the half-life of the reactant is equal to $0.693/K$. If a plot of log concentration versus time is not a straight line, one can conclude that the order of reaction is not the first, but some higher order.

A pseudo–first-order reaction can be defined as second-order reaction involving two reacting species that behave like a first-order reaction. This type of reaction can occur if one reactant is present in greater excess. Under these circumstances, even though there are two reactants in the solution, the reaction rate can be determined by the concentration of only one reactant. Hydroxyl ion–catalyzed ester hydrolysis is an example of a pseudo first-order reaction. For detailed information on kinetics of reactions, the equations involved, and factors affecting rates of reaction, the reader is advised to seek the available references (88,89) and Chapter 11 of this book.

A review of the literature shows that photochemical reactions generally follow first-order kinetics. However, pseudo first-order, zero-order, and even fractional-order kinetics have been reported. Drugs reported to follow first-order kinetics include nifedipine (8) doxepin (41), riboflavin (28), minoxidil (48), adriamycin (52), doxorubicin, daunorubicin and epirubicin (61), folic acid (66), menadione sodium bisulfite (67), tetracycline hydrochloride (71), decarbazine (73), furosemide (74), democlocycline (76) and hydrocortisone, and prednisolone (90). Apparent pseudo first-order rate has been reported for pyridoxine (84). Indomethacin (91), sodium nalidixate (65), methotrexate (92), and metronidazole (68) have been reported to follow zero-order kinetics. The photodecomposition of furnidipine exposed to artificial daylight was reported to follow 0.5-order kinetics (56).

The apparent rate constant for photodegradation has been found in certain cases to be inversely proportional to concentration, i.e. the higher the concentration, the slower the degradation. This observation has been reported for adriamycin (47,52), doxorubicin, daunorubicin and epirubicin (61), furosemide (74), democlocycline hydrochloride (76), and vitamin A (85). This behavior is attributed to the increased absorption of highly concentrated solutions, which act as photon filters (self-filtering effect), thereby decreasing the noted rate of degradation. Figures 10 and 11 are plots showing the first-order and the zero-order photodegradation kinetics for dacarbazine and metronidazole, respectively.

Storage Factors

Source, Intensity, and Wavelength Effects

Several reports have indicated that the nature of photodegradation pathway and rate of photochemical decomposition of drug substances are influenced by the type of the source, intensity, and wavelength. In addition, time of exposure and the distance between the source and the location of the exposed samples within a test chamber are other variables that can affect the noted rate of photodecomposition.

Thoma and Klimek (8) found that the photodegradation of nifedipine in solution was greatly influenced by photon intensity and wavelength. Nifedipine was found to be photolabile at longer wavelengths of UV and in VIS radiation.

The degradation of vitamin A palmitate solution exposed to ultraviolet radiation was found to degrade exponentially during the initial exposure period after which the rate of photodegradation was progressively reduced. This reduction was attributed to the protective absorption effect provided the degradation product. UV radiation provided by two fluorescent tubes was filtered to produce

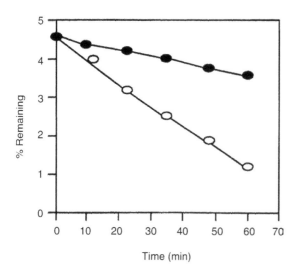

Figure 10   First-order photodegradation of dacarbazine in pH 7 phosphate Buffer. ○ Without PABA; ● with PABA. *Abbreviation*: PABA, *p*-Aminobenzoic acid. *Source*: From Ref. 73.

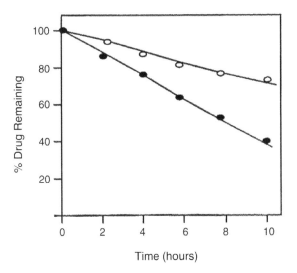

Figure 11 Zero-order photodegradation of metronidazole in phosphate buffer of pH 7.
● without sodium urate; ○ with sodium urate. *Source*: From Ref. 68.

a spectrum range of 254 to 390 nm (85). The maximum rates of degradation for vitamin A palmitate were found to occur between 330 and 350 nm (93).

The effects of exposure of FD&C Red No. 3 solutions to fluorescent, long-wave, and shortwave UV sources were studied by Asker and Jackson (94). They found that exposure to fluorescent lighting (from cool white fluorescent tubes) was more detrimental to the stability of the dye solution than either of the other two UV sources studied. Similarly, the same kind of fluorescent radiation caused a higher degree of degradation of adriamycin solutions than either short wave or long wave UV radiation (47). The results of this study are presented in Table 3.

Habib and Asker (68) showed that metronidazole solutions decompose more under artificial sunlight [from two 40 W fluorescent sunlamps and fluorescent radiation (from cool white tubes)] in a Atlas HPUV Actinic Exposure Cabinet, than under UV-A radiation, as shown in Table 4.

Tetracycline hydrochloride in pH 7 phosphate buffer was found to be the least stable when exposed to radiation from cool white fluorescent tubes when compared to UV-A and UV-B radiation (71). Table 5 shows this effect.

The degradation pattern found for exposed doxepin solutions was quite different, depending on whether the radiation was from a mercury lamp or window-glass filtered daylight (41). Aqueous and ethanolic solutions of ketorolac tromethamine were found to decompose rapidly under a laboratory black light ($\lambda_{max} = 350$ nm) and degradation rate was dependent on the irradiation time (30).

The stability of furnidipine solutions was studied under the influence of UV radiation, artificial daylight, and ambient room daylight. For UV irradiation, a Black-Ray lamp (model B 100 A; 125 W, 366 nm) was used. For artificial daylight irradiation, three 200 W Phillips tungsten lamps were used. Ambient room daylight was a combined mixture of window-glass filtered daylight and fluorescent radiation. Furnidipine solutions exposed to artificial daylight and UV radiation were found to produce two photoproducts (56).

Table 3 Effect of Photon Source and Glutathione (GSH) on the Photodegradation of Adriamycin

Source[a]	Degradation rate constants[b]		Percentage increase[c] in photostability
	Neat K_c	With GSH K_s	
Fluorescent	25.0	6.20	75.2
Shortwave UV	0.36	0.18	50.0
Longwave UV	0.34	0.20	41.2

[a]In pH 7 phosphate buffer and clear glass vials.
[b]$K \times 10^3$ min^{-1}.
[c]$[(K_c - K_s)/K_c] \times 100$.
Source: From Ref. 47.

In a report by Thoma and Kerker (95), the effects of irradiation time and
absolute radiation dose on the decomposition rates and photoproducts of betametha-
zone valerate chloramphenicol, molsidomine, metronidazole, and indomethacin
were studied. Their results show that substances sensitive to UV radiation, such
as the above-listed five drugs, despite their instability in sunlight, have little or no
decomposition upon exposure to irradiation from tubular fluorescent lamps at an
irradiation intensity of $79 \, W/m^2$. In addition, for those compounds that did degrade,
they noted differences in the quantitative relationships of the decomposition products as well as their decomposition rates. The tubular fluorescent lamps used were
found to be unsuitable for simulation of direct daylight.

The photostability of vitamin B_{12} solutions under the influence of various light
sources was studied by DeMerre and Wilson (96). The results of this study indicated
that the rate of photolysis of vitamin B_{12} is influenced by the intensity as well as the
type of light source used. The loss in potency of vitamin B_{12} solution was about 10%
for every 30 minutes of exposure to bright natural daylight of 8000 foot-candles.
When the same solution was irradiated with a 300 W incandescent reflector–type
filament lamp, the losses after two hours of exposure were 11% at 7000 foot-candles
and 45% at 14,000 foot-candles.

Table 4 Effect of Source and Sodium Urate on the
Photodegradation of Metronidazole

Source	Rate constant[a] $K \times 10^2$/hr		
	Neat K_c	With[b] K_w	K_w/K_c
Artificial sunlight	5.0	3.4	0.68
Black light	4.2	4.2	1.00
Fluorescent	6.0	5.0	0.83

[a]Metronidazole $= 9.93 \times 10^{-5}$ M.
[b]Sodium urate, 5.3×10^{-3} M in pH 7.0 phosphate buffer.
Source: From Ref. 68.

Table 5   Effect of Source Type on the Photodegradation of Tetracycline Hydrochloride

Source	Rate constants[a]				Percentage increase in stability $[(K_c-K_s)/K_c]$ $\times 100$
	Neat		With GSH		
	K_c	$t_{1/2}$	K_s	$t_{1/2}$	
Fluorescent	27.0	2.6	8.1	8.6	70.0
UV-A	3.3	21.0	2.1	33.0	36.4
UV-B	3.6	19.3	2.4	28.9	33.3

[a]$K \times 10^2$/hr.
Source: From Ref. 71.

Gu Chiang and Johnson (30) reported that drugs with λ_{max}'s greater than 280 nm have the potential for degradation in sunlight, whereas drugs with λ_{max}'s greater than 400 nm have the potential for degradation in both sunlight and room light. Various light intensities at a wavelength of 363 nm greatly influenced the rate of photodecomposition of riboflavin at 30 (97).

Temperature

Storage temperature can have a pronounced effect on rate of photodegradation. The effect of temperature on the reaction rate is given by the equation first suggested by Arrhenius which is $K = Ae^{-Ea/RT}$ or $\log K = \log A - E_a/2.303\,RT$ in which K is the specific reaction rate, A is the constant known as the Arrhenius or the frequency factor, E_a is the energy of activation for the reaction(s), R is the gas constant, 1.987 calories/deg mole, and T is the absolute temperature (89).

In thermal stability studies, accelerated testing at elevated temperatures is used to minimize the time needed to assess the thermal stability of the substance/product. From the Arrhenius plot, one can predict the shelf life of the product at a proposed storage temperature.

For photostability studies, accelerated tests for liquid preparations may involve the use of some controlled, high intensity photon sources, and should take into account maintenance of a uniform temperature and adequate mixing of the preparation. If the protocol of the photostability study is properly designed, the extent of degradation should be directly proportional to the number of photons absorbed, as reported in the study by Shin et al. (97), and presented in Figure 12.

From a figure plotting the total number of molecules decomposed versus the number of photons absorbed, the amount of riboflavin decomposed for a given number of photons absorbed can be determined.

Sande (98) reported that temperature is not expected to affect the absorption of photons as such, and no additional energy is needed for the reaction to take place. However, the temperature will affect subsequent chemical degradative reactions in the usual manner as described by the Arrhenius equation. If a secondary thermal reaction is involved, a temperature effect on the overall reaction quantum yield would be expected. A change in the viscosity of the liquid as a result of increase of room temperature can influence the rate of photochemical degradation.

The overall reaction of photodegradation of aqueous solutions of chlorpromazine hydrochloride was investigated by Song-ling and Lachman (99) and was

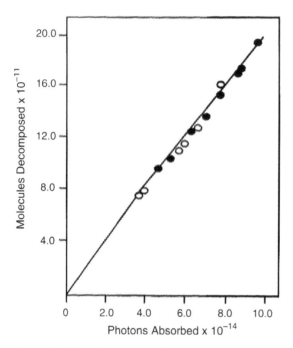

Figure 12   Plot of the number of molecules of riboflavin decomposed as a function of the number of photons absorbed. ○ light of wavelength: 433 nm; ● light of wavelength: 363 nm. *Source*: From Ref. 97.

found to be zero-order kinetics. Song-Ling and Lachman (99) reported that since the solutions were not irradiated at various photon amounts, a prediction of shelf-life photostability of chlorpromazine hydrochloride was not possible for them.

The effect of temperature on the photodegradation rates of solutions of tetra-cycline hydrochloride (71) and adriamycin (77) has been reported. In these reports, solutions of these drugs were stored at different temperatures and exposed to a constant, irradiation intensity. Figures 13 and 14, respectively, present the Arrhenius plots for these two drugs.

The photodegradation of some quinolones used in antimicrobial therapeutics was found to be temperature independent (7).

Type of Packaging

Immediate containers, labels, and carton boxes can play a significant role in protection against photodegradation of drug substances and products. Colored glass containers have been used to protect photolabile preparations. Glass, which is yel-low–green in color, is reported to provide the best protection against UV radiation since it has no transmission below 400 nm. Iron–manganese amber glass has no transmission below 370 nm and therefore gives some protection from UV radiation, but little against the VIS and infrared (IR). Medium-green glass (no details given) has no transmission below 350 nm, but transmits to some extent in the visible range and absorbs somewhat in the IR region (99).

The photon transmissive properties of colored glass containers vary consid-erably with their thickness and absorptivity. Stewart and Tucker (24) considered

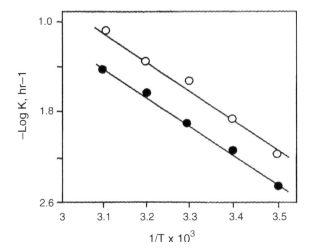

Figure 13 Arrhenius plots of degradation of tetracycline hydrochloride in phosphate buffer at pH 7 under longwave ultraviolet from a black-light tube radiation. ○ without glutathione; ● with glutathione. *Source*: From Ref. 71.

wall thickness less than about 2 mm as unsuitable to be light-resistant containers. Differences in transmission between concave and convex surfaces have been reported to be insignificant (100).

Nifedipine solutions packaged in brown glass ampuls decomposed by 10% in 20 minutes, whereas those in clear glass ampuls showed 50% decomposition after the same exposure (64). No difference in the degree of degradation of vitamin A palmitate exposed to UV radiation of 254 to 390 nm was found, whether packaged in glass or PVC containers (85).

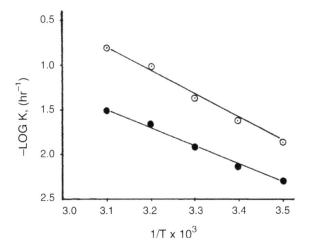

Figure 14 Arrhenius plots for the photodegradation of adriamycin in phosphate buffer at pH 7 exposed to fluorescent radiation. ○ without PABA; ● with PABA. *Abbreviation*: PABA, *p*-Aminobenzoic acid. *Source*: From Ref. 77.

Amber glass is not always protective against photodegradation (76). Wollmann and Gruenert studied the photostability of isoprenaline, adrenaline, and noradrenaline solutions in various containers exposed to VIS radiation (101). They found these drugs to be more stable in plastic than in glass containers, pigmented containers being the most and amber glass ampuls the least. They attributed high decomposition rates in amber glass to the release of alkali and trace amounts of heavy metal ions.

Investigations conducted by Thoma and Klimek (64) showed that commercial amber glass ampuls do not always provide sufficient protection from UV–VIS radiation. Commercial amber glass differs in thickness as well as degree of coloration (64).

Ferdous and Asker (76) found that demeclocycline hydrochloride solution degraded more in clear than in amber glass vials when exposed to fluorescent radiation from cool white fluorescent tubes. Amber glass was also found to provide better protection for solutions of doxorubicin hydrochloride (47), minoxidil (48), and furosemide (74). Figures 15 and 16 illustrate the effect of type of glass on photodegradation of furosemide and democlocycline hydrochloride, respectively.

The photodegradation of doxorubicin, daunorubicin, and epirubicin solutions is more rapid in clear glass than in polyethylene or amber glass containers (61). Colored glass is not preferred for injections since it is difficult to detect discoloration and particulate matter in the liquid preparation. However, a colorless glass capable of cutting off detrimental radiations is still to be discovered.

Porter and Bauwens (102) studied the light transmission properties of amber oral syringes. They found that, of the four brands tested, only one brand consistently met the United States Pharmacopeia (USP) standard for light-resistant packaging.

Infusion solutions of photolabile drugs should be protected since they are exposed to UV–VIS radiation while in their reservoir, passing through the burette, the drip chamber, and the associated tubing of the infusion set. An amber administration set, which filters out light at wavelengths below 470 nm, is as effective as a foil-wrapped set in protecting sodium nitroprusside solution from decomposition when exposed to 750 Lux fluorescent lighting (103).

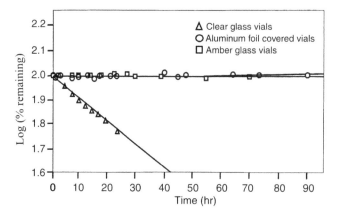

Figure 15 Effect of container type on photodegradation of furosemide solution in phosphate buffer of pH 7 exposed to fluorescent radiation. *Source*: From Ref. 74.

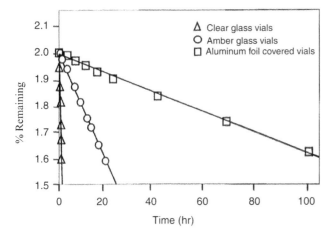

Figure 16 Effect of container type on the photodegradation of demeclocycline hydrochloride in pH 7 phosphate buffer exposed to fluorescent radiation. *Source*: From Ref. 76.

The prevention of the photodecomposition of frusemide solutions in burette administration sets is possible by covering the set with aluminum foil. However, the burette administration sets being opaque, this procedure makes the constant monitoring of the infusion rate more difficult. A more convenient method is to use Amberset burettes coated with a transparent yellow plastic material, which inhibit the transmission of UV radiation in the wavelength range of 220 to 470 nm (104).

A suitable-sized wrap-around label that covers as much as possible of the surface area of the container can also provide some protection against photodegradation. Similarly, cardboard cartons provide photoprotection except when the product is in use. The USP/National Formulary (NF) (105) states that a clear and colorless or a translucent container may be made light-resistant by means of an opaque covering, in which case the label of the container bears a statement indicating that the opaque covering is needed until the contents are to be used or administered. Furthermore, if a container is made light-resistant by means of an opaque covering, a single-use, unit-dose container, or mnemonic pack for dispensing may not be removed from the outer opaque covering prior to dispensing.

PROTECTION TECHNIQUES

In addition to selecting the appropriate solvent system, pH, buffer species and then concentration, ionic strength, dielectric constant, type of container, and storage temperature, the following techniques may be used to enhance the photostability of photolabile drugs.

Addition of Light Absorbers

Photon absorbers are considered photostabilizers by virtue of their ability to preferentially absorb the incident energy responsible for photodecomposition.

To be a suitable photoprotective agent for a drug, the absorber must absorb radiation over the same range that the photolabile drug/product absorbs. Ideally, for the greatest protection, the absorption ranges of the drug and the absorber should overlap as much as possible. The absorber should have high molar extinction coefficients in the desired spectral range, be stable towards oxygen, UV–VIS radiation, and temperature while also being compatible with the drug, soluble in the product and free from other adverse effects, e.g., toxicity (87).

Various UV absorbers have been used to protect photolabile solutions of drugs. p-Aminobenzoic acid absorbs UV radiation in the 360 to 313 nm UV region. Evidence of its photoprotective ability has been reported for solutions of colchicine (70), tetracycline hydrochloride (71), doxorubicin hydrochloride (77), and reserpine (106).

Other benzoic acid derivatives also function as effective photoprotective agents. Methylparaben has a photostabilizing effect on solutions of riboflavin (69). Ethyl aminobenzoate significantly enhanced the photostability of menadione in aqueous solutions (107).

Sodium benzoate was shown by Asker and Harris (108) to be an effective photostabilizer for physostigmine sulfate solutions. The photostability of minoxidil solution is also enhanced by sodium benzoate (48). Figure 17 shows the absorption curves of sodium benzoate and minoxidil. As can be seen in the figure, sodium benzoate absorbs radiation in the same range as minoxidil, thus acting as a photoprotective agent for the drug (109).

Urocanic acid is considered one of the major absorbers of UV radiation in the epidermis with a λ_{max} at 278 nm at pH 7 (110) and therefore functions as a natural sunscreen (111). Habib and Asker (77) found that urocanic acid significantly enhanced the photostability of doxorubicin hydrochloride solutions.

Uric acid in very low concentrations, very strongly absorbs UV radiation and for this reason, has been used to protect various FD&C colors against fading when they will be probably exposed to direct sunlight (112). Uric acid has been found to enhance the photostability of solutions of colchicine (70) and FD&C Blue No. 2 (82,113). Sodium urate, the neutral salt of uric acid, has a photoprotective effect on solutions of metronidazole (68), doxorubicin hydrochloride (77), and physostigmine sulfate (108). In addition to its photon-absorbing property, uric acid has been reported to also possess antioxidant quality (114).

Several dyes have been used as absorbers to stabilize pharmaceuticals. Thoma and Klimek (87) used amaranth, scarlet red GN, and ponceau, to stabilize the red colored solutions of daunorubicin, which are reported to bleach if exposed to a few days of natural unfiltered daylight. Methylene blue solution was stabilized by addition of the blue food dye indigo (87).

Thoma and Klimek (115) also found that the natural food colorant curcumin as well as Fast Yellow, chrysoine, apocarotinol, and, cochineal Red A photostabilized solutions of nifedipine and nitrofurazone. They also found that vanillin and methyl gallate could stabilize solutions of dihydroergotamine. These authors also found that furosemide was stabilized by the addition of vanillin, haloperidol by benzyl alcohol, and vanillin and thiothixene by quinosol and vanillin.

It is interesting to note that the red color produced as a result of the photodecomposition of dacarbazine acts as an effective inner filter inhibiting further photodecomposition (116). Asker and Habib (47) found a decreasing rate of photodegradation for more concentrated solutions of adriamycin. The decease in photodegradation rate-constant obtained at higher concentrations of furosemide is attributable to the yellow coloration formed, which acts as an inner photon

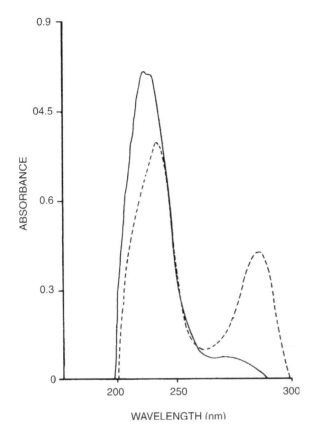

Figure 17 Absorption spectra of sodium benzoate and minoxidil in acetate buffer of pH 4.5.
—— Sodium benzoate, - - - Minoxidil. *Source*: From Ref. 109.

filter (74). This same observation is reported for demeclocycline hydrochloride (76). Allwood and Plane (85) suggested that the nonexponential nature of the degradation curve of vitamin A palmitate might be due to the protective inner-filtering effect of the degradation products.

Judis (117) investigated the possibility of using photoprotective agents such as photostabilizers to guard against the photooxidation of dihydroxyphenylalamine (DOPA). The photooxidation of DOPA in the presence of long wavelength UV radiation and 8-methoxypsoralen is completely inhibited by ascorbic acid, cysteine, cysteamine, β-aminoethylisothiuronium bromide hydrobromide, and partially inhibited by sodium thioglycollate.

Several compounds have been reported to act as radioprotective agents. These include cysteine, cysteamine, EDTA, propyl gallate, glycerin, methionine, thiourea, sodium thiosulfate, dimethylsulfoxide (DMSO), and glutathione (118). These substances include photon absorbers, free radical scavengers, antioxidants, as well as chelating agents. Habib and Asker (77) investigated the use of some of these photoprotective chemicals to photostabilize doxorubicin hydrochloride solution.

The stability of solutions of FD&C Red No. 3 exposed to fluorescent, long, and short wave UV sources was increased by the addition of DMSO (94).

Incorporation of Antioxidants

Many of the photochemical reactions reported are photooxidations. Photooxidation proceeds by a chain mechanism in which the generation of free radicals following photon absorption is the chain initiating mechanism.

The most common form of oxidative degradation in pharmaceutical preparations is autoxidation. Autoxidation involves a free radical chain process. Essentially, autoxidation is the reaction of a compound with molecular oxygen. The product of the combination of radicals and molecules can provide sufficient energy to degrade (oxidize) the molecule. To prevent this degradation, free radicals can be destroyed by the addition of a free radical inhibitor and antioxidant. Well-known free radical scavengers such as thiols, cysteine, and glutathione can retard the photodegradation process by interfering with its chain propagation steps (24).

Only a small amount of oxygen is needed to initiate an oxidative reaction; therefore, the oxygen concentration of a solution is not insignificant (25). Besides autoxidation, oxidation may also occur by the reversible loss of electrons without the addition of oxygen.

A drug's/product's susceptibility to photooxidation can be minimized by the addition of antioxidants and synergists such as chelating agents. Antioxidants are compounds that are more readily oxidized than the active ingredient they protect, and have a lower oxidation potential than the substance/product they are added to protect. Thus, they undergo a preferential degradation or can act as chain inhibitors of free radicals by providing an electron and/or receiving the excess energy possessed by the activated molecule (119). Antioxidants in common use include the following:

1. Inorganic sulfur compounds such as sodium metabisulfite, sulfite, and thiosulfate
2. Organic sulfur compounds such as thioglycerol, thiourea, cysteine dl-methionine, and glutathione
3. Alcohols and enols such as ascorbic acid and propylene glycol
4. Phenols such as hydroquinone and propyl gallate
5. Amino compounds such as glycine and phenylalanine

Sodium metabisulfite is a commonly used antioxidant. It acts by breaking up chains during autoxidation, and/or reacting with oxygen in the solution according to the following equations:

$$Na_2S_2O_5 + H_2O = 2NaHSO_3$$

$$2NaHSO_3 + O_2 = Na_2SO_4 + H_2SO_4$$

Sodium metabisulfite has been used to photostabilize solutions of minoxidil (48) and promethazine hydrochloride (120). Sodium metabisulfite, however, was found to decrease the photostability of chloramphenicol solution (86).

Sodium sulfite stabilizes solutions of daunorubicin hydrochloride (75) and physostigmine sulfate (108). Sodium sulfite, however, was reported by Asker and Canady (83) to be most detrimental to the photostability of sodium nitroprusside solution.

Sodium thiosulfate provides a photostabilizing effect for solutions of minoxidil (48), riboflavin (69), tetracycline hydrochloride (71), and phenobarbital (72).

It is interesting to note here that chlorominoxidil, which is present as an impurity in minoxidil, undergoes photodechlorination in solution. The Cl produced enhances the oxidation of minoxidil. Sodium thiosulfate, which is also known for its "antichlor" activity (removes chlorine, the oxidizer), helps stabilize minoxidil according to the following equation (109):

$$Na_2S_2O_5 + 8Cl + 5H_2O = 2NaHSO_4 + 8HCl$$

Sodium thiosulfate, however, is detrimental to the photostability of solutions of furosemide (74) and doxorubicin hydrochloride (77).

Thiourea and dl-methionine are effective photoprotective agents for solutions of riboflavin (69), tetracycline hydrochloride (71), doxorubicin hydrochloride (77), reserpine (106), and potassium iodide (121). Thiourea enhances the photostability of solutions of minoxidil (48) and furosemide solution (74). Dl-methionine increases the photostability of ascorbic acid solutions (122).

Glutathione, a naturally occurring antioxidant thiol in biological fluids, has been successfully used to stabilize solutions of doxorubicin hydrochloride (47), menadione sodium bisulfite (67), dacarbazine (73), furosemide (74), and demeclocycline (76). However, glutathione was found to increase the rate of photodegradation of chlordiazepoxide solution (123). Propyl gallate is not a very effective antioxidant for stabilizing promethazine hydrochloride solution, but it does inhibit the rate of discoloration of the solution, three times longer than sodium metabisulfite (120).

In addition to dl-methionine, other amino acids also enhance the photostability of some drugs. Glycine was found to increase the photostability of doxorubicin hydrochloride (77) and furosemide (74). Dl-cysteine significantly increases the photostability of tetracycline hydrochloride (71).

Cystine and the aromatic amino acids, e.g., tyrosine, phenylalanine, and tryptophan, absorb radiation at wavelengths greater than 254 nm (124). Tyrosine and tryptophan inhibit the photodecomposition of pyridoxine (84).

Allwood and Plane (85) found that the rate of photodegradation of vitamin A palmitate infusion was reduced by up to 50% in the presence of some amino acids mixtures compared to those prepared in a dextrose infusion. The nature of the protective effect of the amino acids was not elucidated in this report. However, it may be attributable to the antioxidant and the photon absorbing properties of these amino acids.

It is interesting to note that certain antioxidants absorb radiation in a region close to the absorption range of drugs / drug products. Therefore, they may also function as photoprotective agents (filters) by virtue of these two qualities.

Synergists, which increase the efficacy of antioxidants, are generally organic compounds that are capable of complexing small amounts of heavy metal ions. These compounds are referred to as chelating agents and include EDTA derivatives and citric, tartaric, and gluconic acids. EDTA is most commonly used as a chelating agent in liquid preparations.

In selecting a suitable chelating agent, its compatibility with the active drug should be considered. The addition of EDTA to thiomerosal solutions was suggested as a means of removing this most active antimicrobial compound from a solution by chelation (6). EDTA has been reported to enhance the photostability of solutions of minoxidil (48), riboflavin (69), sodium nitroprusside (83), tetracycline hydrochloride (71), demeclocycline hydrochloride (82), reserpine (106), and

Table 6   Effect of Certain Stabilizers on the Photostability of Tetracycline Hydrochloride Solutions

Stabilizer[a]	Rate constant $(K \times 10^2/hr)$	$t_{1/2}(hr)$
None	27.0	2.6
Reduced glutathione	8.1	8.6
p-Aminobenzoic acid	8.5	8.2
Ethylenediamine tetraacetic acid	9.2	7.5
Dl-cysteine	11.5	6.0
Thiourea	14.0	5.0
Sodium thiosulfate	17.6	3.9
DL-Methionine	17.8	3.9

[a]Concentration=0.2% in pH 7 phosphate buffer.
Source: From Ref. 71.

promethazine hydrochloride (120). Citric acid reduced the rate of photodegradation of sodium nitroprusside (83). Tartaric acid is a useful photoprotective agent for physostigmine sulfate solutions (108).

Table 6 lists various photostabilizers and their effects on the photostability of tetracycline hydrochloride solutions exposed to fluorescent radiation from cool white fluorescent tubes.

The effects of a variety of photostabilizers on the photobleaching of riboflavin solutions exposed to fluorescent radiation from cool white tubes are presented in Table 7.

A number of photoprotective agents were found to enhance the photostability of demeclocycline hydrochloride solution exposed to fluorescent radiation from cool white tubes (76). These results are summarized in Table 8.

The photostability of minoxidil solution in the presence of various photostabilizers under fluorescent radiation from cool white fluorescent tubes is presented in Table 9.

It should be pointed out that in selecting a photostabilizer, whether it is a light absorber, antioxidant, or a synergist, one should take into consideration

Table 7   Effect of Various Stabilizers on the Photobleaching of Riboflavin Solution

Stabilizer[a]	Rate constant $(K_c \times 10^2/hr)$	Percentage increase in stability[b]
None	55.8	0
Ethylenediamine tetraacetic acid	2.1	96.2
Thiourea	6.6	88.2
Methylparaben	7.6	86.4
DL-Methionine	13.2	76.3
Sodium thiosulfate	15.1	72.9
Ribonucleic acid	22.2	59.3
Reduced glutathione	41.2	26.2

[a]Concentration=0.3% in pH 7 phosphate buffer.
[b]$[(K_c-K_s)/K_c] \times 100$.
Source: From Ref. 69.

Table 8 Effect of Various Photoprotective Agents on the Photodegradation of Demeclocycline Hydrochloride

Protective[a] agent	Rate constant $K_p \times 10^3$	Percentage increase in photostability[b]
Neat	15.84	–
Glutathione	7.82	50.63
Ethylenediamine tetraacetic acid	11.71	26.07
Sodium thiosulfate	12.08	23.74
Sodium metabisulfite	12.49	21.15
p-Aminobenzoic acid	12.97	18.12
Sodium benzoate	14.17	10.86

[a]$C = 20 \mu g/mL$ for demeclocycline and protective agent, in pH 7; phosphate buffer ($\mu = 0.1$).
[b]$[(K_c - K_p)/(K_c)] \times 100$.
Source: From Ref. 76.

the optimum concentration of the photostabilizer and its compatibility with the drug and other components of the solution. An examination of Table 10 and Figures 18 and 19, clearly demonstrates that there exists an optimum concentration for a photostabilizer.

Complex Formation

Complexes formed as a result of interaction between the drugs and pharmacologically inert substances can enhance the photostability of the drugs. Kowarski and Ghandi (126) found that the presence of the quaternary ammonium compound, cetylethylmorpholinium ethosulfate, in solutions of menadione slowed the rate of photodegradation caused by exposure to 253.7 nm UV radiation. This effect was attributed to complex formation between menadione and the quaternary ammonium compound.

Several electron donors form molecular complexes with menadione, which then suppress its photodecomposition in aqueous solutions (107). DNA-daunomycin complexes exposed to UV irradiation are reported by Gray and Philips (127) to be more stable than daunomycin alone. The photostability of nifedipine increases on the addition of PVP (64,128). However, the rate of photobleaching of riboflavin by VIS radiation is considerably enhanced by PVP (28).

Table 9 Effect of Various Photostabilizers on the Photodegradation of Minoxidil

Photostabilizer[a]	$K \times 10^3/hr$	$t_{1/2}(hr)$	Percentage remaining
Sodium thiosulfate	2.3	301	85
Ethylenediamine tetraacetic acid	4.6	151	62
Sodium benzoate	6.9	100	48
Thiourea	6.9	100	48
Sodium metabisulfite	13.8	50	25
Neat	13.8	50	25

[a]All sample in 0.1 M, pH 4.5 acetate buffer.
Source: From Ref. 48.

Table 10 Effect of Radiation from Cool White Fluorescent Tubes and Dimethyl-sulfoxide on the Absorbance of Sodium Nitroprusside Solutions

Percentage dimethylsulfoxidea[a]	Absorbance				
	0 hr	3 hr	6 hr	9 hr	12 hr
0	0.034	0.231	0.389	0.493	0.570
3	0.036	0.141	0.171	0.171	0.180
5	0.035	0.142	0.151	0.151	0.163
10	0.040	0.105	0.131	0.143	0.155
20	0.039	0.122	0.132	0.161	0.150

[a]All sample 50 mg % sodium nitroprusside in pH 4.65 acetate buffer.
Source: From Ref. 125.

Complex formation between metronidazole and sodium urate is responsible for the photostabilization of metronidazole (68). Asker and Islam (72) found that the enhanced photostability of phenobarbital was due primarily to the complex formation between thiosulfate ion and phenobarbital. The complexation of nimodipine with cyclodextrins was found by Mielcarek (129) to significantly reduce its rate of photodegradation.

CONCLUSION

Photolabile drugs in liquid preparations are more susceptible to photdegradation than any other dosage forms. The simplest and most economical method for guarding against UV–VIS photon induced degradation requires an understanding of the photodegradation processes that take place.

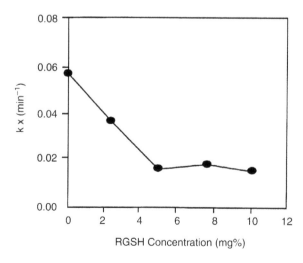

Figure 18 Effect of concentration of reduced glutathione (RGSH) on the rate of photodegradation of dacarbazine in pH 7 phosphate buffer. *Source*: From Ref. 73.

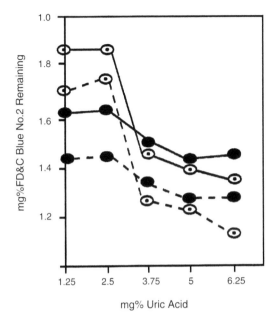

Figure 19 Effect of uric acid concentration on the photostability of FD&C Blue No. 2 aqueous solutions exposed to light from cool white fluorescent tubes. ⊙—⊙ with uric acid exposed 63 hours; ⊙—⊙ without uric acid exposed 63 hours; ●—● with uric acid exposed 111 hours; ●—● without uric acid exposed 111 hours. *Source*: From Ref. 113.

REFERENCES

1. Kaplan EH, Campbell ED, McLaren AD. Photochemistry of proteins. VIII. Inactivation of insulin by ultraviolet light. Biochem Biophys 1950; 4:493–500.
2. Luduena FP, Synder AL, Lands AM. Effect of ultraviolet irradiation of phenylephrine solutions. J Pharm Pharmacol 1963; 15:540–543.
3. Verda DJ, Kneer L, Burge WE. The effect of ultraviolet radiation on the pressor action of epinephrine. J Pharm Expt Therap 1931; 42:383–385.
4. Paiva AC, Paiva TB. The photooxidative inactivation of angiotensinamide. Biochem Bophys Acta 1961; 48:412–414.
5. Tonnesen HH, Devries H, Karlsen J, van Henegouwen GB. Studies on curcumin and curcuminoids IX: Investigation of the photobiological activity of curcumin using bacterial indicator systems. J Pharm Sci 1987; 76(5):371–373.
6. Anthony Y, Meakin BJ, Davies DJ. The antimicrobial action of degraded thiomersal solutions. J Pharm Pharmacol 1981; 33(suppl):73.
7. Tiefenbacher E, Haen E, Prazybilla B, Kurz H. Photodegradation of some quinolones used as antimicrobial therapeutics. J Pharm Sci 1994; 83(4):463–467.
8. Thoma K, Klimek R. Investigations of photoinstability of nifedipine. First communication: decomposition kinetics and reaction mechanism. Pharm Ind 1985; 47(2):207–215.
9. Tarimci N, Agabeyoglu I, Gonul N, Baykara T. Investigation of stability of the pharmaceutical colorants exposed to sunlight. Fabad Farm Bil Derg 1987; 12(4):289–195.
10. Tonnesen HH, Karlsen J, Beijersbergen van Henegouwen G. Studies on curcumin and curcuminoids. VIII. Photochemical stability of curcumin. Zlebensm Unters Forsch 1986; 183(2):116–122.
11. Harmatz D, Blauer G. Reactions of excited methylene blue. Photochem Photobiol 1983; 38(3):385–387.

12. Oster G, Wotherspoon. Photoreduction of methylene blue by ethylenediaminetetraacetic acid. J Am Chem Soc 1957; 794836–4838.
13. Jethwa SA, Stangord JB, Sugden JK. Light stability of vanillin solutions in ethanol. Drug Dev Ind Pharm 1979; 5(1):79–85.
14. Baird GM, Willoughby ML. Photodegradation of decarbazine. Lancet 1978; 2:681.
15. Hartley TF, Philcox JC, Willoughby J. Two methods of monitoring the photodecomposition of sodium nitroprusside in aqueous and glucose solutions. J Pharm Sci 1985; 74(6):668–671.
16. Vargas F, Rivas C, Machado R, Sarabia Z. Photodegradation of benzydamine. Phototoxicity of an isolated photoproduct on erythrocytes. J Pharm Sci 1993; 82(4):371–372.
17. Sax NJ. Dangerous properties of industrial materials. 3rd ed. New York: Van Nostrand Reinhold, 1968:963.
18. Cornellissen PJ, Beijersbergen van Henegouwen G, Gerritsma KW. Photochemical decomposition of 1,4-benzodiazepines. Chlordiazepoxide. Int J Pharm 1979; 3:205–220.
19. Vargas F, Rivas C, Canudas N. Formation of a perbenzoic acid derivative in the photodegradation of fenofibrate: phototoxicity studies on etrythrocytes. J Pharm Sci 1993; 87(6):590–591.
20. Dahl TA, Bilski P, Reszkahj, Chignell CF. Phototoxicity of curcumin. Photochem Photobiol 1994; 59(3):290–294.
21. Bianchi U, Mezzanotte R, Vanni R, Ferrucci L. Do eye drops containing photosensitizers represent therapeutic absurdity? Riv Farmacol 1982; 13(3):233–236.
22. Loveday KS, Bergman CL. Trends in phototoxicity testing. Drug Cosm Ind 1994; 155:30–37.
23. Evans PG, Sugden JK, Vanabbe NJ. Aspects of the photochemical behavior of 1-hydroxypyridine-2-thione. Pharm Acta Helv 1975; 50:94–99.
24. Stewart PJ, Tucker IG. Prediction of drug stability. Part 3: oxidation and photolytic degradation. Aus J Hosp Pharm 1985; 15(2):111–117.
25. Kumar V, Sunder N, Potdar A. Critical factors in developing pharmaceutical formulations—an overview. Part II. Pharm Tech 1992; 16(4):86–92.
26. Mizuno N, Fujiwara A, Morita E. Effect of dyes on the photodecomposition of pyridoxine and pyridoxamine. J Pharm Sci 1981; 33(6):373–376.
27. Baugh R, Calvert RL, Fell JM. Stability of phenylbutazone in the presence of pharmaceutical colors. J Pharm Sci 1977; 66(5):733–735.
28. Kostenbauder HB, Deluca PP, Kowarski CR. Photobinding and photoreactivity of riboflavin in the presence of macromolecules. J Pharm Sci 1965; 54:1243–1251.
29. Moore E, Chappuis PP. A comparative study of the photochemistry of the nonsteroidal anti-inflammatory drugs, naproxen, benoxaprofen and indomethacin. Photochem Photobiol 1988; 47:173–180.
30. Gu L, Chiang H, Johnson D. Light degradation of ketorolac tromethamine. Int J Pharm 1988; 41:105–113.
31. Tamat SR, Moore DE. Photolytic decomposition of hydrochlorothiazide. J Pharm Sci 1983; 72(2):180–183.
32. Moore DE, Tamat SR. photosensitization by drugs: Photolysis of some chlorine–containing drugs. J Pharm Pharmacol 1980; 32:172–177.
33. Castell JV, Gomez MJ, Mirabet V, Miranda MA, Morera IM. Photolytic degradation of benorylate: effects of the photoproducts on cultured hepotocytes. J Pharm Sci 1987; 76(5):374–378.
34. Po LW, Irwin WJ. The photochemical stability of cis-and-trans-isomers of tricyclic drugs. J Pharm Pharmacol 1980; 32:25–29.
35. Fabre H, Ibork H, Lerner DA. Photoisomerization of kinetics of cefuroxime axetil and related compounds. J Pharm Sci 1994; 83(4):553–558.
36. Lerner DA, Bonnefond G, Fabre H, Mandrou B, Simeon De Bouchberg M. Photodegradation paths of cefotaxime. J Pharm Sci 1988; 77:699–703.

37. Vargas F, Rivas C, Machado R. Photodegradation of nifedipine under aerobic conditions. Evidence of formation of singlet oxygen and radical intermediate. J Pharm Sci 1992; 8(4):399–400.
38. Iwaoka T, Kondo M. Mechanistic studies on the photooxidation of chlorpromazine in water and ethanol. Bull Chem Soc Jap 1974; 47(4):980–986.
39. Shih IK. Photodegradation products of chloramphenicol in aqueous solution. J Pharm Sci 1971; 60(12):189–190.
40. Mubarak IM, Stanford JB, Sudgen JK. Some aspects of the photochemical degradation of chloramphenicol. Pharm Acta Helv 1981; 57:226–230.
41. Tammilehto S, Heikkinen L, Jarvela P. Glass capillary gas chromatographic determination of doxepin photodecomposition in aqueous solutions. J Chromatog 1982; 246:308–312.
42. Beijersbergen van Henegouwen GM, Van de Zijde HJ, Van de Griend J, De Vries H. Photochemical decomposition of diphenhydramine in water. Int J Pharm 1987; 35(3):259–262.
43. Zanocco AL, Diaz I, Lopez M, Nunez-Vergara LJ, Squella JA. Polarographic study of the decomposition of nimodipine. J Pharm Sci 1992; 8(9);920–924.
44. Moore DE, Sithipitaks V. Photolytic degradation of frusemide. J Pharm Pharmacol 1982; 35:489–493.
45. Seno S, Kessler WV, Christian JB. Thin layer chromatographic study of prochlorperazine photodeterioration. J Pharm Sci 1964; 53(9):1101–1103.
46. Mokrosz J, Bojarski J. Photochemical degradation of barbituric acid derivatives. Part 3: rate constants of photolysis of barbituric acid and thiobarbituric acid derivatives. Pharmazie 1987; 37(11):768–773.
47. Asker AF, Habib MJ. Effect of glutathione on photolytic degradation of doxorubicin hydrochloride. J Parent Sci Tech 1988; 42(5):153–156.
48. Chinnian D, Asker AF. Photostability profiles of minoxidil solutions: PDA. J Pharm Sci Tech 1996; 50(2):94–98.
49. Smith EC, Metzler DE. The photochemical degradation of riboflavin. J Am Chem Soc 1963; 85:3285–3288.
50. Sanniez WH, Pilpel N. Photodecomposition of sulfonamides and tetracycline at oil-water interfaces. J Pharm Sci 1980; 69(1):5–7.
51. Stavchansky S, Wallace JE, Wu P. Thermal and photochemical studies of promethazine hydrochloride. A stability-indicating assay. J Pharm Sci 1993; 72(5):546–548.
52. Tavoloni N, Guarino AM, Berk PD. Photolytic degradation of adriamycin. J Pharm Pharmacol 1980; 32:860–862.
53. Morkrosz J, Zurowska A, Bojarski J. Photochemical degradation of barbituric acid derivatives. Part 4: kinetics and TLC investigations of photolysis of proxibarbal. Pharmazie 1982; 37:832–835.
54. Cornelissen PJ, Beijersbergen van Henegouwen GM, Gerritisma KW. Photochemical decomposition of 1,4-benzodiazepine. Diazepam. Int J Pharm 1978; 1:173–181.
55. Moore ED, Fallen M. Burt CD. Photooxidation of tetracycline—a differential pulse polarographic study. J Pharm 1983; 14:133–142.
56. Nunez-Vergara L, Sunkel C, Squella JA. Photodecomposition of a new 1,4-dihydropyridine: furnidipine. J Pharm Sci 1994; 83(4):502–507.
57. Pawelczyk E, Knitter B. Kinetics of drug decomposition. Part 51: kinetics of indomethacin photodegradation. Pharmazie 1977; 32(11):608–609.
58. Piechocki J, Wolters RJ. Use of actinometry in light-stability studies. Pharm Tech 1993; 17(6):46–52.
59. Sandeep N, Washkuhn RJ, Beussink DR. Photostability Testing: an overview. Pharm Tech 1995; 19(3):170–185.
60. Hochberg M, Melnick D, Siegel L, Oser BL. Destruction of vitamin B_6 (pryidoxine) by light. J Biol Chem 1943; 148:253–254.

61. Wood MJ, Irwin WJ, Scott DK. Photodegradation of doxorubicin, daunorubicin and epirubicin measured by high-performance liquid chromatography. J Clin Pharm Therap 1990; 15:290–300.

62. Bundgaard H, Nargaard T, Nielson NM. Photodegradation and hydrolysis of furosemide and furosemide ester in aqueous solution. Int J Pharm 1988; 42:217–224.

63. Taugir A, Igbal A. Effect of pH on the photostability of sulfacetamide eye-drop solutions. Pak J Pharm Sci 1988; 1(1):1–4.

64. Thoma K, Klimek R. Investigations of photoinstability of nifedipine. Part 2: influence of medium conditions. Pharm Ind 1985; 47(3):319–326.

65. Pawelczyk E, Plotkowiak Z, Ponitka E. Kinetics of drug decomposition. Part XIII. Photodegradation of sodium nalidixate in alkaline solutions. Dissert Pharm Pharmacol 1970; 22(6):449–453.

66. Akhtar MJ, Khan MA, Ahmad I. Photodegradation of folic acid in aqueous solution. J Pharm Biomed Anal 1999; 19(3–4):269–275.

67. Asker AF, Habib MJ. Photostabilization of menadione sodium bisulfite by glutathione. J Parent Sci Tech 1989; 43(5):204–207.

68. Habib MJ, Asker AF. Complex formation between metronidazole and sodium urate: effect on photodegradation of metronidazole. Pharm Res 1989; 6(1):58–61.

69. Asker, AF, Habib MJ. Effect of certain stabilizers on photobleaching of riboflavin solutions. Drug Dev Ind Pharm 1990; 16(1):149–156.

70. Habib MJ, Asker AF. Influence of certain additives on the photo-stability of colchicine solutions. Drug Dev Ind Pharm 1989; 15(5):845–849.

71. Asker AF, Habib MJ. Effect of certain additives on photodegradation of tetracycline hydrochloride solutions. J Parent Sci Tech 1991; 45(2):113–115.

72. Asker AF, Islam MS. Effect of sodium thiosulfate on the photolysis of phenobarbital: evidence of complex formation. PDA J Pharm Sci Technol 1994; 48(4):205–210.

73. Islam MS, Asker AF. Photostabilization of dacarbazine with reduced glutathione. PDA J Pharm Sci Technol 1994; 48(1):38–40.

74. Asker AF, Ferdous AJ. Photodegradation of furosemide solutions. PDA J Pharm Sci Technol 1996; 50(3):158–162.

75. Islam MS, Asker AF. Photoprotection of daunorubicin hydrochloride with sodium sulfite. PDA J Pharm Sci Technol 1995; 49(3):122–126.

76. Ferdous AJ, Asker AF. Photostabilization of demeclocycline hydrochloride with reduced glutathione. Drug Dev Ind Pharm 1996; 22(2):119–124.

77. Habib MJ, Asker AF. Photostabilization of doxorubicin hydrochloride with radioprotective and photoprotective agents: Potential mechanism for enhancing chemotherapy during radiotherapy. J Parent Sci Tech 1989; 43(6):259–261.

78. Lachman L, Deluca P, Akers MJ. Kinetics principles and stability testing. In: Lachman L, Lieberman HA, Kanig JL, eds. Theory and Practice of Industrial Pharmacy. Philadelphia: Lea & Febigers, 1986:760–803.

79. West GB, Whitlet TD. A note on the stability of solutions of phenylephrine. J Pharm Pharmacol 1960; 12(suppl):113–115.

80. Chen TM, Chafetz L. Kinetics of procaterol autoxidation in buffered acid solutions. J Pharm Sci 1987; 76(9):703–706.

81. Kowarski CR. Effect of polysorbate 80 and sodium lauryl sulfate on the photodegradation of flavin mononucleotide: a kinetic study with electron spin resonance. J Pharm Sci 1969; 58(3):360–361.

82. Asker AF, Colbert DY. Influence of certain additives on the photostabilizing effect of uric acid for solutions of FD&C Blue No. 2. Drug Dev Ind Pharm 1982; 8(5): 759–773.

83. Asker AF, Canady D. Influence of certain additives on the photostabilizing effect of dimethyl sulfoxide for sodium nitroprusside solutions. Drug Dev Ind Pharm 1984; 10(7):1025–1039.

84. Mizuno N, Nakayama F, Morita E. Photodecomposition of pyridoxine accelerated by flavine nucleotide. Vitamin 1979; 53(5–6):213–219.

85. Allwood MC, Plane JH. The degradation of vitamin A exposed to ultraviolet radiation. Int J Pharm 1984; 19(2):207–213
86. Mubarak IM, Stanford JB, Sugden JK. Effect of additives on the photostability of chloramphenicol in Clark and Lubs' borate buffer at pH 7.8. Pharm Acta Helv 1985; 60(7):187–192.
87. Thoma K, Klimek R. Photoinstability and stabilization of drugs. Possibilities of an applicable stabilizing principle for general use. Pharm Ind 1991; 53(5):504–507.
88. Kostenbauder HB. Reaction kinetics. In: Gennaro AR, ed. Remington: The Science and Practice of Pharmacy. Easton: Mack Publishing Company, 1995:231–240.
89. Martin A. Physical Pharmacy. 4th ed. Philadelphia: Lea and Febiger, 1993.
90. Hamlin WE, Chulski T, Johnson RH, Wagner JG. A note on the photolytic degradation of antiinflammatory steroids. J Am Pharm Assoc (Sci Ed) 1960; 49(4):253–255.
91. Pawelczyk E, Knitter B, Knitter K. Kinetics of drug decomposition. Part 50: graphic method for calculation of zero-order rate constants of indomethacin photodegradation. Pharmazie 1977; 32(8–9):483–485.
92. Chatterji DC, Gallelli JF. Thermal and photolytic decomposition of methotrexate in aqueous solution. J Pharm Sci 1978; 67(4):526–531.
93. Allwood MC, Plane JH. The wavelength-dependent degradation of vitamin A exposed to ultraviolet radiation. Int J Pharm 1986; 31:1–7.
94. Asker AF, Jackson D. Photoprotective effect of dimethyl sulfoxide for FD&C Red No. 3 solutions. Drug Dev Ind Pharm 1986; 12(3):385–396.
95. Thoma K, Kerker R. Photoinstability of drugs. Communication 1. Report of the behavior of substances absorbing only in the UV range during daylight simulation. Pharm Ind 1992; 54(2):169–177.
96. DeMerre LJ, Wilson C. Photolysis of vitamin B12. J Am Pharm Assoc (Sci Ed) 1956; 45:129–134.
97. Shin CT, Sciarrone BJ, Discher CA. Effect of certain additives on the photochemistry of riboflavin. J Pharm Sci 1970; 59(3):297–302.
98. Sande SA. Mathematical models for studies of photochemical reactions. In: Tonnesen HH, ed. Photostability of Drugs and Drug Formulations. Bristol, Pennsylvania: Taylor and Francis, 1996:323–339.
99. Song-Ling L, Lachman L. Photochemical considerations of parenteral products. Bull Parent Drug Assoc 1969; 23(4):149–165.
100. Krogerus VE, Nieminen Elna, Savolainen UM. Light transmission of glass containers used in pharmacy. Farm Aikak 1970; 79:101–107.
101. Wollmann H, Gruenert R. Influence of visible light on stability of isoprenaline, epinephrine and levarterenol solutions in different containers. Part 86: contributions about the problem concerning the use of plastic containers for liquid medicinal preparations. Part 21: stability of pharmaceuticals and preparations. Pharmazie 1984; 39(3):161–163.
102. Porter WR, Bauwens SF. Variation in light transmission properties of amber oral syringes. Am J Hosp Pharm 1986; 43:913–916.
103. Davidson SW, Iyall D. Sodium nitroprusside stability in light-protective administration sets. Pharm J 1987; 239:599–601.
104. Yahya AM, McElnay JC, D'Arcy PF. Photodegradation of frusemide during storage in burette administration sets. Int J Pharm 1986; 31(1–2):65–68.
105. United States Pharmacopeia XXIII-National Formulary XVIII. The United States Pharmacopeial Convention, Inc; Rockville, MD, 1995.
106. Asker AF, Helal MA, Motawi MM. Light stability of some parenteral solutions of reserpine. Pharmazie 1970; (26):90–92.
107. Shun-ichi H, Mizuno K, Tomioka S. Effects of electron donors on the photodecomposition of menadione in aqueous solution II. Stability of menadione in aqueous solution of electron donors. Chem Pharm Bull 1967; 15(11):1796–1799.
108. Asker AF, Harris CW. Influence of certain additives on the photostability of physostigmine sulfate solutions. Drug Dev Ind Pharm 1988; 14(5):733–746.

109. Chinnian DM. Photostability of Minoxidil Solutions. Master Thesis, Florida A&M University, Tallahassee, FL, 1995.
110. Morrison H, Deibel RM. Photochemistry and photobiology of urocanic acid. Photochem Photobiol 1986; 43:663–665.
111. Anglin JH. Urocanic acid, a natural sunscreen. Cosmet Toilet 1976; 91:47–49.
112. Mecca SB. Uric acid UV absorber. Drug Cosmet Ind 1965; 1:97.
113. Asker AF, Collier A. Influence of uric acid on photostability of FD&C Blue No. 2. Drug Dev Ind Pharm 1981; 7(5):563–586.
114. Djerassi D, Machlin LJ, Nocka C. Vitamin E: biochemical function and its role in cosmetics. Drug Cosmet Ind 1986; 3:46–77.
115. Thoma K, Klimek R. Photostabilization of drugs in dosage forms without protection from packaging materials. Int J Pharm 1991; 67:169–175.
116. Horton JK, Stevens MF. A new light on the photodecomposition of the anti-tumor drug DTIC. J Pharm Pharmacol 1981; 33:808–811.
117. Judis J. The photooxidation of dihydroxyphenylalanine in the presence of 8-methoxypsoralen. J Am Pharm Assoc 1960; 49(7):447–450.
118. Foye WO. Radiation-protective agents in mammals. J Pharm Sci 1969; 58(3):283–300.
119. Vadas EB. Stability of pharmaceutical products. In: Gennaro AR, ed. Remington's Pharmaceutical Sciences. Easton Pennsylvania: Mack Publishing Company, 1990:1507.
120. Cox N, Meakin BJ, Davies DJ. The influence of some antioxidants on the photochemical decomposition kinetics of promethazine hydrochloride. J Pharm Pharmacol 1976; 28(suppl):45.
121. Asker AF, Gragg R. DL-methionine and thiourea as photostabilizers for saturated solution of potassium iodide. Drug Dev Ind Pharm 1988; 14(1):165–169.
122. Asker AF, Canady D, Cobb C. Influence of dl-methionine on the photostability of ascorbic acid solutions. Drug Dev Ind Pharm 1985; 11(12):2109–2125.
123. Cornelissen PJ, Beuersbergen van Henegouwen GM. Photodecomposition of 1,4-benzodiazepine. Quantitative analysis of decomposed solutions of chlordiazepoxide and diazepam. Pharm Weekbl 1980; 2:547–555.
124. Pavlichko JP. Sunscreens in hair care products. Drug Cosmet Ind 1985; 12:35–81.
125. Asker AF, Gragg R. Dimethyl sulfoxide as a photoprotective agent for sodium nitroprusside solutions. Drug Dev Ind Pharm 1983; 9(5):837–848.
126. Kowarski CR, Ghandi HI. Complex formation between menadione and cetylethylmorpholinium ethosulfate: effect of UV photodegradation of menadions. J Pharm Sci 1975; 64(4):696–698.
127. Gray PJ, Philips DR. Ultraviolet photoirradiation of daunomycin and DNA-daunomycin complexes. Photochem Photobiol 1981; (33):297–303.
128. Ali SL. Nifedipine. In: Florey K, ed. Analytical Profiles of Drug Substances. New York: Academic Press, 1989:221–288.
129. Mielcarek J. Analytical study of photodegradation of inclusion complexes of nimodipine with a, g-cyclodextrin, methyl-b-cyclodextrin and hydroxypropyl-b-cyclodextrin. Drug Dev Ind Pharm 1998; 24(2):197–200.

18 Photostability of Cosmetic Sunscreens

Hans U. Gonzenbach

Roche Vitamins Ltd., Geneva, Switzerland

INTRODUCTION

Photostability is a term ambiguously defined. In the context of drugs, it is used with a meaning analogous to material testing known from areas such as paints, polymers, and textiles, etc. Some of these materials have chemical structures sensitive to sunlight or at least a part of the solar spectrum—especially ultraviolet radiation.

Exposure to solar irradiation (sunlight) may alter their chemical integrity and in due course some physical properties such as mechanical strength (e.g. brittleness) or color (e.g. fading) as well. Additives like industrial ultra violet (UV)-absorbers, hindered amine light stabilizers, radical scavengers and antioxidants are incorporated or applied in these products to effectively reduce or delay such radiation induced deteriorations.

Industries testing protocols focus on the quantification of radiation-induced changes in the protected material but not on the additive used as a protector. Drug substances (actives) sometimes have chemical structures, which are very sensitive to electromagnetic radiation in the UV and/or visible (VIS) range.

The International Conference on Harmonization (ICH) guideline for the photostability testing of drug substances and drug products (1) deals with methodologies of exposure of samples to simulated radiation sources and subsequent analysis, in order to assess the amount of change induced by the exposure. The aim is to make sure that a drug substance/product at the moment of administration has not suffered an "unacceptable change" (1).

The product passing this protocol should thus be "safe and effective." Again, the focus is on the product itself, and like with the other materials discussed before, not on the protection that (if necessary) could be provided by handling instructions (e.g., subdued lighting) or packaging (opaque). Cosmetic sunscreens and their active screening ingredients are a special case of photostability testing and can be discussed in various ways, most often with a meaning different from the foregoing one.

In the United States, sunscreens are classified as over-the-counter (OTC) drugs (2) hence, the ICH guideline is applicable. The requirement should then be to demonstrate that the product has the potential to protect human skin prior to application. This can be achieved analytically, e.g., utilizing UV absorption measurements or high-performance liquid chromatography (HPLC) (3), or more appropriately by an in vitro determination of its protective efficacy, e.g., sun-protection factor (SPF), UV-A protection factor (PF_{UV-A}) or critical wavelength λ_{crit} (4–6).

In real life, this is a nonissue. Substances used in sunscreen products are designed to absorb UV radiation. The energy to which each molecule might be exposed during manufacturing, compounding and storage is very small compared to the dose of radiation they will be exposed to when applied as a thin layer to human skin. Furthermore, the active ingredients are supplied in opaque packaging with the recommendation to use opaque containers for the finished products (i.e., not transparent to VIS or UV wavelengths).

Very rarely has a problem of efficacy or safety been observed when a manufacturer developed a product containing certain sunscreens packaged in a clear transparent pack because such products are normally stopped in the development phase, before they reach the marketplace. In those instances where it has occurred, a discoloration was noted adjacent to the walls of the package. This poses an esthetic problem, generally, analytically not detectable or affecting either product efficacy or safety (7).

Another way of looking at cosmetic sunscreens is to consider them for what they are—a protection system for human skin against harmful solar or artificial UV radiation. The focus then is on "use conditions," to which the ICH guidelines (1) do not apply.

An analogy to the previously mentioned material tests, where the efficacy is evaluated in terms of amount of photodegradation of the protected material, is the case of sunscreens where changes occurring on protected versus unprotected human skin should be considered. This is exactly what is being done when an in vivo test such as SPF (2,6,8) is carried out.

At the time of this writing, the debate on UV-A protection methodology is still ongoing. In view of the lack of an official UV-A (in vivo) method, some manufacturers have tried to substitute photostability measurements. Various protocols have been used, none of which has been validated nor even generally accepted.

Finally, yet importantly, cosmetic sunscreen substances can be looked at as materials being exposed to solar radiation. Their photostability can then be evaluated as change of protective potential based on a transmission measurement or loss (or survival rate) of individual UV filters upon exposure to simulated solar radiation. It is in this sense that the term "photostability" is normally used in the context of cosmetic sunscreens. It is also in this sense that the "Photostability of Cosmetic Sunscreens" will be discussed in this chapter (and not in the meaning of drug/ICH guidelines).

One question which remains to be discussed is, "Are there means of protection or stabilization for more photolabile but usable UV filters?" Protection from UV radiation, using a UV filter which is supposed to absorb the radiation by another UV filter, would be similar to one protecting an umbrella from getting wet by deploying a second umbrella over it. If an umbrella does not repel rain, it needs to be improved otherwise. Stabilization of UV filters in this sense will be discussed. Examples will be given to demonstrate how the in-use lifetime of UV filters can be extended.

Photostability and the Sun-Protection Factor

The first photostability measurements performed on cosmetic UV filters were carried out, probably in the 1930s, by Kunz (9). He prepared dilute solutions of UV filters in (cosmetic) solvents, exposed them to artificial UV radiation and recorded the changes in their UV absorption profile before and after exposure. At that time, the term, "sunprotection factor," was unknown and there was no standard methodology to evaluate the protective potential of a sunscreen on human skin. Hence, the behavior of the filter was used as a surrogate for a protection assessment.

This situation changed in the 1950s with the work of Schulze (10), which subsequently lead to the publication of several defined protocols: U.S. Food and Drug Administration (FDA) (2) (first published in 1978), Deutsches Institut für Normung (DIN) in 1984 (11), and the European Cosmetic, Toiletry, and Perfumery Association

(COLIPA) (8) in 1994. After the publication of these protocols, it became possible for the first time to quantify the protective effectiveness of sunscreens by observing changes induced by simulated sunlight on human skin with and without sunscreen. The focus of these methods is clearly on the skin protection and the performance of the sunscreen is expressed by its ability to inhibit sunburn.

Confusion subsequently arose when people started to ask questions about SPF of products that are not photostable. In other words, "can a UV filter which is altered by UV radiation be protective?" To answer this question, it is necessary to analyze how SPF is determined.

In practice, human skin without sunscreen is exposed to UV solar-simulated radiation until erythema is observed. This dose, called the minimal erythemal dose (MED), varies with skin color as well as from volunteer to volunteer of the same skin color. In practice, the same volunteers have had adjacent skin areas exposed, which are protected with the test sunscreen until erythema could be observed. The ratio of the protected to the unprotected dose is the SPF.

$$SPF = \frac{MED \text{ of protected skin}}{MED \text{ of unprotected skin}}$$

Let us now consider Figure 1, which compares two hypothetical sunscreens with a nominal SPF 15, one being 100% photostable and the other having a reduced photostability. For both, the endpoint is reached when the protected skin has received a dose of 15 times the dose of unprotected skin. The stable preparation would show no change, equivalent to a horizontal line, with a constant SPF value of 15 over time throughout the measurement. The less stable preparation must necessarily have an initial value well above 15 and end up below 15 so that its average value will correspond 15.

It is inherent in this technique of SPF measurement that we do not have access to any intermediate values. Once the end point is reached, but not before, we observe an erythema. Thus, this in vivo SPF measurement cannot determine which of one of the two preparations is undergoing changes. Any changes in the absorption of the UV filter(s) are not detectable but nevertheless built into the experiment and taken care of in the end result (9).

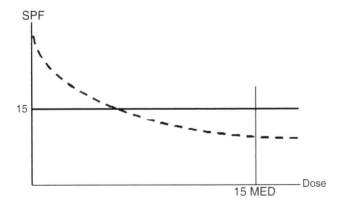

Figure 1 Comparison of two hypothetical sunscreen preparations with a nominal SPF of 15: (———) 100% photostable; (- - - - -) less photostable.

Similar comments regarding this problem were also submitted by the Cosmetics, Toiletry and Fragrance Association and Nonprescription Drug Manufacturers Association, both of the United States, to the FDA, on December 6, 1996: to quote, "...formulations which do not achieve or provide the desired level of protection for any reason can be screened out. If the photochemistry or absorbance characteristics of the active ingredients are significantly altered rendering them incapable of providing UV protection or..., the formulation would fail to meet its expected SPF level."

Two additional comments are worth considering in the context of Figure 1:

1. Skin which is exposed beyond the end point will burn and it is obvious that the consequences will be more severe for the less stable product. This applies to low-SPF products because with high-SPF products the end point is practically not reached under normal use conditions.
2. On the other hand, if the user follows the advice of most manufacturers, which is to "reapply frequently," the user will restore the initial SPF dose long before its end point is reached, especially with high-SPF products. As a consequence, with the less-stable product, the user will have the benefit of an SPF value higher than the average of the SPF test, which means higher than the nominally determined SPF.

PHOTOSTABILITY OF SUNSCREENS

The photostability of sunscreens has been, since the 1990s, primarily a UV-A issue due to the fact that no agreed method was (and still isn't at this time) available to assess the protective capacity of a sunscreen preparation towards UV-A radiation. As in the early days, the photostability of the UV filter or the sunscreen preparation itself was tentatively used instead of a skin-related end point.

The term "photostability" in the context of UV filters and sunscreen preparations is most often used in either of two ways:

1. Analytically: related to actual quantity before and after exposure to a given dose of simulated solar radiation. This implies that analytical techniques being used will allow both a separation and a quantification of the UV filter(s) and any potential isomers. This allows one to take into account reversible transformations, e.g., *cis* Δ *trans* isomerizations, and in the case of complex blends, to obtain information as to which filter shows what amount of loss or the "survival rate" of each of the compounds in a mixture.
2. Spectrally: related to anticipated protection. In practice, the spectral profiles before and after exposure are compared and any changes are interpreted as photoinstability. This is a very crude analysis, not allowing one to identify reversible transformations, which after an equilibration phase may lead to stable systems (e.g., ethylhexyl methoxycinnamate [EHMC, formerly OMC (octyl methoxycinnamate)]. Because it gives an overall profile, it is also not possible, in the case of complex mixtures, to know which filter is doing what.

Consequently, this kind of analysis may often lead to the premature disqualification of UV filters or UV filter combinations that are often found in sunscreen preparations.

Optimally, the "analytical" and the "spectral" approaches should be combined. The two are complementary and together provide the most insight into the behavior of a UV filter or a sunscreen preparation.

UV FILTERS AND THEIR PHOTOCHEMISTRY

When studying UV filters or sunscreen preparations, it is useful and desirable to have some knowledge about the photochemical behavior of the components under study. It is the absence of this knowledge which, over the last two decades, has led to many misinterpretations and erroneous conclusions.

Organic UV filters are designed to absorb UV radiation or, on a molecular level, UV photons. The absorbed energy produces an excited state of the molecule, giving it higher energy content and a very short lifetime (micro to milliseconds). Such an excited UV filter molecule can stabilize itself by dissipating this energy via one or more of three possible physical processes:

A. emitting radiation of a wavelength longer than the excitation wavelength— fluorescence or phosphorescence
B. transmitting its energy as vibrational energy to surrounding molecules, thus producing heat
C. transferring its energy to another molecule, and producing an excited state in another molecule

All these processes are physical and compete against each other. They allow the excited UV filter molecule to free itself of its excitation energy and return to the ground state, where it can absorb another photon.

Besides these physical pathways, some molecules may also lose their excitation energy through chemical transformations, some of which are even desirable. When considering the photochemical reactions of UV filters, it is important to distinguish between:

a) reversible transformations = stable system
b) nonreversible transformations = photodegradation

Reversible Transformations

Reversible transformations are by no means rare and often lead to confusion because they are not easy to detect using simple analytical means. The most prominent example is that of the *cis* Δ *trans* isomerization, of a double bond. The best-documented example of this transformation is EHMC as outlined in Figure 2.

The normal pure synthetic material contains more than 98% the *trans*-isomer. Upon exposure to UV radiation, it will undergo isomerization to the *cis* form which itself is a UV filter with a very similar shape of absorption curve. Hence, the *cis*-isomer will also absorb UV radiation and isomerize back to *trans* form. Neither of the two isomers are stable and their back and forth isomerization will lead to a photostationary equilibrium (60% *cis* + 40% *trans*); the two isomers together form a stable system. This equilibration is fast and accompanied by a drop in overall absorbance. This decrease is due to the fact that, although the two isomers have absorption curves very similar in shape, they differ in magnitude of their individual extinction coefficients as shown below.

Isomer	Solvent	Molar extinction coefficient ε
trans EHMC	EtOH	24,300
cis EHMC	EtOH	14,500

The equilibrated mixture will exhibit an absorbance lower by 20% to 25%, depending on the medium, compared to the same sample before exposure (Fig. 3).

This equilibrium problem is a deficiency that has been discussed in many published papers. Many authors have exposed samples (under various conditions) containing EHMC, then compared the absorbance at time zero with one other point at time X and found a drop in absorbance, which they then interpreted as photoinstability. They have not performed a kinetic analysis, which would have revealed that once the system is equilibrated it is virtually stable as shown in Figure 3. One paper even describes such a study but uses a nonlinear scale for the kinetics and completely misses this important point (12). This most important point was brought to the attention of the public, by Gonzenbach (13). Use of another analytical approach has confirmed this. Figure 4 shows the same experiment previously illustrated in Figure 3, but this time the analysis was performed using a kinetic approach based on HPLC analysis, with the results plotted on an expanded time scale (14).

Figure 4 shows the very rapid drop in concentration of the *trans* form to about 40% of the initial value—the remainder (not shown) being the *cis* isomer. In parallel, a sample of pure *cis* isomer also exposed under identical conditions produced about 40% of the *trans* isomer. As can be seen, the two curves meet at the same level, i.e., when the equilibrium has been reached. EHMC is not an exception. The UV filters shown in Figure 5 show a similar behavior (15).

In 1981, Beck, Deflandre, Arnaud, and Lemaire (16) stated that E Δ Z, *cis* Δ *trans* isomerization is a desirable process for a UV filter, i.e., "Isomerization E Δ Z is thus a choice process for a sunscreen agent. In practice, it is this mixture of E and Z forms, which acts as a UV filter and it is therefore the mixture's stability which should be considered rather than that of the initial compound."

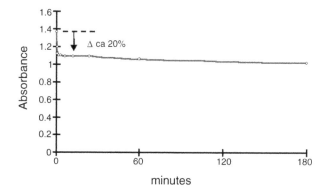

Figure 2   Isomerization of ethylhexyl methoxycinnamate. Synthetic material is over 98% *trans*, upon exposure rapid isomerization leads to a photostationary state: about 40% *trans* and 60% *cis*.

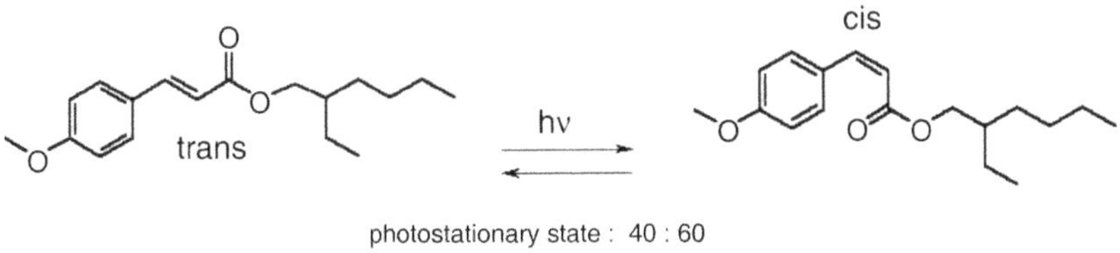

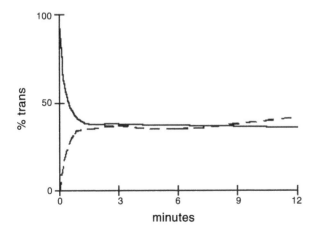

photostationary state : 40 : 60

Figure 3 Parsol MCX in polymer film (amount/surface area equivalent to 2% in a cream spread at 2 mg/cm²) exposed to sun at Jungfraujoch, Switzerland, 11,500 feet a.s.l./3580 m, ü.M. Irradiance: 4.2 SU/hr ≈ 4.2 MED/hr.

Other reversible transformations involve tautomeric structures. This is, for instance, the case for 1,3-dicarbonyl compounds particularly dibenzoylmethane (DBM), the parent compound of the UV A filter butylmethoxydibenzoylmethane (BM-DBM). The photochemistry of DBM (Fig. 6) has been studied by Veierov (17), Markov (18,19), Yankov (20), Moriyasu (21) and others.

DBM is almost fully enolized and its exposure to UV radiation leads to the diketo form. The diketo form is not stable and will revert back to its keto-enol tautomer. This back reaction occurs in the absence of radiation (dark reaction) and is influenced by the polarity of the medium, temperature and pH. This produces an equilibrium in which the forward reaction is driven by UV radiation and the back reaction is not. Therefore, one would expect that an increase of irradiance would result in a displacement of the equilibrium. This expectation is in agreement with the findings of an effectiveness study by Gonzenbach et al. (22) in which the protective properties of BM-DBM (Parsol® 1789, Avobenzone) were assessed on human skin using a ninefold variation of irradiance. The enolization of the diketo form, back reaction, is a relatively slow process. Therefore, it can be observed analytically by HPLC (3) but may also be observable in photostability studies monitored by simple, nonspecific spectral means. It is not unusual to find a low extinction value

Figure 4 Same as Figure 3 but analyzed by high-performance liquid chromatography. Start with 100% *trans* isomer leads rapidly to about 40% *trans* while the difference (60%) is the *cis* isomer (not shown). Start with 100% *cis* isomer produces, in a parallel experiment, about 40% *trans* isomer.

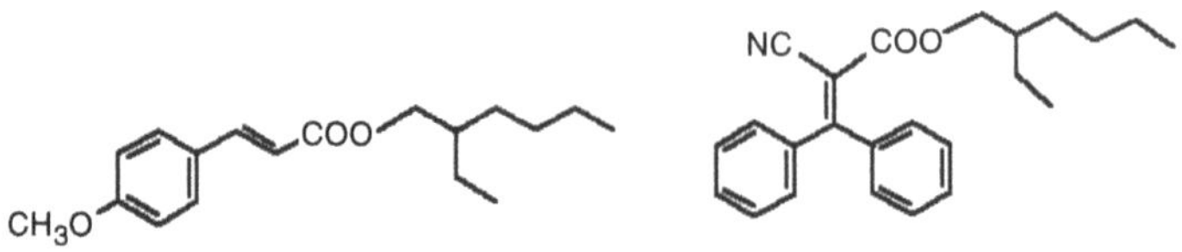

NC COO

isomerization degenerate

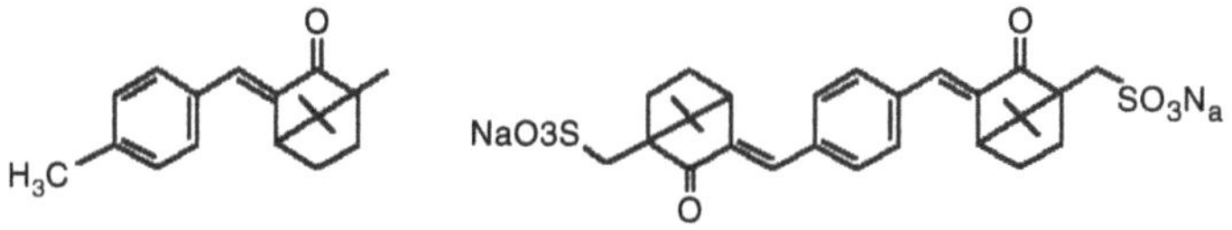

H. Gonzenbach, R.Schwarzenbach 1986
Photostationary state: E/Z ratio = 40 : 60 [13]

NaO3S   SO3Na

A. Deflandare, G. Lang 1988 [14]

Figure 5 Cosmetic ultraviolet-filters using E/Z isomerization as preferred mechanism to dissipate absorbed radiation energy.

in the 360 nm region in measurements immediately after exposure, and a (at least) partial recovery in the following minutes to hours later, depending on the system.

A different case is the photoenolization of ortho-hydroxybenzoyl derivatives, as outlined in Fig. 7. Here we see there is practically a total reconversion back to starting material at a very fast rate, leading one to conclude that there is total photostability. The same conclusion would be reached whether one is using either a normal specific analytical (e.g., HPLC) or the simple spectral approach. Special photophysical equipment is needed to observe these very short-lived photo-enol transformations.

Nonreversible Transformations

All previously discussed pathways, physical and chemical; regenerate the UV filter in its ground state (or an isomer thereof) where it is again capable of absorbing a photon. The same molecule can thus go through very many of these cycles. Occasionally, it may take a different (nonreversible) pathway, leading to its consumption

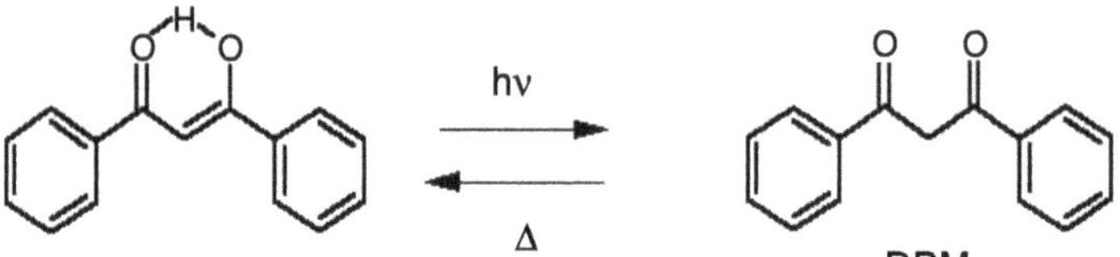

hν

Δ

DBM

Keto-enol from
strong UV-A absorber

Diketo from
weak UV-B absorber

Figure 6 Tautomeric equilibrium of dibenzoylmethane. Keto-enol dominates in solution. Diketo is formed upon exposure and reverts back to keto-enol mix in the absence of radiation (dark reaction).

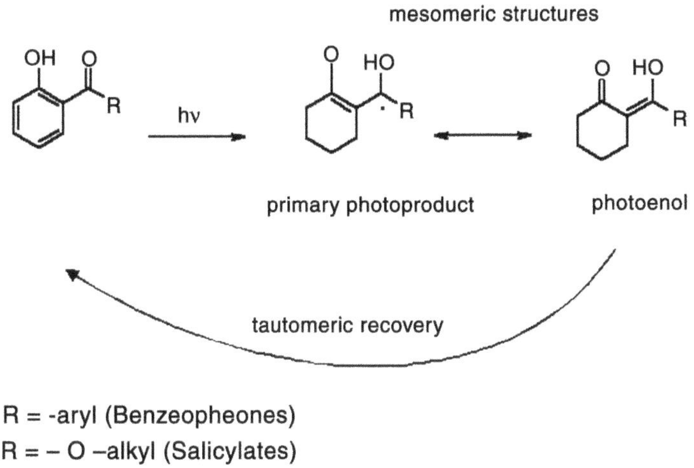

Figure 7 Intramolecular hydrogen transfer in *ortho*-hydroxy-benzoyl filters leading to an unstable photoenol which can stabilize itself by a tautomeric proton transfer.

in the process. The most prominent examples of these nonreversible reactions are given in the following sections.

Dealkylation of Dialkyl-p-Amino-Benzoic Esters

Berge reported that ethylhexyldimethyl-PABA is demethylated via a photo oxidation/decarbonylation sequence (23,24). The photoproducts produced are N-mono-methyl substituted or possess a free amino-group. These products have UV absorption spectra very similar to that of the starting dimethylamino compound. Therefore, this transformation cannot be detected, utilizing a simple spectral measurement (12,13). This is an example of a nonreversible photodegradation, which, in the spectral-only measurement, produces a result, which falsely indicates that the test substance/product has a higher photostability than is actually the case.

2+2 Addition of Cinnamates

Cinnamates are known to undergo 2+2 addition reactions, either with themselves or with other olefins. The structures of the main isomers of the dimerization of EHMC are shown in Figure 8 (Thorel P.J. Personal communication 1989).

The 2+2 addition to olefins, or e.g., the double bond of the keto-enol form of BM-DBM can also occur, leading to intermediate cyclobutanols which then undergo ring opening (De Mayo reaction) to form structures such as those shown in Figure 9. (Bringhen A. Personal communication 1993).

Alpha-Cleavage (Norrish I) of Dibenzoylmethane

Alpha-cleavage (Norrish I) is a frequently encountered photoreaction for aromatic carbonyls. In the case of DBM and BM-DBM, it can be assumed to occur in the diketo form producing two reactive (and therefore short-lived) radicals, which produce a variety of secondary photoproducts depending on the medium (25).

An important role in this cascade is that played by the recombination of radicals or radical pairs. This leads, in the case of BM-DBM, to the production original

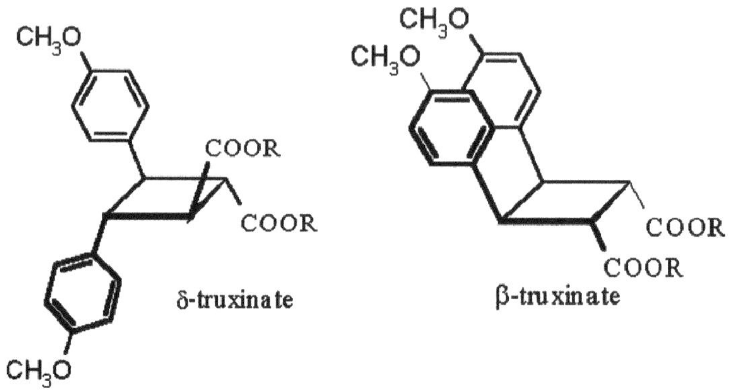

Figure 8 Structures of the main isomers resulting from 2+2 addition of ethylhexyl methoxycinnamate to itself. Note that only the *trans* isomers are involved.

starting material or other symmetrically substituted DBMs, both having UV-A filter properties. One of the products, dianisoylmethane, produced in this process has even been used commercially (Parsol DAM) as a cosmetic UV-A filter prior to the use of BM-DBM.

DETERMINATION OF PHOTOSTABILITY

Single Ultraviolet Filters

At the request of COLIPA, a task force mainly composed of experts from the cosmetic UV filter producing industry worked to produce a protocol for the determination of the photostability of UV-filters (26). It was their objective to define an in vitro model,

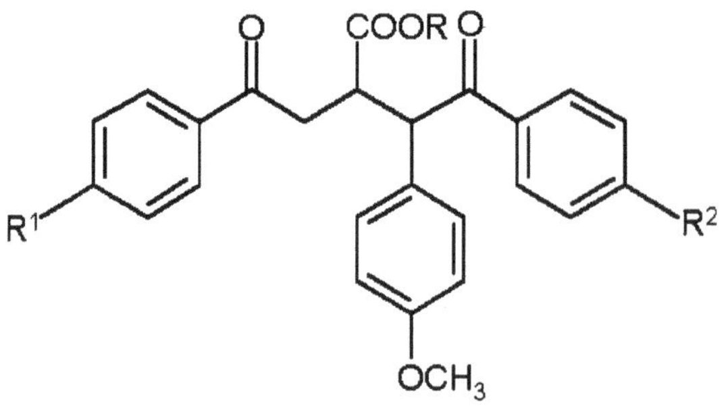

$R^1 = -OCH_3$ when $R^2 = $ t-Butyl

$R^1 = $ t-Butyl when $R^2 = -OCH_3$

Figure 9 Structures resulting from 2+2 addition of ethylhexyl methoxycinnamate to butylmethoxydibenzoylmethane followed by subsequent ring opening of the intermediate cyclobutanol.

simulating conditions of actual use, which is simple and reliable. Key points of this method are as follows:

- Light source: The UV part of the solar spectrum should be correctly simulated, using e.g., a long-arc xenon lamp filtered through a Schott WG 320/2 mm or a spectrally equivalent setup. The irradiance level should not exceed 4 MED/hr at sample level in order to avoid potential problems due to nonreciprocity of UV filters, e.g., DBM filters.
- Sample: The product should be irradiated as a liquid film in a cosmetically acceptable solvent. For that purpose, the product is dissolved in a 70:30 mixture of two solvents, 70% volatile (e.g., alcohol or water) and 30% nonvolatile (e.g., cosmetic esters or glycerin, etc.). The concentration should be the (max) use concentration. A 20 µL portion of this solution is then spread on a surface of 10 cm² glass plate and then left to dry. The volatile part evaporates leaving behind a film, which can be considered an adequate model for a sun protection product applied at 2 mg/cm², after drying. It is not advisable to cover the film with a quartz plate because this would hinder free access of ambient oxygen. A model requiring the use of a cover plate would be unrealistic in view of the fact that oxygen may interact with excited states and therefore a result in the absence of oxygen would not be representative.
- Exposure: The samples, thus prepared, are then exposed for 5, 10 MEDs. This dose is considered high enough for a single filter. In high-SPF products, blends of filters are used and the share of energy each filter is exposed to does not exceed such a dose.

After exposure, the samples are immersed in a suitable solvent, their container immersed in an ultrasonic bath, sonicated and the resultant solution analyzed. It is strongly recommended that each sample be run in triplicate and the results compared to those of three similar unexposed controls kept in the dark for the same exposure duration. This protocol is outlined in Figure 10.

Although the analytical protocol was rigorously defined for the round robin test, it was deliberately left open in the "Proposed Protocol" so that one can use the analytical means, which are most appropriate for the product under study. On the other hand, it is strongly recommended that the solution of each "washed sample" be analyzed by using two different analytical techniques, e.g.,

> direct spectrophotometric analysis
> chromatographic analysis (e.g., HPLC) for taking into account individual isomers (e.g., EHMC).

The two results obtained are often not identical. The UV absorption profile contains information as to the anticipated protection after exposure whereas the chromatographic result indicates quantitatively a recovery rate or "survival rate" of each of the components including their isomers (which may have a lower extinction, as previously discussed, e.g., EHMC).

Sunscreen Preparations

The previously described protocol for single UV filters is easy to use and has been successfully applied to study the photostability of blends of UV filters and adapted

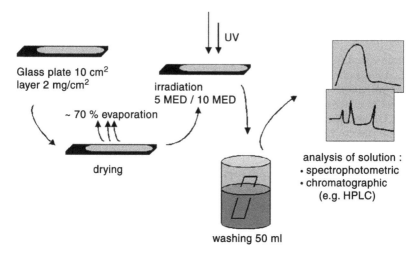

Figure 10 Scheme of the proposed protocol for determination of photostability of cosmetic ultraviolet filters (25).

to work with sunscreen preparations (emulsions). Nevertheless, COLIPA has asked a second task force to deal with the determination of the photostability of sunscreen preparations. This group was composed of representatives from the industry manufacturing finished preparations. Their experimental work has been completed but not published, to date. Starting with the previous protocol, the key points of difference in this new protocol are as follows:

• Dosimetry: An improved and more rigorous monitoring of irradiance over the entire UV spectrum and a more careful control of the ratio of the UV A part to the UV-B part of the spectrum.
• Sample: The dosage and spreading of an emulsion on a flat support is difficult and requires special attention. A roughened support surface is recommended.

   Both these points are essential if acceptable intra- and interlaboratory reproducibility is to be achieved. These were lessons learned from round robin tests in which a very severe analytical protocol, based on HPLC, was followed.

PHOTOSTABILIZATION
The Inner Filter Effect
It has been suggested that stable UV filters be used to stabilize (shield) other less stable UV filter compounds (e.g., of the DBM type). If the two filters of these two types are in a test blend and their two absorption curves do overlap at several wavelengths, they will share the incident energy in proportion to their respective extinction coefficients and amount of overlap. The larger the amount of overlap and the higher the extinction coefficient of the stable filter, the less the amount of energy available to interact with the unstable component. The system will be more stable, even though there is no direct (molecular) interaction between the two substances. This shielding effect is due to what is known as "inner filter effect."

If in the extreme case where the unstable absorption curve overlaps completely with that of the stable one and this latter compound has a much higher extinction coefficient, it could essentially absorb all incident photons. The unstable compound would therefore appear to be stable when in actuality it would be useless as a protecting substance. Under these conditions, it would absorb an insignificant amount of energy and hence not make any significant contribution to the overall protection of the product (27).

One could compare this to an umbrella made out of paper that is not resistant to rain. If, a second umbrella, made out of strong and water-resistant material, is deployed over it, thus providing protection, the first and lower umbrella is shielded. Thus, the more fragile paper umbrella appears to provide some degree of protection only because of the action of the upper second umbrella. The paper umbrella by itself would be quickly destroyed in the rain. What is then the role and use of the paper umbrella?

Triplet–Triplet Energy Transfer

It is well known that electronically excited molecules can transfer their excitation energy to other molecules under certain conditions. First, a donor molecule, D, capable of being excited must be exposed to radiation and reach an excited state D*. This molecule D*, must have a sufficiently long-lived excited state, usually a triplet state. During this lifetime, it must encounter an acceptor molecule A. If an energy transfer does occur, the donor molecule then returns to its original ground state and the acceptor molecule, A*, is now in its excited (triplet) state. This process is called a triplet–triplet energy transfer.

$$D^* + A \rightarrow D + A^*$$

A second requirement of such an energy transfer is that molecule A be capable of existing in an excited (triplet) state at an energy level similar to that of the D* molecule—ideally slightly below that of the donor molecule because of energy transfer losses. The third and last condition is that molecule A be present at a concentration, sufficient to ensure successful collisions with D* during its short lifetime. If all these conditions are fulfilled and D* has no other option by which it can dissipate its excitation energy than through irreversible chemical reaction or energy transfer and A can successfully take over this energy, we have a photostable system. This assumes that A* can dissipate this trapped energy in a safe and efficient way such as phosphorescence, internal conversion, or E/Z isomerization, etc.

BM-DBM has a triplet energy of 59.5 kcal/mol (28) and therefore a suitable quencher for BM-DBM should have a triplet energy in the order of 55 to 60 kcal/mol. Many organic molecules fulfill this requirement but not all are usable because they do not have an acceptable regulatory status which would allow their use in cosmetic preparations nor are they sufficiently photostable, per se, for this purpose. Figure 11 presents three such substances, which are all efficient quenchers for BM-DBM (29).

The requirement that a molecule possess an appropriate triplet energy level implies the presence of a chromophore. All three of the compounds presented in Figure 12 are UV-B absorbers and have the possibility to dissipate excitation energy via an E/Z isomerization. This means they are capable of performing this highly efficient process and are not consumed in the process. Figure 12 illustrates the

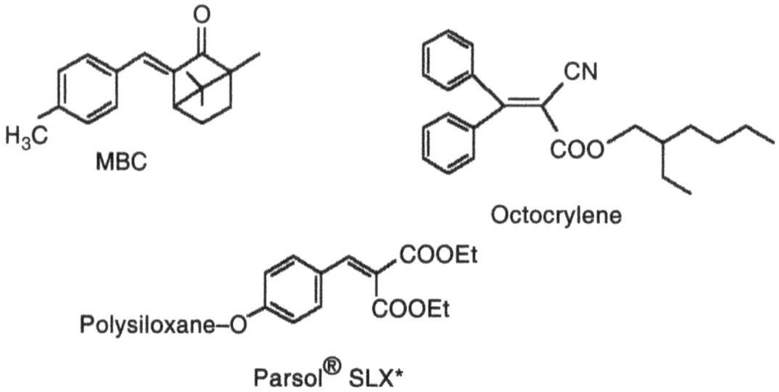

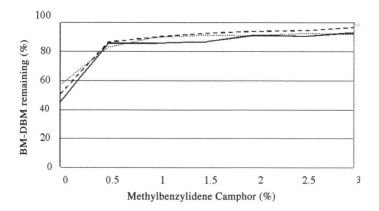

*INCI name : not yet defined.

Figure 11 Structures of suitable triplet energy quenchers for butylmethoxy-dibenzoylmethane. Note that all have a double bond allowing for E/Z isomerization. In octocrylene and Parsol® SLX, this isomerization is for symmetry reasons degenerate and therefore not detectable.

behavior of BM-DBM at different concentrations for methylbenzylidene camphor (MBC), one of these compounds.

Please note that the stabilizing effect is not a question of a defined ratio as would be expected if it were a filter effect. Note also that the same curves are obtained for different concentrations of BM-DBM, namely 1%, 2%, and 3%. A patent (30) claims that a ratio of at least 3:1 MBC:BM-DBM is required to achieve protection of the more photolabile BM-DBM. The curve presented in Figure 12, however suggests a threshold concentration of MBC in the order of 0.5% to 1% is sufficient for an energy transfer mechanism. Figure 13 demonstrates a similar behavior obtained using octocrylene as stabilizer.

Considering the similarity of the curves presented in Figures 12 and 13 and the fact that octocrylene has filtering characteristics (which are very different from MBC both in wavelength and in extinction capacity) it is obvious that the

Figure 12 BD-DBM, (·······) 1%, (- - - - -) 2%, (——) 3%; tested according to the presence of increasing amounts of methylbenzylidene camphor, 0% to 3%. The analysis of remaining BM-DBM after exposure is based on high-performance liquid chromatography. *Abbreviation*: Butylmethoxy-dibenzoylmethane. *Source*: From Ref. 25.

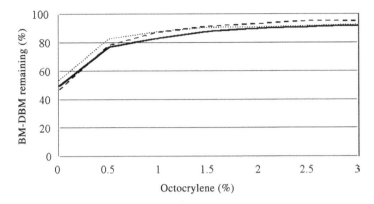

Figure 13 BM-DBM, (········) 1%, (- - - - -) 2%, (——) 3%; tested according to Schwak and Rudolph in the presence of increasing amounts of octocrylene, 0% to 3%. The analysis of remaining BM-DBM after exposure is based on high-performance liquid chromatography. *Abbreviation*: BM-DBM, Butylmethoxy-dibenzoylmethane. *Source*: From Ref. 25.

stabilization is not due to solely a filtering or shielding effect. The combination of BM-DBM and octocrylene has also been the object of patents (31,32).

Finally, Figure 14 shows the results obtained with Parsol SLX. This is a new polysiloxane, UV-B filter with a statistical, average chain length of 64 Silicon units, and an average of four chromophores grafted to each chain (33). At the time of this writing, the International Nomenclature Cosmetic Ingredients name for this new substance is still pending but regulatory approval in the European Union (EU) has been given.

If one takes into account that the siloxane backbone is not active and considers only at the equivalent weight of the chromophores, the situation is very much the same as with octocrylene and with MBC. A 5% preparation of Parsol SLX (total weight) gives the same stabilization as 1% of octocrylene (OC) or MBC (34).

One may ask if EHMC is suitable for stabilizing BM-DBM. It has molecular features similar to the previously discussed three UV-B sunscreens and an acceptable triplet energy of 57 kcal/mol (27). In fact, EHMC does stabilize BM-DBM when the mixture is exposed to UV-A radiation only (>340 nm). This is another proof that the stabilization is not due to a shielding effect. EHMC does not absorb any radiation under these conditions (Gonzenbach H and Berset G, personal communication, 1999).

A stabilization effect has also been reported for full spectrum exposures (UV-B + UV-A) when EHMC is present in relatively low amounts (1–2%) in a system (35). This is in agreement with our own findings but at higher concentrations of EHMC. The 2 + 2 addition of EHMC to BM-DBM yields the photoproduct shown in Figure 9, which overrides its stabilization effect (Gonzenbach H and Berset G. personal communication, 1999).

SYNOPSIS

The photostability of cosmetic sunscreens in the sense of the ICH guidelines (before use) is not an issue. In contrast to some drugs, which may be very sensitive to UV and/or VIS radiation, sunscreen agents are designed to absorb solar UV radiation.

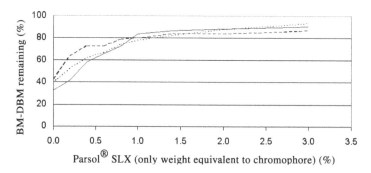

Figure 14 BM-DBM, (······) 1%, (- - - - -) 2%, (——) 3%; tested according to Schwak and Rudolph in the presence of increasing amounts of Parsol SLX, 0% to 3% weight equivalent to chromophore corresponding to a total weight of 0% to 15%. The analysis of remaining BM-DBM after exposure is based on high-performance liquid chromatography. *Abbreviation*: Butylmethoxy-dibenzoylmethane. *Source*: From Ref. 25.

Further, sunscreen preparations are usually packaged in materials, which are opaque to UV. The integrity of a sunscreen preparation can be checked by an in vitro SPF determination or more preferably and accurately by chemical analysis, e.g., HPLC (3) or spectrophotometrically.

The photostability of cosmetic sunscreens has been and still is a subject of discussion during use where the ICH guidelines do not apply. The major concern being that an important breakdown (photodegradation) could lead to a significant reduction of photoprotection and could potentially have fatal consequences.

In the UV-B spectral region, this aspect is adequately covered by an in vivo SPF determination, which also does imply an exposure to UV-A radiation. In the UV-A spectral region, this will also be the case if the method to assess UV-A protection, to be adopted in a near future, is based on an in vivo model. If this is not the case and the adopted method would be based on a in vitro determination, it should incorporate a (pre-) irradiation step in order to take into account photochemical changes which may occur during exposure and also during use, as suggested by Gers-Barlag et al. (36).

The photostability of a UV filter can be influenced by the presence of another filter in a formulation. This interaction can also be destabilizing, e.g., EHMC adding to BM-DBM, or provide a significant stabilization, as seen by the action of MBC, OC or Parsol SLX on BM-DBM.

The photostability of sunscreen formulations does not require regulation because SPF assessment covers this aspect. Consumers want to buy protection and not be bothered by technical items, which, they cannot understand.

UV filter and finished product photostability are a topic for the synthetic and formulation chemists. Their use of reliable and specific assessment protocols provides an excellent screening tool for development of new ingredients and formulations. Insight into the photophysical and photochemical behavior of the substances/products under study will help to avoid errors in the design of experiments. It is also essential to a meaningful interpretation of data and helps prevent premature disqualification of substances/products, which for instance undergo equilibration when exposed.

REFERENCES

1. International Conference on Harmonization. Guidelines for the Photostability Testing of New Drug Substances and Products, Federal Register. May 16, 1997; 62(95).
2. Sunscreen Drug Products for Over-the-Counter Human Use, Final Monograph. Federal Register. May 21, 1999; 64(98).
3. Schneider P, Bringhen A, Gonzenbach HU. Single step HPLC analysis of sunscreen products containing complex UV-filter systems. Drug Cosmet Ind 1996; 32–38, 77.
4. Diffey BL, Robson J. A new substrate to measure sunscreen protection factors throughout the ultraviolet spectrum. J Soc Cosmet Chem 1989; 40:127–133.
5. Diffey BL. A method for broad-spectrum classification of sunscreens. Int J Cosmet Sci 1994; 16:47–52.
6. Australian/New Zealand Standard A5/NZs Sunscreen Products-Evaluation and Classifiaction 2604; 1997.
7. Johncock W. Sunscreen interactions in formulations. Cosmet Toil 1999; 9(114):75–82.
8. COLIPA Sun Protection Factor Test Method. The European Cosmetic Toiletry and Perfumery Association, Bruxelles, 1994.
9. Kunz EC. An Advance in the Field of Sunburn Preventive Chemicals. Givaudanian, January 1937.
10. Schulze R. Gewöung und umstimmung der menschlichem haut bei Brestrahlung mit sonne und ultrvitaluxlampe. Strahlentherapie 1952; 86:51–61.
11. Deutsches Institut für Normung (DIN). Normen ausschuss Lichttechnik, Berlin, Ger. 1984.
12. Kammeyer A, Westerhof W, Bolhuis PA, Ris AJ, Hische EA. The spectral stability of several sunscreening agents on stratum corneum sheets. Int J Cosmet Sci 1987; 9: 125–136.
13. Gonzenbach HU, Schudel P. Letter to the editor. Int J Cosmet Sci 1988; 9:287–292.
14. Schwarzenbach R, Gonzenbach HU, Klecak G. New approach for testing UV-filters. 14th Int Congress of the IFSCC, Barcelona, 1986.
15. Deflandre A, Lang G. Photostability assessment of sunscreens. Benzylidene and dibenzoylmethane derivatives. Int J Cosmet Sci 1988; 10:53–62.
16. Beck I, Deflandre A, Lange G, Arnaud R, Lemaire J. Study of the photochemical behavior of sunscreens-benzylidene camphor and derivatives. Int J Cosmet Sci 1981; 3:139–152.
17. Vireo D, Bercovici T, Mazur Y, Fischer E. Effect of additives and solvents on the fate of the primary photoproduct of 1,3-dicarbonyl compounds. J Org Chem 1978; 43: 2006–2010.
18. Markov P, Petkov I. On the photosensitivity of dibenzoylmethane, benzoylacetone and ethyl benzoylacetate in solution. Tetrahedron 1977; 33:1013–1015.
19. Markov P. Light-induced tautomerism of β-dicarbonyl compounds. Chem Soc Rev 1984; 13:69–97.
20. Yankov P, Saltiel S. Photoketonization of dibenzoylmethane in polar solvents. Chem Phys Lett 1986; 128:517–520.
21. Moriyasu M, Kato A, Hashimoto YJ. Kinetic studies of fast equilibrium by means of high-performance liquid chromatography. Part 11. keto-enol tautomerism of some β-dicarbonyl compounds. Chem Soc Perkin Trans 1986; II:515–520.
22. Gonzenbach HU, Mascotto RE, Cesarini JP. UV-A sunscreen in vivo effectiveness measurements. Cosmet Toil 1991; 106:79–84.
23. Ten Berge CDM, Bruins HP, Faber JS. Die photochemie von sonnenschutzmitteln. I. Über die photochemie von methyl-*p*-dimethylaminobenzoat. J Soc Cosm Chem 1972; 23:289–299.
24. Ten Berge CDM, Bruins JP. Die photochemie von sonnenschutzmitteln. II. Über die photochemie von methyl-*p*-dimethylaminobenzoat. J Soc Cosm Chem 1974; 25:263–269.
25. Schwack W, Rudolph T. Photochemistry of dibenzoylmethane UV-A filters. J Photochem Photobiol B: Biology 1995; 28:229–234.
26. Berset G, Gonzenbach H, Christ R, et al. Proposed protocol for determination of photostability Part I: Cosmetic UV-filters. Int J Cosmet Sci 1996; 18:167–177.

27. Gonzenbach HU, Pittet GH. Photostability a must? Conference: broadspectrum sunprotection, the issues and status, London, U.K., March 11–12, 1997: Summit Events Ltd., London, U.K.
28. Gonzenbach H, Hill TJ, Truscott TG. The triplet energy levels of UV-A and UV-B sunscreens. J Photochem Photobiol B: Biology 1992; 16:377–379.
29. Berset G, Bringhen A, Gonzenbach H. Photostable UV-filter combinations containing butyl methoxy dibenzoylmethane. Poster presented at the 7th Congress of the European Society for Photobiology, Stresa, Italy, 1997.
30. LU 86 703, priority 8.12.86, Luxemburg, L'Oréal S.A, Paris, France.
31. EP 0514 491, priority 14.2.90, France, L'Oréal S.A, Paris, France.
32. EP 0780 119, priority 18.12.95, EU, Givaudan-Roure (International) S.A, Vernier, Geneva, Switzerland.
33. EP 0538 431, priority 10.5.91, Great Britain, Givaudan-Roure (International) S.A, Vernier, Geneva, Switzerland.
34. EP 0709 080, priority 14.10.94, Givaudan-Roure (International) S.A, Vernier, Geneva, Switzerland.
35. US 6 071 501, priority 5.2.99, U.S.A., Procter & Gamble, Cincinnati, Ohio, USA.
36. Gers-Barlag H, Klette R, Springob C, et al. In vitro testing to assess the UVA protection performance of sun care products. Int J Cosmet Sci 2001; 23:3–14.

19 Photodegradation and Photoprotection in Clinical Practice

Irene Krämer

Department of Pharmacy, Johannes Gutenberg-University Hospital, Mainz, Germany

INTRODUCTION

Ultraviolet (UV) and visible (VIS) radiation–induced degradation may take place during manufacturing, storage, and/or administration of medicinal products. The International Conference on Harmonization (ICH) Tripartite guideline (1), covering the Stability Testing of New Active Substances and Medicinal Products, notes that photostability testing should be an integral part of stress testing. The systematic approach to photostability testing includes tests on the active substance, the product outside of its immediate packaging and, if necessary, tests on the product in its immediate packaging and/or in its marketed package. The resulting stability information is an important element in the quality assurance of drug therapy, e.g., in determining the expiration date (shelf life), storage instructions, and use life (i.e. stability in an opened container).

The ICH document does not make any recommendations regarding how to test for the UV–VIS proper protection of medicinal products during administration. In addition, the guideline does not cover the testing of the photostability of medicinal products during and after administration, i.e., under conditions of actual use (1). The guideline only states that it may be appropriate to test certain products such as infusion liquids, dermal creams, etc., to determine their photostability in use. The extent of this testing should depend on and relate to the directions for use, and is left to the applicant's discretion (1).

Valid drug use instructions are indispensable if the dispenser/user is to ensure the quality, efficacy, and safety of drug therapy. Otherwise, e.g., the infusion reservoirs of photolabile infusion solutions may be exposed for hours to UV–VIS radiation during use, or prefilled syringes may be stored under the wrong illumination conditions. From our own experience with photolabile drug products in hospital use, there is a lack of valid photostability information in current labeling, regarding many medicinal products during administration. This leads to uncertainty, excessive precautions, or simple ignorance on the part of doctors or nurses administering the drugs. The problem of deciding what photoprotection may be applicable during the administration of a certain drug, in clinical practice by use of package inserts, is illustrated in Table 1. This table presents the results of a questionnaire distributed to 15 wards in our hospital, before implementing a guideline on photoprotection during drug administration.

The lack of photoprotection for the drug product during administration may lead to increased/decreased efficacy or increased toxicity of photolabile drugs. These episodes may be categorized as preventable medication errors, in particular, as wrong administration-technique errors (2). On the other hand, superfluous photoprotection measurements represent an unproductive use of nurses' time and a waste of materials.

Table 1 Results of a Questionnaire on Protection of Light-Sensitive Drugs During Administration; 15 Wards Were Surveyed Before Implementing Guidelines

		Photoprotected	
Drug	Wards using	Yes	No
Amhotericin B	5	5	0
Adrenaline/epinepherine	3	1	2
Cisplatin	8	6	2
Dacarbazine	6	4	2[a]
Etoposide	5	3	2
Farmorubicin	1	0	1
Methotrexate	5	4	1
Mitomycin C	3	2	1
Nifedipine	3	3	0
Nimodipine	2	2	0
Norepinepherine	3	1	2
Plicamycin	2	1	1
Teniposide	2	1	1

[a]Administered as a short-term infusion.
Source: From Ref. 3.

In hospital practice, several different photoprotective shielding wraps are commonly used to cover infusion reservoirs or sets frequently, a time-consuming procedure. The wrapping materials may include the use of aluminium foil, plaster dressings, disposable opaque plastic, commonly polyolefin, paper bags, specially sewn washable bags, or even baby diapers in pediatric wards. Photoprotection during administration is rarely ensured by infusion devices supplied by the drug manufacturer.

One certainly needs to reevaluate photostability information by searching the literature, in order to appropriately apply the right means of photoprotection in clinical practice. From a safety and economic perspective, valid recommendations concerning photoprotection measures are important to improve the outcome of drug therapy. However, it remains difficult to decide whether particular exposure conditions will produce unacceptable changes of medicinal products or not. In the following sections, drug dosage forms and approved drugs sensitive to photo deterioration are reviewed.

PHOTODEGRADATION AND PHOTOPROTECTION DURING ADMINISTRATION

The photodegradation of drugs, during administration, depends on drug and illumination-related parameters. The photolysis of a drug can be viewed as the absorption of UV–VIS radiant energy by drug substances/components, yielding a subsequent physical/chemical reaction. The amount of these photodegradants depends on the intensity, wavelength of the irradiating source, and time of exposure. Information on wavelength ranges of various photon sources is briefly summarized in Table 2.

The proper use of standardized terms is important to the usefulness of photostability information. Geographical and seasonal differences in UV–VIS daylight composition and intensity make the interpretation and comparison of the results of photolysis studies using natural illumination, quite difficult. Even the distance of the

Table 2　Wavelength Ranges of Various Light Sources

UV range	185–380 nm
UV-C	185–290 nm
UV-B	290–320 nm (2% of all light reaching the earth)
UV-A	320–380 nm
Visible range	380–780 nm
Sunlight	290–1000 + nm
Fluorescent light	320–780 nm
Incandescent light	390–780 nm
Blue light or bilirubin light	400–500 nm
D 65, CIE No. 85	Outdoor daylight standard
ID 65 (ISO 10977) (1993)	Window-glass–filtered daylight standard

drug product from a window has to be considered. The closer to the window or a fluorescent ceiling fixtures that the uncovered admixtures are placed, the faster will the photolabile drugs decompose. To get reproducible results, photolysis studies should be performed using artificial sources operated under standardized conditions.

Intravenous

Special illuminating conditions need to be considered during the blue-lamp phototherapy for neonatal jaundice (hyperbilirubinemia). In practice, the neonate is placed under a blue fluorescent lamp having a λ_{max} at about 450 nm, at an irradiance level between 4 and 6 μW/cm². This procedure is very effective in decreasing serum bilirubin concentrations. The high irradiance levels used in phototherapy may also enhance the degradation of drugs in concurrent use.

Comparisons of the results of photostability studies are complicated by the variations in the overall illumination level used. The ICH guideline recommends the use of a standardized radiation intensity and sample preparation in confirmatory studies, to allow for the direct comparisons of results (1). The overall illumination is, in general, given in lux hours for visible illumination or W hr/m² in the UV range. The ICH Guideline (1) recommends an overall illumination of not less than 1.2 million lux hours for the VIS and an integrated UV energy of not less than 200 W hr/m². However, there are no recommendations for overall illumination in studies concerning photostability during drug administration. Moreover, the photostability testing of medicinal products during and after administration should simulate actual clinical practice (radiation conditions, and administration conditions), as closely as possible. Studies should be done on the final drug formulation after admixing and not just on the drug substance solution. Drug-related parameters, which need to be considered, include the type of dosage form, concentrations of the active substances, excipients, type of administration, and duration of administration.

Infusions

Intravenous (IV) drug administration is the most critical type of administration for photolabile drugs, because of the dosage form and administration-related factors. In general, photosensitive drug substances degrade more rapidly in solution, the prevalent dosage form for IV administration, than in their solid state. Emulsions, however, do provide a degree of photoprotection for photosensitive drugs.

Administration-related factors, influencing the extent and rate of photodegradation, are discussed, below:

Infusion Period
While the exposure for bolus injections is negligible, photoexposure during intermittent or especially continuous infusions may be crucial. For instance, admixtures used in total parenteral nutrition (TPN) therapy are administered over 12 to 24 hours, and the infusion reservoirs are exposed to UV–VIS illumination for prolonged periods.

Type of Infusion Reservoir
Colourless glass does not appreciably transmit radiation wavelengths below about 310 nm, and amber glass does not appreciably transmit wavelengths below 400 nm. Clear glass is moreover, used to facilitate visual inspection of the infusion solutions during administration. For this same reason, in clinical practice, clear infusion bags are often preferred to opalescent polyethylene (PE) or polypropylene bottles.

Type of Infusion Set
Infusion sets components (dripping chamber, and infusion line) are produced using different plastic materials, which may vary in their photon absorbing ability [e.g., clear polyvinylchloride (PVC) tubing or opalescent PE tubing]. Additionally, the thickness of the tubing walls also influences the transmission of incident radiation.

Length and Diameter of Intraluminal Tubing
The length of the infusion line determines, in combination with the infusion rate, the overall exposure time of the infusion solution. In addition, solutions in infusion lines are exposed to greater irradiance than in the reservoirs because they expose a thinner layer of the liquid solution, and therefore show increased photodegradation.

If the photodegradation of photosensitive drugs during (continuous) infusion or injection cannot be avoided by changing the drug delivery system, additional photoprotection is indicated. Infusion reservoirs, whether bottles or bags, can be protected by the use of photon absorbing over wraps, especially colored plastic bags.

Martinelli and Mühlebach (4) studied the effectiveness of three different plastic coverings used in Swiss hospitals to photoprotect infusion reservoirs. The coverings used were made from polyolefines of different colors, amounts of pigmentation, and thicknesses. Their study showed that only some of these over wrap plastic coverings were effective. The lowest protection was achieved using a covering, marketed as a "light-tight" bag. The conclusion to be drawn from their study is, that the evaluation of similar coverings requires experimental studies and documentation.

Aluminium foil is also a safe alternative. When administering photosensitive drugs by continuous injection, perfusor syringes with black (e.g., manufactured by BlBraun Melsungen) or amber barrels (e.g., manufactured by Avon Medicals, Baxter, and Becton Dickinson) may be used as infusion reservoirs. The use of opaque or amber infusion sets or infusion lines may provide additional photon protection during administration. The opaqueness of these devices is achieved by incorporating carbon black, titanium dioxide, or zinc oxide into the plastic. Two of these devices, currently available on the market, are presented in Figures 1 and 2.

In contrast to opaque syringes and tubing, amber syringes and infusion sets allow the visual inspection of the solutions for particulates/precipitates, during administration. According to the United States Pharmacopeia (USP) standard, 90% of the radiation, at any wavelength between 290 and 450 nm, should be absorbed by the amber glass used.

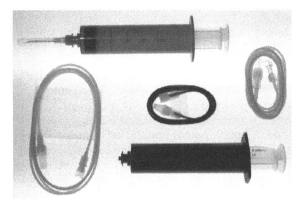

Figure 1   Photoprotective syringe and infusion line.

No generally valid recommendations for the use of opaque or amber infusion devices can be given. Opaque or amber syringes and administration sets should be used in accordance with published results of experimental studies or drug information leaflets.

Oral
Solid dosage forms, during oral drug administration, are usually not compromised by UV–VIS radiation because the drug substance is in the solid state and the exposure times are very short. The stabilization of photolabile products during storage is accomplished by the use of special coatings (e.g., colored or pigmented coatings) and/or special packaging.

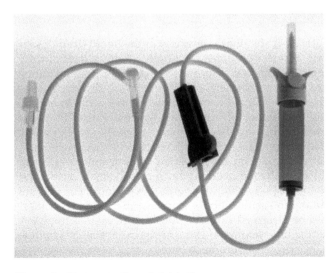

Figure 2   Photoprotective administration.

Tablets and Powders

The effectiveness of photon protection, e.g., by packaging for nifedipine tablets, depends on the color used, as well as the degree of pigmentation and thickness of the blister used. Nifedipine preparations (tablets, capsules) approved in Germany showed varying degrees of photodegradation, when stored without photoprotection (5). Nurses and patients should be taught not to store the photosensitive molsidomine and nifedipine tablets outside their primary packaging.

Powders are an important dosage form, which can be easily added to a small amount of any pleasant tasting food or liquid. Nifedipine, 1.0 in 500 mg of a powder, carefully compounded extemporaneously from nifedipine extended-release tablets, was found to be stable for up to one year when stored in waxed, sealed papers, and then placed in black plastic bags. When stored without the photoprotective black plastic bags, losses of the active ingredient, exceeded 20% within three hours and 40% within eight hours of exposure (6). This study shows that the storage of nifedipine powder is possible. Parents and other health-care providers should be educated about the correct storage and handling of this product.

Liquids

Oral liquids of very photosensitive drugs such as molsidomine and nifedipine are compromised by photodegradation during administration even in short exposure. For safety reasons and in view of its high photosensitivity, molsidomine oral liquid was withdrawn from the market.

Nifedipine aqueous solutions are also extremely photosensitive. Studies performed using a commercially available nifedipine oral liquid, (Nifedicor), produced 5% degradation after only 10 minutes exposure under ambient illumination, even though the product contained a photostabilizing surfactant, PE glycol (7). In another study, 10% degradation was found after one-minute exposure to sunlight. Less than 5% degradation was observed immediately after dosage by dropping in ambient illumination or sunlight (5). Therefore, nifedipine oral liquid should be administered immediately after dropping.

Topicals

During topical drug administration, the photosensitivity of the liquid drug dosage forms used (e.g., eye drops), and the prolonged exposure of other dermal drug dosage forms (e.g., ointments, creams) on unprotected skin after administration need to be considered. Eye drops containing chloramphenicol, dexamethasone phosphate, or tetracycline have been shown to be photolabile in use (8). In natural PE, used as a primary packaging material for many eye drops, more than 20% of the chloramphenicol decomposed during unprotected storage over a six-week period. No degradation was detected when white, opaque PE containers were used (8).

On exposure to sunlight, under in vivo–related circumstances relevant to the eye under treatment, *p*-nitrobenzaldehyde, *p*-nitrobenzoic acid, and *p*-nitrosobenzoic acid were found (9) in chloramphenicol eye drops. The amount of *p*-nitrosobenzoic acid found, represented 45 mol% of the initial amount of chloramphenicol. Patients should be informed if photoprotective storage of these eye drops is necessary, and that the intense photon exposure of the eye after administration of chloramphenicol eye drops should be avoided.

Photoexposure of unprotected skin after the administration of dermal preparations containing corticosteroids or antimycotics may cause photodegradation of the drug substance, on and in the skin. Irradiation (Novasol test chamber, sunlight) of the unprotected skin, after the administration of a betamethasone-containing gel, produced, after a 20 minute exposure, a 17% loss of active substance, when compared to a duplicate sample stored in the dark (10), Figure 3.

The extent and rate of photodegradation can be related to the water content of the dosage form, solubility of the active substance in the ointment base, and the excipients used. Solubilized drug substances degrade more rapidly than other dosage forms, e.g., suspensions. Nystatin was shown to degrade more rapidly in different ointment preparations than when suspended in cream preparations (11). As shown for a nystatin ointment containing zinc oxide, the addition of opacifying pigments can decrease the photodegradation of drug substance in semisolid preparations.

The photodegradation of drug preparations after administration on the skin is very complex and is influenced by absorption and photon–skin interactions. Loss of activity seems to be more significant than the occurrence of toxic degradation products.

In contrast, phototoxic reactions associated with systemic antimicrobial therapy, e.g., quinolones, sulphonamides, and tetracyclines are well known. Patients undergoing quinolone therapy may develop erythema on surfaces exposed to the sun. The photoreactivity and phototoxicity is mostly influenced by the halogen-substituent at the C8-position of the quinolone nucleus. In volunteers, with lomefloxacin, a correlation was found between the concentration of the drug in the skin and the degree of the phototoxic reactions (12). The simple avoidance of UV–VIS radiation or the use of appropriate photoprotective measure will minimize the degree of phototoxicity.

PHOTODEGRADATION OF SELECTED DRUGS AND MANAGEMENT OF PHOTOPROTECTION DURING ADMINISTRATION

Anti-Infectives

Amphotericin B

Amphotericin B is an amphoteric heptaene macrolide with antifungal activity. Conventional amphotericin B for injection contains amphotericin B, sodium deoxycholate, and sodium phosphates. The drug product must be reconstituted with water for

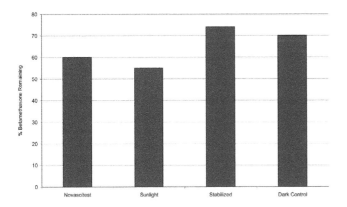

Figure 3 Loss of betamethasone-17 valerate (in an isopropanol-containing polyacrylate gel) after a 20-minute irradiation. *Source*: From Ref. 10.

injection (10 mL) and further diluted, usually, with 500 mL of 5% dextrose injection. Reconstituted solutions should be photoprotected.

Normally, reconstituted amphotericin B solutions are administered by slow infusion over a period of six hours (minimum one to two hours) to avoid potentially serious adverse effects. The manufacturers state that during administration the IV infusion solutions should be "protected from light." Photoprotection during adminis-tration is not required for ampothericin B cholesteryl sulfate complex, amphotericin B lipid complex, or liposomal formulations of amphotericin B. The administration of amphotericin B in parenteral fat emulsions is no longer an accepted practice.

A methanolic solution of the active drug substance, amphotericin B, was shown to be very photosensitive. After 100 minutes of irradiation (Suntest CPS with UV filter), the solution of amphotericin B no longer absorbed UV radiation because of the complete degradation of the heptaene moiety (11). In contrast, Utz (13) reported that little decomposition occurred during the infusion period of amphotericin B in colloidal dispersion, even though the infusion bottle was not photoprotected. When compared to solutions, this result is congruent with the generally recognized rela-tive photostability of dispersions.

Using a colorimetric and microbiological assay, the stability of the ready-to-use 0.1 mg/mL amphotericin B infusion solutions (conventional colloidal dispersions) over long exposure periods, were investigated. On exposure to the fluorescent light and storage at 25°C, the solutions lost approximately 26% of the active and 10% of their microbiological activity in three days (14). However, short-term exposure, 8 or 24 hours of fluorescent illumination, did not significantly affect the bioactivity of the amphotericin B infusion solutions tested (15).

The photostability of some amphotericin B infusion solutions has also been demonstrated using high performance liquid chromatography (HPLC) assays. Kintzel et al. (16) studied the stability of amphotericin B solutions, packaged in polyolefin containers, at concentrations >0.1 mg/mL, used for the administration via a central venous line. Amphotericin B infusion solutions (0.5, 0.7, and 0.8 mg/mL), without any protection from room irradiation, were found to be chemically stable when stored for up to 24 hours at 25°C. In addition, amphotericin B in 5% dextrose injection packaged in PVC containers at concentrations of 0.2, 0.5, and 1.0 mg/mL, was chemically stable when stored at 25°C for up to 120 hours, under 40 W fluorescent irradiation (17).

Based on these data, it is apparent that the potency of conventional amphotericin B is not significantly diminished if the infusion solution is exposed to UV–VIS radiation only during administration. The exposure time for this procedure is usually not more than 8 hours, at a maximum, 24 hours. Protection, in our experience, is not needed.

Doxycycline

Doxycycline is a semisynthetic, tetracycline, antibiotic drug. Doxycycline hyclate (doxycycline hydrochloride hemiethanolate hemihydrate) injection solutions (100 mg/5 mL) may be administered by slow IV injection or by IV infusion (maxi-mum four hours). Reconstituted solutions, in either 0.9% sodium chloride injection or 5% dextrose injection, are stable for 48 hours under fluorescent lamps (Product Information Sheet, Vibramycin®, 1995).

No significant loss of doxycycline was found during a seven hour simulated infusion using a glass infusion bottle and plastic infusion set (18). It is recom-mended that doxycycline infusion solutions be protected from direct sunlight during storage and administration (Product Information Sheet, Vibramycin, 1995).

Based upon the manufacturers package insert and our own experience, we suggest that doxycycline infusion solutions may be safely administered without photoprotection, if diffuse daylight or fluorescent irradiation is used. In the case of intensive sunshine, the slow infusion of doxycycline should be performed with photoprotection of the infusion container.

Metronidazole

Metronidazole, a synthetic nitroimidazole derivative with antibacterial and anti-protozoal activity, is available for oral and parenteral use. Commercially available ready-to-use infusion solutions of metronidazole (100 mL, 5 mg/mL) are administered by intermittent IV infusion over a 20- to 30-minute period. The degradation of metronidazole depends on temperature, pH, and radiation conditions.

Metronidazole tablets and infusion solutions darken following prolonged exposure to UV–VIS radiation (Fig. 4).

The photosensitivity of this active ingredient is attributable to its imidazole ring system and nitro moiety. Photolysis results in increased concentrations of nitrite. Studies have shown that the effects of temperature and pH are more significant than those of UV–VIS radiation (19,20).

As a result of a questionnaire distributed to five manufacturers regarding the packaging and storage of metronidazole infusion solutions, it was concluded that protecting metronidazole injection from UV–VIS radiation should not be of particular concern, during dosing, in a typical hospital setting. Short-term exposure to fluorescent and diffuse daylight (i.e., daylight passing through a glass window) during dosing should not adversely affect the stability of metronidazole (21).

Quinolones

A major group of synthetic antibiotics used orally and IV (e.g., ciprofloxacin, levoflox-acin, and ofloxacin) are the 8-Fluoroquinolones. Tablets containing fluoroquinolones should be protected from intense UV radiation. In general, fluoroquinolone injec-tions are infused undiluted (2 mg/mL) over a period of 30 to 60 minutes. Injection solutions should be stored, protected from UV–VIS radiation. Ciprofloxacin and ofloxacin/levofloxacin are sensitive to UV illumination, losing potency if exposed to radiation of wavelengths between 300 to 400 nm. Some of the fluoroquinolone

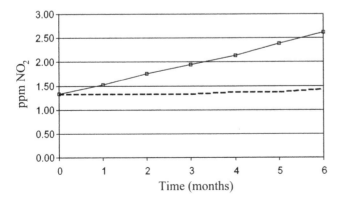

Figure 4 Nitrite content of a metronidazole infusion solution (pH 3.2); unprotected, (– –) photo protected. *Source*: From Ref. 20.

photodegradation products have been reported to exert phototoxic or photoallergic effects in vitro and in vivo (22,23).

During radiation studies (radiation source: Suntest CPS with UV filter) of ciprofloxacin and ofloxacin infusion solutions, after eight hours of exposure, 82% of ciprofloxacin and 96% of ofloxacin remained undegraded (24). Based on this study, the photoprotection of these infusion solutions during administration does not seem to be necessary. Diluted ciprofloxacin infusion solutions (10^{-5} mol/L), stored in transparent glass bottles, showed photodegradation, which could be related to the radiation dose. After 4 and 96 hours of exposure to ambient room irradiation, 1.3% and 8.8% of the ciprofloxacin, respectively, were degraded. Numerous new photodegradants were formed in the process (23). No significant drug loss was observed during one hour simulated infusions of 250 mL of pefloxacin, ciprofloxacin, and ofloxacin, using regular PVC infusion bags and administration sets (25).

In another experiment, infusion bags containing 1.6 mg/mL of pefloxacin or 0.8 mg/mL of ofloxacin or ciprofloxacin solutions were stored for eight hours without photoprotection. No significant losses of the parent drug were detected (25). No loss of diluted ciprofloxacin, 1.5 mg/mL, was found by HPLC assay after 48 hours exposure at 25°C. This study used samples prepared in 5% glucose injection, or 0.9% sodium chloride injection, and all samples were exposed to fluorescent radiation (3).

In common clinical practice, in our hospital, the protection of fluoroquinolone injections during administration does not appear necessary. Visible radiation sources did not reduce the antibiotic potency of fluoroquinolones, under the conditions of our experimental studies. No experimental studies on the exposure of infusion solutions to sunlight were performed.

Antineoplastic Agents

Anthracyclines

The anthracycline antibiotic drugs approved for antitumour drug therapy include doxorubicin (adriamycin), epirubicin, daunorubicin, idarubicin, and aclarubicin. These drug substances are structurally similar and characterized by a tetracyclic aglycone linked to amino sugar moieties. Particularly in solutions, the anthracyclines are photolabile. The formation of inactive aglycones and polymeric substances as a result of the irradiation of anthracycline solutions has been reported.

The extent of photolytic decomposition depends on the pH, nature of the solvent(s) used, and the initial drug concentration (27). Degradation is faster in diluted solutions and at higher pHs. The photolysis is characterized in terms of first-order kinetics.

Walker et al. (26) reported that doxorubicin, 2 mg/mL in glass vials or plastic syringes and also 1 mg/mL in plastic syringes, is stable for 124 days when stored at 23°C without photoprotection. However, the photodegradation of anthracycline drug solutions (daunorubicin, doxorubicin, epirubicin) may be substantial at concentrations below 0.1 mg/mL, if the solutions are exposed to UV–VIS radiation for longer periods (28,29). Above concentrations of 0.1 mg/mL, little or no photodegradation was observed.

In samples exposed to ambient radiation, rapid photolysis occurred when doxorubicin, dissolved in distilled water at concentrations ranging from 0.01 to 0.05 mg/mL, were studied. The amount of photodegradation was found to be a function of the intensity of the radiation (28). In the studies performed by Wood et al. (29), using ambient illumination (mixed window-glass–filtered daylight and fluorescent radiation), the rates of photodegradation of daunorubicin, doxorubicin, and epirubicin were inversely proportional to the drug concentration and accelerated by increases in the pH of the vehicle. For 0.1 mg/mL doxorubicin solutions packaged in clear glass vials, the remaining concentration was >90% after 168 hours, and in amber glass vials or PE bags, >95% after 188 hours. Diluted idarubicin solutions (0.01 mg/mL) have been reported to undergo degradation after exposure to radiation for periods greater than six hours (30).

For concentrations of at least 0.1 mg/mL, these studies indicated that no special precautions are needed to protect anthracycline infusion solutions from UV–VIS radiation during IV administration. When infusion solutions are to be exposed to intense UV–VIS radiation, it is essential that a diluent with, as low a pH as practical, be used.

Carmustine

Carmustine infusion solutions are administered over a period of one to two hours by slow infusion. The stability of carmustine, a nitrosourea derivative, after reconstitution and dilution depends on pH, temperature, radiation exposure, and sorption of delivery devices. Because of the pH factor, diluted solutions are more stable in 5% dextrose injection than in 0.9% sodium chloride injection. Different degradation mechanisms exist for the photoexposed and dark reactions.

The manufacturer states that carmustine infusion solutions, packaged and stored in glass containers, are stable at room temperature (25°C) for eight hours when protected from intense UV–VIS radiation or normal fluorescent illumination (60 ft. candles = 645 lux), as determined by spectrophotometric and HPLC assays.

Frederiksson et al. (31) studied the influence of UV–VIS radiation on carmustine test solutions in a testing cabinet. The drug solutions were placed in glass petri dishes with lids. The exposure at 1000 lux, produced 10% degradation of the active ingredient in two hours. However, it is difficult to transfer the results to clinical practice where photon-induced degradation rates are likely to be reduced by the glass or plastic infusion containers used and higher volumes of solutions.

In view of these conflicting study results, precautionary photoprotection for carmustine and the photoprotection of infusion reservoirs, during administration, are recommended.

Cisplatin

The approved platinum complexes, with antineoplastic activity, differ in their photochemical stability. Among these, cisplatin, in particular, undergoes a photolysis reaction. Early studies proposed that this degradation was due to the photoaquation of the chloride ligands, and its subsequent photon-induced isomerization to transplatin (32,33). Later studies found that the main photolysis-induced degradation product, was trichloroammineplatinate II (TCAP) (34,35). This decomposition product is limited to 1% in cisplatin for injection by the USP.

Cisplatin injection solutions (1 mg/mL), prepared in PVC bags and exposed to fluorescent radiation of approximately 1700 Lux, degraded substantially. After 24 hours, the TCAP content of cisplatin solutions reached 12% (34). The absorption spectrum of cisplatin, which has a UV maximum at 300 nm (35), overlaps the spectral power distributions of all UV-A and UV-B lamps. Therefore, cisplatin will rapidly degrade when exposed to radiation in the UV spectral region. As expected, cisplatin solutions have been shown to be more sensitive to fluorescent than to normal incandescent radiation, which is UV deficient.

Various types of containers containing cisplatin solution were exposed for four hours to 1500 Lux of fluorescent radiation; the TCAP content of cisplatin solutions, in PVC containers and PE syringes used, did not exceed the USP limit. In long-term studies, with solutions protected from UV–VIS radiation, TCAP content was found to depend strongly on pH. Degradation gives rise to an increase in pH because of the release of ammonia.

In older studies, where only the content of undegraded cisplatin was measured and the TCAP content not determined, similar results were found. The degradation rate was found to be dependent on the pH of the cisplatin injection solution, composition of the container used, and the source and intensity of the UV–VIS radiation.

In summary, available reports confirm that the cisplatin is not adversely affected by normal room irradiation for 24 hours or up to 72 hours, after dilution in 0.9% sodium chloride solution (36–38). The accelerated decomposition of cisplatin by UV–VIS radiation has been demonstrated by the studies of Cubbels et al. (39). Solutions exposed to UV–VIS radiation (300–35,000 lux) for two weeks, degraded more rapidly than solutions protected from radiation. The observed degradation rate constant, obtained from log concentration–time curves of these samples, was determined to be 0.385 per day (t_{90} = 6.5 hours). The manufacturer states that a cisplatin solution removed from its amber vial should be protected from UV–VIS radiation if it is not used within six hours, or if it is exposed to direct sunlight (Technical Information, Platinol, Bristol-Myers Squibb).

Under normal ambient illumination conditions used in experimental studies, cisplatin injection solutions do not need photoprotection when administered over several hours, as indicated in the manufacturers literature. However, intense sunlight should be avoided during administration. The photoprotective storage of ready-to-use cisplatin infusion solutions is strongly recommended.

Dacarbazine

Dacarbazine (DTIC) is chemically characterized by a dimethyl-triazeno structure. Alkylation, via a methyldiazonium ion metabolite is the most likely mechanism involved in its anti-tumour activity. Reconstituted solutions of DTIC are administered as slow IV bolus injection or the infusion of diluted solutions. DTIC is sensitive to photonic energy and heat.

Reconstituted solutions (10 mg/mL, pH 3.5–4) are reported to be stable for at least 24 hours at room temperature and at least 96 hours when refrigerated, and when photoprotected (40). Solutions, further diluted with 5% dextrose or 0.9% sodium chloride diluent, remain stable for 24 hours under refrigeration and 8 hours when stored at room temperature and photoprotected (Product Information, DTIC-Dome, 1998).

A color change from pale yellow to pink indicates the photodecomposition of DTIC. The photolysis of DTIC initially produces dimethylamine and 5-diazoimdazole-4-carboxamide (Diazo-IC), and the latter decomposition product, subsequently cyclizes to 2-azahypoxanthine (41). 2-Azahypoxanthine has been identified as the main degradation product of DTIC hydrolysis and DTIC photolysis, under all study conditions (42). This specific decomposition product is limited to 1% in DTIC for injection, by the USP.

The further degradation of Diazo-IC produces conjugated polymers, which give rise to a pink coloration (42). When studying the photosensitivity of IV DTIC solutions, Kirk (43) found that solutions stored in unprotected infusion containers became pink in color after 30 to 40 minutes exposure to window-glass–filtered daylight. This was associated with a 4% loss of DTIC content (Fig. 5).

In addition, Kirk also reported a 2% loss in DTIC content for a solution passed through a nonphotoprotected IV administration set, which had been exposed for 30 minutes to window-glass–filtered daylight. Amberset (Avon Medical Ltd.) protected solutions showed no sign of discoloration, even after 24 hours of storage, exposed to daylight of the same photon intensity. Samples of these solutions contained 92% undegraded DTIC. Infusion solutions exposed to artificial radiation, from a fluorescent source for 24 hours, and solutions stored in the dark for 24 hours, exhibited similar degrees of stability. A 4% loss of DTIC occurred, in the reservoir containers, over the 24-hour period.

Direct sunlight ($4000\,\mu W\,cm^{-2}$ UV light intensity at 365 nm) was extremely deleterious, producing a 12% loss in DTIC content after 30 minutes of exposure. Unfortunately, experimental results related to the extent of 2-azahypoxanthine formation, during DTIC administration under various illuminating conditions, are currently not available. Moreover, the pharmacological significance of the presence of the degradation products on the overall efficacy and toxicity of DTIC remains unclear. It has been postulated that 2-azahypoxanthine explains some of the drug's activity and toxicity, including local pain along the cannulated vein (44,45).

The photostabilization of DTIC, using reduced glutathione in different buffer systems, has been reported (46). The results of this study are difficult to interpret.

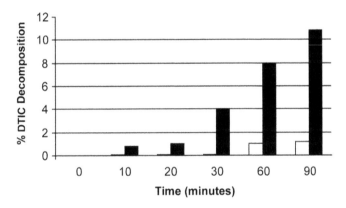

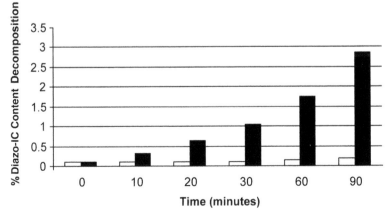

Figure 5 Percent degradation of Dacarbazine intravenous solution and increase of Diazo-IC in unprotected polyvinylchloride infusion containers and protected by Amberset A90, exposed to indirect sunlight (~300 micro W cm^{-2} UV light intensity at 365 nm); (□) protected, (■) unprotected. *Source*: From Ref. 43.

DTIC itself, not the formulated drug product, was used and only solutions exposed to fluorescent radiation were studied. In spite of these results, the photostabilization has not been generally implemented in clinical practice.

From these studies it is apparent that VIS radiation has little to no effect on DTIC photostability. But, DTIC is very sensitive to UV radiation. Exposure to direct sunlight should be strictly avoided and exposure to daylight should be minimized. During administration, the reservoirs and administration sets used for infusion should be photoprotected. Sets capable of excluding the UV component of all incident radiation are a convenient method of photoprotection in clinical practice. The utility of Amberset products, for this purpose has been experimentally proven.

Etoposide

Etoposide injection must be diluted prior to use and should be administered over a 30 to 60-minute period. Experimental studies have shown that the photostability of etoposide infusion solutions is not affected by exposure to normal room fluorescent radiation (47). The photostability of etoposide has also been demonstrated in IV admixtures containing cisplatin.

Fluorescent radiation (900–1000 lux) has no significant effect on the stability of etoposide over 24 hours (48). A ready-to-use etoposide infusion solution was found to be stable for 48 hours, when stored at room temperature unprotected from UV–VIS radiation. In an admixture of etoposide, cytarabine, and daunorubicin for infusion, the content of etoposide declined more rapidly when exposed to UV–VIS radiation (9% after 12 hours), than when photoprotected (3% after 12 hours) (49). Etoposide injection, diluted with water (10 mg/mL) and filled into 5 mL plastic oral syringes, was stable for 22 days at ambient temperature and found to be stable under fluorescent radiation (50).

Etoposide infusion solutions, in our experience, may be administered without photoprotection. However, experimental data on use in intense or bright window-glass–filtered daylight are not available.

Fluorouracil

Both, the thermal and photochemical decomposition of 5-fluorouracil (5-FU), result in the opening of the pyrimidine ring and the production of urea. Therefore, 5-FU injection solutions need to be protected from strong window-glass–filtered daylight and elevated temperatures during storage. The solutions are administered undiluted, by IV push, or diluted and administered by IV infusion.

The storage of undiluted 5-FU injection in clear glass vials or plastic syringes for up to seven days was found to produce less than 2% degradation (51). To investigate the photosensitivity of 5-FU, various concentrations of the solution were placed in glass vials and opaque or translucent perfusor sets, which were then exposed to intense window-glass–filtered daylight. No photodegradation was observed in samples stored in these packagings and exposed for up to two weeks (52).

Furthermore, resistance to photodegradation has also been demonstrated over a wide range of conditions, for periods up to two months. No changes in the photodegradation of 5-FU solutions were found when various diluents or different containers (plastic bags or glass bottles) were studied. Exposure to fluorescent radiation, 24 hours a day, under ambient temperatures had no significant effect on 5-FU stability (53).

Based on these studies, it would appear that the 5-FU injection solutions might be administered without photoprotection, even in intense window-glass–filtered daylight.

Fotemustine

Fotemustine is a nitrosourea derivative, approved in France for the treatment of melanoma. Diluted fotemustine solutions are to be administered, by infusion, over a one-hour period. Fotemustine infusion solutions are known to be very photosensitive. Infusion solutions containing 0.5 to 2 mg/mL fotemustine, in a 5% dextrose diluent, are stable for eight hours at room temperature, and 72 hours if refrigerated and photoprotected. A 10% or 30% decline in the fotemustine concentration occurs when these solutions are exposed to ambient radiation at room temperature for one hour or eight hours, respectively (54). In sunlight, the fotemustine concentration was found to decrease more than 75% after two hours storage in PVC infusion bags.

These studies indicate that the fotemustine reservoirs, administration sets, and infusion solutions should be photoprotected from UV–VIS radiation during administration.

Irinotecan, Topotecan

Irinotecan and topotecan are recently approved topoisomerase 1 inhibitors with antineoplastic activity. Both are administered by IV infusion. Diluted irinotecan and topotecan infusion solutions proved to be photostable for a minimum of four weeks, when stored at room temperature and protected from UV–VIS radiation (55).

Topotecan infusion solutions, packaged in PVC bags and stored under mixed window-glass–filtered daylight and normal laboratory fluorescent radiation conditions, remained stable (90% of the initial concentration) for 17 days. The degradation is accompanied by an increase in pH.

Irinotecan infusion solutions, packaged in PVC bags, and stored under mixed window-glass–filtered daylight and normal laboratory fluorescent radiation conditions remained stable (90% of the initial concentration) for at least seven days (Fig. 6).

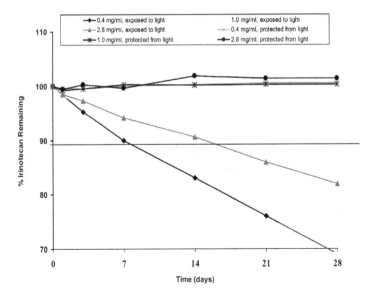

Figure 6 Dependence of the degree of photodegradation on irinotecan infusion solution concentration. Concentration 0.4 mg/mL; (x) protected; (♦) exposed. Concentration 1.0 mg/mL; (✳) protected; (■) exposed. Concentration 2.8 mg/mL; (△) protected; (▲) exposed.

The increased rate of photodegradation correlates with the concurrent increase in the pH of the infusion solution and decrease in irinotecan concentration.

Based on our results we conclude that the topotecan and irinotecan infusion solutions may be administered safely without any photoprotection.

Methotrexate

The manufacturers recommend that methotrexate injection solutions be stored at controlled room temperature and protected from light. Injection solutions are administered undiluted (25 mg/mL) by IV bolus injection or infusion after dilution. Methotrexate solutions are susceptible to hydrolytic and photolytic degradation. On prolonged exposure to window-glass–filtered daylight, a yellow precipitate is formed.

When undiluted methotrexate injection (50 mg/mL) in plastic syringes was stored under controlled UV–VIS radiation conditions (21,200 lux and 25°C), no change in the methotrexate concentration was observed after 28 days of exposure (56). Dyvik et al. (57) also reported no significant loss of methotrexate following UV–VIS exposure for four hours (laboratory illumination and in front of a continuous radiation source (200–700 nm). The solutions tested contained 0.1, 1, or 20 mg/mL of methotrexate and were packaged in PVC containers.

Chatterji and Galleli (58) determined that the photodegradation reaction of methotrexate after an initial lag phase follows zero-order kinetics. The photolability of the drug increases at higher dilutions. Bicarbonate ion has also been identified as a catalyst for the photolysis of methotrexate. No photodegradation products were observed, in an infusion solution containing 2 mg/mL methotrexate and 0.05 mol sodium bicarbonate, after 12 hours exposure to room radiation.

Studies were performed utilizing aqueous solutions of the drug substance (0.5, 1, and 2.5 mg/mL), adjusted to pH 8.3 and packaged in glass vials, which were then exposed to different levels of illumination (4000, 8000 or 12,000 lux). The worst case, methotrexate concentration 0.5 mg/mL, showed 20% degradation when exposed for 72,000 lux hours. The 1 mg/mL solution had less than 10% degradation after 72,000 lux hours, and the 2.5 mg/mL solution had 10% degradation after 216,000 lux hours exposure (59).

Methotrexate solutions (1 mg/mL) stored in translucent syringes and exposed to intense window-glass–filtered daylight showed more than 10% degradation after eight hours exposure. Similarly, 5 mg/mL solutions had less than 5% degradation after 24 hours exposure (52).

McElnay et al. (60) studied the photodegradation of methotrexate (1 mg/mL in 0.9% sodium chloride solutions) in the drip chambers of three different burette administration systems. Storage of these samples, under normal radiation conditions (window-glass–filtered daylight plus fluorescent tube room illumination) produced little change in the drug concentration in the first 24 hours. Static storage, in standard administration tubes, under normal illumination conditions produced more than 10% degradation within 12 hours. When the drug solution was exposed to intense window-glass–filtered daylight (seven hours of continuous sunlight via the laboratory window), an 11% decline in drug concentration occurred. The use of Ambersets or tinfoil prevented photodegradation over this same period of time.

From the cited studies we conclude that the undiluted methotrexate injection is minimally photosensitive and may be administered without protection, even in intense window-glass–filtered daylight. Diluted (<1 mg/mL) methotrexate solutions should be protected from intense window-glass–filtered daylight during administration.

Mitomycin C

Mitomycin C is intended for IV, intravesical, or ophthalmic use. The stability of reconstituted and diluted mitomycin solutions is pH dependent and is most stable between pH 7 and 8. According to the studies of Beijnen et al. (61) and Quebbeman et al. (62), the stability of mitomycin solutions are not markedly influenced by the container composition or room fluorescent illumination, even if exposed for several days.

Based on this information it would appear that the Mitomycin C injection solutions may be administered without any need for photoprotection. Experimental data concerning the influence of intense UV–VIS radiation is not available.

Miscellaneous Antineoplastic Drugs

On occasion, the photostability of certain older cytotoxic drugs have been open to question, including amsacrine, bleomycin, cyclophosphamide, cytarabine, dactinomycin, mitozantrone, nimustine, teniposide, thiotepa, and the vinca alkaloids. UV–VIS photon–induced degradation is not a significant factor during injection or infusion of these agents. In normal clinical practice, photoprotection is not necessarily based on our present state of knowledge.

Cardiovascular Drugs

Molsidomine

Molsidomine is very photosensitive and undergoes rapid photodegradation to form the toxic product morpholin. It is strongly recommended that oral preparations (tablets, liquids) and injection solutions be stored with photoprotection. Without their appropriate packaging materials, the preparations are not adequately protected from UV–VIS radiation (63). The tablets should not be stored outside their packaging. For safety reasons and in view of its high photosensitivity, molsidomine oral liquid was withdrawn from the market.

The photolability of molsidomine infusion solutions in administration sets was investigated using both static and dynamic conditions (64). Molsidomine injection was diluted to a concentration of 0.08 mg/mL using 0.9% sodium chloride and packaged in PVC bags. After 24 hours exposure of the bags to fluorescent or incandescent illumination, no degradation was detected. On exposure to window-glass–filtered daylight, an inverse relation between the half-life and the radiation intensity was observed (Fig. 7).

Dynamic experiments also demonstrated photostability of molsidomine infusion under condition of artificial source illumination. A flow rate of 12.5 mL/hr was used, resulting in a residence time of one hour in the infusion tubing. Simulated infusions exposed to window-glass–filtered daylight or simulated sunlight (Ultra-Vitalux lamps) showed that protection of the infusion bag alone, results in a degradation rate similar to that found for unprotected infusion bags (Fig. 8).

The results of these experiments clearly demonstrate that the photoprotection of the infusion bag alone is insufficient to prevent degradation of molsidomine. Under varying weather conditions the half-lives of solutions, T_{50}, infused through unprotected infusion sets exposed to window-glass–filtered daylight varied between 17 and 50 minutes. Photoprotection of the infusion bag and the infusion tubing with a UV filter (for the bag: aluminium foil or Sureset A 261 infusion tubing (Avon Medicals, Rehau, Germany) increases the half-life of the infusion to several days. These results clearly indicate that, during the continuous molsidomine administration, both the infusion reservoir and administration set should be photoprotected.

Nifedipine

On exposure to UV–VIS radiation nifedipine, a nitrophenyldihydropyridine compound decomposes very rapidly to form, primarily, the inactive nitroso-phenylpyridine analog. Because of its photosensitivity to UV–VIS radiation and most artificial radiation sources, nifedipine drug substance and formulations are manufactured only under red ambient illumination conditions. The oral liquid and ready-to-use infusion solutions are protected from UV–VIS radiation by using special primary packaging (brown glass bottles painted black or brown glass vials with a yellow plastic over-wrap). The infusion vial should not be removed from any of its packaging until immediately before use. In its plastic over-wrap, nifedipine infusion retains its complete potency for one hour under window-glass–filtered daylight and for six hours under artificial illumination (Fachinfo, Adalat® pro infusione).

Thoma and Kerker (5) found 5% degradation in samples exposed for eight hours in a Novasoltest cabinet, and 3% degradation for those exposed to window-glass–filtered daylight. Removal of the infusion solution from its bottle resulted in a 45% degradation of the nifedipine after only a 30 seconds exposure.

During the administration of nifedipine infusion solution the opaque syringes and tubing provided by the manufacturer of nifedipine infusion solution should be used to guarantee sufficient photoprotection (Fig. 9).

Nifedipine solutions are extremely photosensitive (65). Oral solutions should be taken immediately after dispensing. Ready-to-use infusion solutions should be admin-istered using the opaque syringes and tubing sets provided by the manufacturer. The use of Y-site injection via translucent administration sets should be avoided.

Nimodipine

IV formulations of nimodipine may be administered via continuous injection or intracisternally during surgery. Nimodipine infusion solutions are less photosen-sitive than nifedipine solutions. Degradation half-lives of 16 and 56 hours were

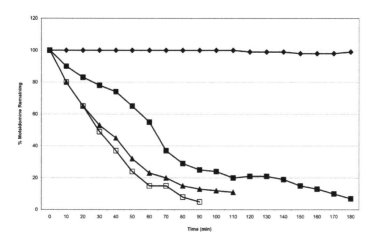

Figure 7 Mean molsidomine concentrations of unprotected infusion bags of product stored on a windowsill at three different irradiation levels (each $n = 3$); (♦) protected from daylight; (■) 3.9 lux; (▲) 5.5 lux; (□) 12.5 lux. *Source*: From Ref. 64.

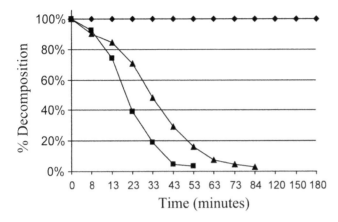

Figure 8 Mean molsidomine concentrations (% initial) as a function of time for dynamic experiments, performed in simulated sunlight with an unprotected infusion bag and unprotected infusion tubing (■), with an infusion bag, covered with aluminium foil and an unprotected infusion tubing (▲), with an infusion bag, covered with a ultraviolet (UV)-cover and an infusion tubing with UV-filter (◆) (*n* = 1). *Source*: From Ref. 64.

found following exposure of nimodipine infusion solutions (0.2 mg/mL) to UV radiation [360 nm (λ_{max}), 300 lux] and daylight, respectively. Exposure of the solutions to bright microscopic radiation for up to 30 minutes, following intracranial instillation during surgery, did not appear to produce substantial photodegradation of the drug (66). In the presence of diffuse window-glass–filtered daylight or fluorescent illumination nimodipine infusion, solutions may be administered without photoprotection for a period of up to 48 hours. If the solutions may be exposed to window-glass–filtered sunlight, protection by means of opaque syringes and administration sets is recommended (Product Information, Nimotop® Infusionslösung, Germany).

Figure 9 Nifedipine infusion solution in photoprotective packaging as well as photoprotective, blackened perfusor syringe and infusion line.

From this information it appears the nimodipine infusion solutions may be administered over periods of up to 10 hours without the use of photoprotection. Intense window-glass–filtered daylight irradiation should be avoided.

Nitroglycerin

Exposure to even high-intensity UV–VIS radiation, does not adversely affect the photostability of nitroglycerin. No difference in the photostability of nitroglycerin injection (0.5, 1 mg/mL) was found, whether stored in room light or in the dark (67) for at least three years at room temperature. When the stability of diluted nitroglycerin infusion solutions (0.2, 0.4 mg/mL) packaged in polyolefin and glass containers was studied, even solutions exposed to intense VIS radiation (15,000–21,500 lux) proved stable for up to 24 hours of exposure; and a 4% loss of potency was produced. Solutions stored at ambient room temperatures with normal room illumination proved to be stable after 28 days of exposure (68). According to the literature, nitroglycerin infusion solutions require no photoprotection during administration.

Sodium Nitroprusside

Sodium nitroprusside (sodium nitrosyl pentacyanoferrate dihydrate) is administered in hypertensive crisis by IV infusion, using a controlled infusion device. The drug is commercially available as reddish-brown lyophilized powder. Prior to infusion, it is reconstituted and diluted with 5% dextrose injection.

Sodium nitroprusside in solution is extremely photosensitive, undergoing rapid and numerous photodegradation reactions (69). The deterioration of the product is evidenced by a color change from brown to blue, resulting from the reduction of the ferric ion to the ferrous ion. Hydrogen cyanide is one of the toxic degradation products formed. Therefore, reconstituted solutions should be stored protected from UV–VIS radiation by wrapping the container with aluminium foil or some other opaque material. Solutions with adequate photoprotection are stable for up to 24 hours (70–72).

Sewell et al. (73) evaluated the stability of sodium nitroprusside (0.5 and 1.67 mg/ mL in 5% dextrose) injection packaged in polypropylene syringes and administered by a syringe pump system. When exposed to artificial radiation and window-glass– filtered daylight, the time to 10% decomposition, $t_{90\%}$, was about four hours. After 24 hours of exposure, the level of free cyanide exceeded 2 mg/L.

The photodegradation of sodium nitroprusside is wavelength, intensity, temperature, and pH dependent. Studies by Vesey and Batistoni (74) showed that sunlight behind a window is more deleterious than less intense daylight, and window-glass–filtered daylight is more deleterious than artificial illumination (75 W tungsten filament bulb or 80 W fluorescent tube). A 45% decrease in sodium nitroprusside concentration and a hydrogen cyanide concentration of 6 mg/L were found after two hours of exposure to sunlight, for solutions packaged in bottles.

The amount of loss occurring in the administration sets, during sodium nitroprusside infusion, depends on the concentration of the drug used, composition and thickness of the tubing used, and the flow rate of the solution. In view of all these variables, it is no wonder that the results of different studies vary. Sewell et al. (73) reported 10.3% and 3.7% losses from 0.5 and 1.67 mg/mL samples, respectively, when they were delivered by pumps at 3 mL/hr through tubing exposed to ambient radiation. In another study, over a period of six hours, using a sodium nitroprusside concentration of 0.1 mg/mL and a flow rate of 30 mL/hr, the maximum potency difference found was only 3.5%, between the photoprotected and nonphotoprotected administration sets (70). The simulated infusion experiments of Mahony et al. (71) did not find any potency loss when sodium nitroprusside infusion solutions were infused through a 3 m long PVC infusion set tubing without photoprotection.

Because of the toxic degradation products formed, reservoirs containing sodium nitroprusside infusion solutions should be photoprotected during administration. As an additional measure of precaution, solutions in administration sets should also be photoprotected. Amber administration sets are an effective method of photoprotection for sodium nitroprusside infusion solutions (75).

Catecholamines

In intensive care settings, sympathomimetic catecholamines [e.g., dobutamine, dopamine, epinephrine (adrenaline), isoprenaline (isoproterenol), norepinephrine (noradrenaline, and levarterenol] are often administered via continuous infusion. In clinical practice, reservoirs and administration sets of these drugs are routinely changed every 12 or 24 hours. As the pharmacological efficacy of catecholamines is directly related to their intact phenolic groups, their stability over these dosing periods is questionable.

Catecholamines degrade rapidly via oxidative reactions and are catalyzed by oxygen, pH's >6, heavy metal ions, heat, and UV–VIS radiation. UV–VIS radiation is more deleterious than temperature, and UV is 15 times more deleterious than VIS radiation. The degradation rates, of individual catecholamines, vary and are dependent on the position of their phenolic groups and the type of N-substituents in the aminoalkyl side chain. For instance, norepinephrine is more stable than epinephrine, and epinephrine is more stable than isoprenaline.

The oxidation products of catecholamines are often colored (quinones, adrenochroms). However, the extent of discoloration does not truly indicate the extent of degradation. Therefore, visual inspection based on color changes is an inadequate means of assessing the compatibility and stability of catecholamines (76). Except for dobutamine, colored solutions of catecholamines should not be used. The photoprotected storage of catecholamines is imperative.

According to the manufacturer dobutamine injection, diluted with 5% dextrose solutions, should be used within 24 hours of preparation. Solutions may exhibit a pink coloration due to slight oxidation of the drug, but there is no significant loss of potency during this period of time. In addition, diluted dobutamine infusion solutions (5 mg/mL), packaged and stored in polypropylene syringes, have been shown to be stable for periods of up to 48 hours, when stored at ambient temperature under ambient irradiation conditions (72,77,78). Solutions continuously administered through PE lines or PVC administration sets have also been shown to be stable for up to 48 hours (77). Highly concentrated dobutamine solutions (10 mg/mL), packaged in 50 mL syringes, have been reported to be stable for up to 24 hours, when stored at room temperature and not photoprotected. Samples stored in direct sunlight were not stable (79).

No unacceptable potency losses have been reported for dopamine infusion solutions, packaged in syringes and stored under ambient illumination conditions for up to 48 hours (72,77,78). The exposure of dopamine infusion solutions, packaged in glass bottles (1 mg/mL in 5% dextrose), to normal fluorescent light and blue lamp phototherapy (wavelength 400–500 nm, four F40B Sylvania blue bulbs) for 36 hours at 25°C, static or flowing through tubing at 2 mL/hr, resulted in no significant drug loss compared to controls stored in the dark. The phototherapy irradiation levels ranged from 5.1 and 6.6 µW/cm² during the experiment (80).

The stability of dopamine, epinephrine HCL (4 mg/L), isoprenaline HCL (2 mg/L), and norepinephrine bitartrate (8 mg/L) injections, packaged in a 250 mL, type 1 glass bottles containing 5% dextrose injection, were studied using fluorescent irradiation (76). The samples were stored upright, 0.9 m from the continuously illuminated fluorescent bulb (30 W), at ambient temperature in a laminar flow hood. The results showed that all of the drugs tested were stable for at least 24 hours. The degradation time to 90% of the original potency ($t_{90\%}$) was 24.5 hours for isoprenaline, 50 hours for epinephrine, 142 hours for dopamine, and 2500 hours for norepinephrine. The pH values of the solutions studied were in a range of 3 to 4.4, which is optimal for their stability.

Adrenaline acid tartrate (0.4 mg/L in normal saline and packaged in polypropylene syringes) exposed for 24 hours to ambient illuminating conditions also showed no loss of potency (78). However, various epinephrine preparations may vary in their stability, depending on the form of epinephrine as well as the preservatives and packaging used. Hoechst (today, Sanofi-Aventis) claims that, according to their experimental studies, diluted solutions (0.2, 1, 2 mg/L) of Arterenol® (norepinephrine HCL) and Suprarenin® (epinephrine HCL) are stable for up to 24 hours in polypropylene syringes, if stored without photoprotection.

From this data we conclude that the catecholamine infusion solutions may be administered over periods of up to 12 or 24 hours without any need for photoprotection. Furthermore, it does not appear necessary to protect dopamine solutions from blue-light irradiance to ensure clinically acceptable stability, during infusion. Intense window-glass–filtered daylight should be avoided.

Diuretics

Among diuretics, furosemide, also known as frusemide, is recognized as photolabile. The injection solution is available in amber glass ampouls and the tablets are packaged in photoresistant foil, affording good protection during storage. Exposure to UV–VIS radiation may cause discoloration of the drug dosage form.

Solutions should not be used if a yellow color has developed. Storage of furosemide (1.0 mg/mL in 0.9% sodium chloride packaged in a glass infusion bottle) solution at room temperature while exposed to ambient illumination for 70 days resulted in about an 80% loss of the active substance and the formation of a yellow–orange precipitate (81). The UV irradiation of furosemide solutions under different testing conditions is reported to cause photooxidation, photohydrolysis, and dechlorination of the active substance, resulting in the formation of various degradation products (82).

Furosemide injection may be administered by slow IV injection or by intermittent or continuous IV infusion at a rate not exceeding 4 mL/min. Therefore, the question arises whether photoprotection of the drug reservoir is necessary during parenteral administration. Furosemide injection contains the sodium salt of furosemide formed in situ by the addition of sodium hydroxide during the manufacturing process. The injection solution has a pH of 8 to 9.3. In use, it can be mixed with weakly alkaline and neutral infusion solutions (e.g., 0.9% sodium chloride or Ringer's solution) or with some weakly acidic solutions having a low buffer capacity (e.g., dextrose 5% in water).

The solubility and photostability of furosemide are pH dependent. Alkaline solutions of furosemide are very photostable. According to its pH/photodegradation profile, the photolabile species of furosemide is the unionized acidic form (83). Undiluted injection solutions, 10 mg/mL packaged in 25 mL polypropylene syringes (Becton Dickinson) and stored at 25°C, were exposed to ambient illuminating conditions or in the dark for 24 hours. No detectable change in the furosemide content occurred under either these storage conditions (81). The same results were obtained when furosemide injection was mixed with sodium lactate injection BP or 0.9% sodium chloride injection British Pharmacopeia, packed in "Viaflex" PVC bags. At a concentration of 1 mg/mL, no degradation was found after 24 hours of exposure to ambient illumination or storage in the dark, both at 25°C (81).

Studies have also demonstrated that the short-term stability of furosemide (2 mg/mL in 5% or 10% glucose) infusions stored at room temperature under fluorescent illumination. No significant loss of furosemide was measured after a 24-hour exposure period and no degradation products were detected (84). Moreover, no furosemide degradation products were detected after 48 hours exposure of furosemide solutions (1 mg/mL in 0.9% sodium chloride), whether packaged in normal or Amberset burette sets (85). The solutions were exposed to a mixture of artificial (822 lux fluorescent light intensity, Thorn fluorescent White 85 Lamp) and indirect window-glass–filtered daylight (total irradiation 395 mW hr/cm^2). However, solutions exposed to intense window-glass–filtered daylight (beside the laboratory window, 700 mW hr/cm^2) showed that furosemide is labile when exposed to indoor daylight, with detectable breakdown occurring after only 30 minutes of exposure (85).

The studies cited from the literature argue that furosemide solutions may be slowly injected without having to use any photoprotection. In a mixture of artificial and window-glass–filtered daylight (room light), furosemide infusion solutions may be administered by intermittent or continuous infusion without photoprotection of the reservoir or administration set. Intense window-glass–filtered daylight should be avoided during the infusion of furosemide solutions. In countries where daylight is most frequent and very intense, photoprotection from all sources of UV–VIS irradiation should be instituted, e.g., by the use of aluminium foil (reusable) or Amberset (Avon Medicals) administration sets and burette sets.

TOTAL PARENTERAL NUTRITION SOLUTIONS

TPN requires the IV administration of carbohydrates (usually glucose), amino acids, lipids, electrolytes, trace elements, and vitamins. In clinical practice, these components are usually administered as admixtures of water-soluble and lipid-soluble components or even by all-in-one admixtures. Parenteral nutrition solutions are administered by a 24-hour continuous infusion, or intermittent infusions of 12 to 18 hours each day. These admixtures are very complex and the stability of each depends on numerous factors. During administration the reservoirs, and solutions passing through the administration sets are exposed to a variety of UV–VIS irradiation conditions. The photodegradation of certain amino acids and vitamins are a matter of concern, especially, during long-term TPN (vitamin deficiencies are becoming more clinically relevant) and in neonatal medicine (very small infusion rates resulting in long exposure times of the solutions).

Amino Acids

Amino acid IV solutions may be administered without photoprotection, although the photooxidation of some amino acids may occur. The most susceptible amino acids are cystine, histidine, methionine, tryptophan, and tyrosine. The photoprotected storage of amino acid injection solutions is strongly recommended.

When amino acids in parenteral solutions are exposed to relatively intense illumination for 24 hours, simulating phototherapy in neonatal nurseries, most individual amino acids decrease only slightly. Decreases in the concentrations of methionine, proline, tryptophan, and tyrosine are significantly greater in the presence of riboflavine. The observed decreases in amino acid concentrations are unlikely to be nutritionally important. However, in view of the possibility that photooxidation products may exert toxic effects, it is best to shield amino acid solutions containing vitamins from strong sources of UV–VIS irradiation (86).

Vitamins

Certain vitamins if not photoprotected are very much degraded during administration. The photostability of parenteral vitamins is enhanced by their administration in fat emulsions. The stability of fat-soluble and water-soluble vitamins (Vitalipid® N infant and Soluvit® infant) in a soybean fat emulsion (Intralipid® 10%), under conditions simulating the IV feeding of neonates and children without having to use photoprotection has been demonstrated.

Multivitamins

Multivitamin products for parenteral administration are available in a variety of compositions from different manufacturers. As different formulations are available generally, valid stability information cannot be provided. Stability data obtained on vitamins, derived from studies of a single vitamin, cannot be accurately extrapolated to all multivitamin preparations because of possible vitamin, preservative, and excipient interactions.

According to their package inserts, fat emulsions containing Vitalipid Adult, Infant (fat-soluble vitamins) and Soluvit N (water-soluble vitamins) may be administered without any photoprotection. A sufficient degree of photoprotection, against a mixture of artificial and window-glass–filtered daylight during administration (self-absorption), is provided by the fat emulsion. Only intense window-glass–filtered daylight should be avoided. However, for Soluvit N concentrate diluted in glucose infusion solutions, photoprotection during administration is recommended. On the other hand, photoprotection during administration of Multibionta® or Cernevit® infusion solutions, which are also multivitamin preparations marketed in Germany, is not recommended by the manufacturer.

A high degree of vitamin availability to the patients has been demonstrated in experiments which simulated the delivery of Vitalipid N or Vitalipid N Infant (fat-soluble vitamins) and Soluvit N or Soluvit N Infant (water-soluble vitamins), prepared in a TPN admixture or fat emulsion, using various conditions of irradiation (87,88). The fat emulsion admixtures contained the recommended daily doses of vitamins and fat normally required by a neonate or an 18 kg child (88). The admixtures for the neonate were exposed to window-glass–filtered daylight (200–400 lux) and bilirubin light (1400 lux). The admixtures for a child were exposed to a mixture of artificial and window-glass–filtered daylight irradiation for 24 hours. Only the amounts of delivered riboflavin and total ascorbate were reduced. The amount of ascorbate was even more reduced in the experiments simulating administration to a young child. This greater loss is attributed to the introduction of air, by the peristaltic pump used, into the set during the simulated infusion. Comparable vitamin stability results were obtained for the all-in-one admixtures compounded for adult patients (87).

There are some vitamins, which are adversely effected by exposure to UV–VIS radiation during administration if not photoprotected. Among the fat-soluble vitamins, vitamin A (retinol, retinyl palmitate) photodegradation is of clinical significance.

Individual Vitamins

Vitamin A
Vitamin A is extremely sensitive to UV photonic energy. The photosensitivity of vitamin A makes it unsuitable for use in bags for 24-hour infusions, if the bag is not photoprotected. Only 10% of the labeled amount of vitamin A (10 mL of Multibionta added to a complex aqueous nutritional admixture) was found after a eight-hour exposure to ambient radiation and temperature (89). Protection from UV radiation is possible. Use of suitable UV-absorbing covers, or the presence of a fat emulsion (opacification) in the admixture, can reduce the loss of vitamin A considerably.

In a simulated 24-hour infusion of multivitamins, admixed in a fat emulsion, losses of vitamin A, in the presence of UV–VIS ambient radiation, were negligible (87). When the administration of vitamin A in 3 L TPN regimens were simulated, in an area entirely illuminated by "warm white light" (Thorne Electrical Ltd., only 0.4 W UV radiation in the range 300–400 nm, none below 300 nm), no detectable effect on vitamin A photostability was found. In contrast, only 7% of the added vitamin A was delivered from the bags exposed to window-glass–filtered daylight (90). When exposed to pure fluorescent radiation, vitamin A degraded within acceptable limits, when administered through an unprotected set. In contrast to the rapid degradation of vitamin A when exposed to ambient room radiation experienced when employing conventional unprotected IV sets, photodegradation can be reduced by using Amberset IV administration sets (91).

Depending on the type of ester used, losses of vitamin A by binding to the administration systems can be considerable and complicate the interpretation of the study results (91). Special emphasis should be put on delivering adequate doses and the photoprotection of vitamin A, when this active is employed in neonatal care. The effect of phototherapy radiation on vitamin A photostability, present in TPN admixtures, has been investigated (92). The addition of Intralipid fat emulsion to TPN admixtures significantly reduces losses of vitamin A (Fig. 10).

Vitamin K₁ (Phytomenadione)
When phytomenadione is added to a fat-free TPN admixture and exposed to artificial daylight, it shows less than 10% decomposition after 4.5 hours of exposure. Decomposition is effectively prevented by photoprotection, achieved by the addition of a lipid emulsion to the TPN admixtures (93). It has also been reported that the addition of lipids to this same mixture had no effect on the photostability of phytomenadione. After three hours of exposure to window-glass–filtered sunlight, a 50% loss of the phytomenadione has been reported (94). Therefore, exposure to

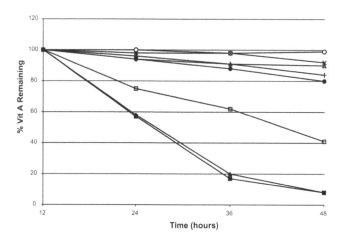

Figure 10 Effect of phototherapy (*blue*) light, Intralipid fat emulsion, and storage container composition on the vitamin A levels in admixtures. (○) Glass, Intralipid; (x) plastic, Intralipid; (●) glass, Intralipid + light; (+) plastic, Intralipid + light; (△) glass; (□) plastic; (▲) glass + light; (■) plastic + light. *Source*: From Ref. 92.

intense window-glass–filtered daylight should be avoided during administration of phytomenadione liquid preparations by the use of photoprotective measures.

Vitamin E
The photostability of Vitamin E exposed to window-glass–filtered sunlight, has been confirmed (94).

Thiamine
Among the water-soluble vitamins subject to photodegradation during administration, thiamine, ascorbic acid, and riboflavine must be considered. A multivitamin product containing all of these vitamins was added to both 0.9% NaCl and 5% dextrose infusion solutions packaged in PVC and Clearflex containers. These admixtures were then exposed to photonic energy (2000 lux) for 24 hours and showed a rapid degradation of both riboflavine and ascorbic acid (95).

Thiamine under these conditions exhibited maximum losses of 11% in 24 hours. Thiamine is reported to be stable at room temperature for up to eight-hours, under fluorescent light in a windowless room or if exposed to indirect window-glass–filtered daylight, when admixed in a TPN solution (96). Direct window-glass–filtered daylight produced a 28% thiamine loss, as determined by microbiological assay. Immediate decreases of thiamine and folic acid levels, sometimes reported, may actually reflect adsorption of these vitamins on to the surface of the bags.

Ascorbic Acid
The stability of ascorbic acid in TPN solutions, neat and in the presence of a fat emulsion has been studied and produced unclear results. Ascorbic acid's stability has been shown to be strongly effected by the oxygen content of the solution, presence of copper, and composition of the amino acid injection used (97).

Some studies have shown that photoprotection does not affect the rate of decomposition significantly (87,88). A 50 mg/mL ascorbic acid solution has been reported to be stable for up to 2.16 mlux hours (59).

Riboflavine

Riboflavine solutions are extremely photosensitive. In the presence of UV–VIS radiation, riboflavine is reported to be the stability-limiting water-soluble vitamin component in multivitamin preparations (95,96,98). Potency levels fell as much as 100% in 24 hours. Studies have indicated that the photon-induced losses of 40% for riboflavine after an eight-hour exposure, and 55% after a typical 24-hour administration period. Losses during passage through the administration set can lead to a further 2% loss (97). Losses caused by exposure to window-glass–filtered daylight can be reduced by the use of protective covers or by starting infusions in the evening, to ensure that the patient receives greater amounts of this vitamin.

CONCLUSION

The ICH guideline fails to supply the clinician with any clear guidelines as to how to test for the photostability of drug products during their administration to patients. The published literature abounds with many conflicting reports attributable to special locality variations, e.g., how much daylight illumination is used in the design of care facilities.

Our experiences and recommended procedures put into use are reported as a guide to others who are working in a hospital environment and are faced with similar situations.

REFERENCES

1. The European Agency for the Evaluation of Medicinal Products. ICH Topic Q1B, Photostability testing of new active substances and medicinal products (CPMP/ICH/279/95), ICH Harmonised Tripartite Guideline, 5 November 1996.
2. American Society of Health-System Pharmacists Reports. Draft guidelines on preventable medication errors. Am J Hosp Pharm 1992; 49:640–648.
3. Goodwin SD, Nix DE, Heyd A, Wilton JH. Compatibility of ciprofloxacin injection with selected drugs and solutions. Am J Hosp Pharm 1991; 48:2166–2171.
4. Martinelli E, Mühlebach S. Kunststoffumbeutel als Lichtschutz für Infusionen. Ergebnisse einer praxisnahen Untersuchung mit Vitamin K_1 als Testsubstanz. Krankenhauspharmazie 1997; 16:286–289.
5. Thoma K, Kerker R. Photoinstabilität von Arzneimitteln. 3. Mitteilung: Photoinstabilität und Stabilisierung von Nifedipin in Arzneiformen. Pharm Ind 1992; 54:359–365.
6. Helin MM, Kontra KM, Naaranlahti TJ, Wallenius KJ. Content uniformity and stability of nifedipine in extemporaneously compounded oral powders. Am J Health-Syst Pharm 1998; 55:1299–1301.
7. Gambaro V, Caligara M, Pesce E. Studio della stabilita alla luce della nifedipina in una formulazione in gocce nelle condizioni d'uso terapeutico. Boll Chim Farm 1985; 124:13–18.
8. Thoma K, Marschall M, Schubert OE. Augentropfen. Physikalische und chemische Instabilität nach Anbruch. Dtsch Apoth Ztg 1991; 31:1739–1743.
9. Vries HDE, Beijersbergen van Henegouwen GM, Huf FA. Photoinstability of chloramphenicol. Pharm Weekbl Sci Ed 1984; 6:54.
10. Thoma K, Kerker R. Photoinstabilität von Arzneimitteln. 5. Mitteilung: Untersuchungen zur Photostabilisierung von Glukokortikoiden. Pharm Ind 1992; 54:551–554.
11. Thoma K, Kübler N. Untersuchung der Photostabilität von Antimykotika. 2. Mitteilung: Photostabilität von Polyenantibiotika. Pharmazie 1997; 52:294–302.

12. Stahlmann R, Lode H. Safety overview. In: Andriole VT, ed. The Quinolones. San Diego: Academic Press, 1998:375,384–385.

13. Utz JP. Amphotericin B toxicity. Ann Inter Med 1964; 61:334–354.

14. Gallelli JF. Assay and stability of amphotericin B in aqueous solutions. Drug Intell 1967; 1:103–105.

15. Block ER, Bennett JE. Stability of amphotericin B in infusion bottles. Antimicrob Agents Chemother 1973; 4:648–649.

16. Kintzel PE, Kennedy PE. Stability of amphotericin B in 5% dextrose injection at 25°C. Am J Hosp Pharm 1991; 48:1681.

17. Lee MD, Hess MM, Boucher BA, Apple AM. Stability of amphotericin B in 5% dextrose injection stored at 4 or 25°C for 120 hours. Am J Hosp Pharm 1994; 51:394–396.

18. Kowaluk EA, Roberts MS, Polack AE. Interactions between drugs and intravenous delivery systems. Am J Hosp Pharm 1982; 39:460–467.

19. Kraus JJ, Vermeji P. Composition and stability of a metronidazole infusion fluid. Pharm Weekblad 1984; 119:1097–1098.

20. Theuer H. Der Einfluss von Herstellungs und Lagerungsbedingungen auf die Stabilität der Metronidazol-Infusionslösungen. Pharm Ztg 1983; 128:2919–2923.

21. Cano SB, Glogiewiewicz FL. Storage requirements for metronidazole injection. Am J Hosp Pharm 1986; 43:2983–2984.

22. Matsumoto M, Kojima K, Nagano H, Matsubara S, Yokota T. Photostability and biological activity of fluoroquinolones substituted at the 8-position after UV irradiation. Antimicrob Agents Chemother 1992; 36:1715–1719.

23. Tiefenbacher EM, Haen E, Przybilla B, Kurz H. Photodegradation of some quinolones used as antimicrobial therapeutics. J Pharm Sci 1994; 83:463–467.

24. Thoma K, Kübler N. Untersuchungen zur Photostabilität von Gyrasehemmern. Pharmazie 1997; 52:519–529.

25. Faouzi MA, Dine T, Luyckx M, et al. Stability and compatibility of pefloxacin, ofloxacin, and ciprofloxacin with PVC infusion bags. Int J Pharm 1993; 89:125–131.

26. Walker S, Lau D, DeAngelis C, lazetta J, Coons C. Doxorubicin stability in syringes and glass vials and evaluation of chemical contamination. Can J Hosp Pharm 1991; 44:71–78, 88.

27. Bouma J, Beijnen HJ, Bult A, Underberg WJM. Anthracycline antitumour agents. A review of physicochemical, analytical and stability properties. Pharm Weekbl Sci 186; 8:109–133.

28. Tavaloni N, Guarino M, Berk PD. Photolytic degradation of adriamycin. J Pharm Pharmacol 1980; 32:860–862.

29. Wood MJ, Irwin WJ, Scott DK. Photodegradation of doxorubicin, daunorubicin, and epirubicin measured by high performance liquid chromatography. J Clin Pharm Ther 1990; 15:291–300.

30. Trissel LA. Handbook of Injectable Drugs. 8th ed. Bethesda: American Society of Hospital Pharmacists, 1994.

31. Frederiksson K, Lundgren P, Landersjo L. Stability of carmustine—kinetics and compatibility during administration. Acta Pharm Suec 1986; 23:115–124.

32. Perumareddi JR, Adamson AW. Photochemistry of complex ions V. The photochemistry of some square-planar platinum (ll) complexes. J Phys Chem 1968; 72:414–420.

33. Thomson AJ, Williams RJP, Reslova S. The chemistry of complexes related to cis-Pt(NH3)₂Cl₂. An anti-tumour drug. In: Dunitz JD, Hemmerich P, Ibers JA, et al., eds. Structure and Bonding. Berlin, Heidelberg, New York: Springer-Verlag, 1972:11:28–29.

34. Zieske PA, Koberda M, Hines JL, et al. Characterization of cisplatin degradation as affected pH and light. Am J Hosp Pharm 1991; 48:1500–1506.

35. Macka M, Borak J, Semenkova L, Kiss F. Decomposition of cisplatin in aqueous solutions containing chlorides by ultrasonic energy and light. J Pharm Sci 1994; 83:815–823.

36. Greene RF, Chatterji DC, Hiranaka PK, Galleli JF. Stability of cisplatin in aqueous solution. Am J Hosp Pharm 1979; 39:38–43.

37. Cheung Y, Cradock JC, Vishuvajjala BR, Flora KP. Stability of cisplatin, iproplatin, carboplatin and tetraplatin in commonly used intravenous solutions. Am J Hosp Pharm 1987; 44:124–130.

38. Hincal AA, Long DF, Repta AJ. Cis-platin stability in aqueous parenteral vehicles. J Parent Drug Assoc 1979; 33:107–116.

39. Cubbels PM, Aixela PJ, Brumos GV, Pou DS, de Bolos Capdevilla J. Cis-platinum stability in physiological saline solutions (0.9% NaCl) in glass and PVC containers in environmental light and temperature conditions. Farm Clint Spain 1992; 9:768–773.

40. KIeinman LM, Davignon JP, Cradock JC, Penta JS, Slavik M. Investigational drug information. Drug Intell Clin Pharm 1976; 10:48–49.

41. Stevens MFG, Peaty L. Photodegradation of solutions of the antitumour drug DTIC. J Pharm Pharmacol 1978; 30:47P.

42. Shetty BV, Schowen RL, Slavik M, Riley CM. Degradation of dacarbazine in aqueous solution. J Pharm Biomed Anal 1992; 10:675–683.

43. Kirk B. The evaluation of a light protecting giving set. Intensive Ther Clin Monitor 1987; 8:78–86.

44. Montgomery JA. Experimental studies at Southern Research Institute with DTIC (NSC-45388). Cancer Treat Rep 1976; 60:125.

45. Koriech OM, Shukla VS. Dacarbazine (DTIC) in malignant melanoma: reduced toxicity with protection from light. Clin Radiol 1981; 32:53–55.

46. Islam MS, Asker AF. Photostabilization of dacarbazine with reduced glutathione. PDA J Pharm Sci Tech 1994; 48:38–40.

47. Beijnen JH, Beijnen-Bandhoe AU, Dubbleman AC, van Gijn R, Underberg WJM. Chemical and physical stability of etoposide and teniposide in commonly used infusion fluids. J Parent Sci Tech 1991; 45:108–112.

48. Stewart CF, Hampton EM. Stability of cisplatin and etoposide in intravenous admixtures. Am J Hosp Pharm 1989; 46:1400–1404.

49. Chevrier R, Sautou V, Pinon V, Demecocq F, Chopineau J. Stability and compatibility of a mixture of the anti-cancer drugs etoposide, cytarabine and daunorubicin for infusion. Pharm Acta Helv 1995; 70:141–148.

50. McLeod HL, Relling MV. Stability of etoposide for oral use. Am J Hosp Pharm 1992; 49:2784–2785.

51. Sesin GP, Milette LA, Weiner B. Stability study of 5-fluorouracil following repackaging in plastic disposable syringes. Am J IV Ther Clin Nut 1982; 9:23–30.

52. Lorillon P, Coberl JC, Mordelet MF, Basle B, Guesnier LR. Photosensitivity of 5-fluorouracil and methotrexate in translucent or opaque perfusors. J Pharm Clin 1992; 11:285–295.

53. Biondi L, Nairn JG. Stability of 5-fluorouracil and flucytosine in parenteral solutions. Can J Hosp Pharm 1986; 39:60–63,66.

54. Dine T, Khalfi F, Gressier B, et al. Stability study of fotemustine in PVC infusion bags and sets under various conditions using a stability-indicating high-performance liquid chromatographic assay. J Pharm Biomed Anal 1998; 18:373–381.

55. Krämer I, Thiesen J. Stability of topotecan infusion solutions in polyvinylchloride bags and elastomeric portable infusion devices. J Oncol Pharm Pract 1999; 5:75–82.

56. Wright MP, Newton JM. Stability of methotrexate injection in prefilled plastic disposable syringes Int J Pharm 1988; 45:237–244.

57. Dyvik O, Grislingaas AL, Tønnesen HH, Karlsen J. Methotrexate infusion solutions - a stability test for the hospital pharmacists. Clin Hosp Pharm 1986; 11:343–348.

58. Chatterji DC, Gallelli JF. Thermal and photolytic decomposition of methotrexate in aqueous solutions. J Pharm Sci 1961; 67:526–531.

59. Battelli G, Marra M, Romagnoli E, Tagliapietra L, Tassara A. Fotosensibilita di soluzione acquose di methotrexate e di acido ascorbico. Boll Chim Farm 1983; 123:149–157.

60. McElnay JC, Elliott DS, Cartwright-Shamoon J, D'Arcy PF. Stability of methotrexate and vinblastine in burette administration sets. Int J Pharm 1988; 47:239–247.

61. Bejnen JH, van Gijn R, Underberg WJM. Chemical stability of the antitumour drug mitomycin C in solutions for intravesical instillation. J Parent Sci Tech 1990; 44:332–335.

62. Quebbeman EJ, Hoffman NE, Ausman RK, Hamid AAR. Stability of mitomycin admixtures. Am J Hosp Pharm 1985; 42:1750–1754.

63. Thoma K, Kerker R. Photoinstability of drugs. Part 6. Investigations on the photostability of molsidomine. Pharm Ind 1992; 54:630–638.
64. Vandenbossche GM, De Muynck C, Colardyn F, Remon JP. Light stability of molsidomine in infusion fluids. J Pharm Pharmacol 1993; 45:486–488.
65. Thoma K, Klimeh R. Untersuchungen der Photoinstabilität von Nifedipin. 1. Mitt.: Zersetzungshinctih und Reahtions mechanismus. Pharm Ind 1985; 47:207–215.
66. Jakobsen P, Mikkelsen EO. Determination of nimodipine by gas chromatography using electron-capture detection; external factors influencing nimodipine concentrations during intravenous administration. J Chromatogr 1986; 374:383–387.
67. Hanff PAJM, van den Biggelaar JPFA. Stability determination of nitroglycerin solutions for parenteral use. Ziekenhuisfarmacie 1994; 10:134–138.
68. Wagenknecht DM, Baaske DM, Alam AS, Carter JE, Shah J. Stability of nitroglycerin solutions in polyolefin and glass containers. Am J Hosp Pharm 1984; 41:1807–1811.
69. Leeuwenkamp OR, van Bennekom WP, van der Mark EJ, Bult A. Nitroprusside, antihypertensive drug and analytical reagent. Review of (photo)stability, pharmacology, and analytical procedures. Pharm Weekbl Sci 1984; 6:129–140.
70. Baaske DM, Smith MD, Karnatz N, Carter JE. High-performance liquid chromatographic determination of sodium nitroprusside. J Chromatog 1981; 212:339–346.
71. Mahony C, Brown JE, Stargel WW, Verghese CP, Bjornsson TD. In vitro stability of sodium nitroprusside solutions for intravenous administration. J Pharm Sci 1984; 73:838–839.
72. Pramar Y, Gupta VD, Gardner SN, Yau B. Stabilities of dobutamine, dopamine, nitroglycerin and sodium nitroprusside in disposable plastic syringes. J Clin Pharm Ther 1991; 16:203–207.
73. Sewell GJ, Forbes DR, Munton TJ. Stability of sodium nitroprusside infusion during administration by motorized syringe pump. J Clin Hosp Pharm 1985; 10:351–360.
74. Vesey CJ, Batistoni GA. The determination and stability of sodium nitroprusside in aqueous solutions. J Clin Pharm 1977; 2:105–117.
75. Davidson SW, Lyall D. Sodium nitroprusside stability in light-protective administration sets. Pharm J 1987; 14:599–601.
76. Newton DW, Fung EYY, Wiliams DA. Stability of five catecholamines and terbutaline sulfate in 5% dextrose injection in the absence and presence of aminophylline. Am J Hosp Pharm 1981; 38:1314–1319.
77. Sautou-Miranda V, Gremeau I, Chamard I, Cassagnes J, Chopineau J. Stability of dopamine hydrochloride and of dobutamine hydrochloride in plastic syringes and administration sets. Am J Health-Syst Pharm 1998; 53:186–187.
78. The stability of drugs in syringes and bags: a collaborative study by hospital quality control staff. Pharm J 1991; 246:H525.
79. Ritter C. Stabilität hochkonzentrierter Dobutaminlösungen. Krankenhauspharmazie 1996; 17:398–399.
80. Dandurand KR, Stennett DJ. Stability of dopamine hydrochloride exposed to blue-light phototherapy. Am J Hosp Pharm 1985; 42:595–597.
81. Neil JM, Fell AF, Smith G. Evaluation of the stability of frusemide in intravenous infusion by reversed-phase high-performance liquid chromatography. Int J Pharmaceutics 1984; 22:105–126.
82. Rowbotham PC, Stanford JB, Sugden JK. Some aspects of the photochemical degradation of frusemide. Pharm Acta Helv 1976; 51:304–307.
83. Bundgaard H, Norgaard T, Nielsen NM. Photodegradation and hydrolysis of furosemide and furosemide esters in aqueous solutions. Int J Pharmaceutics 1988; 42:217–224.
84. Murdoch JM. Short-term stability of frusemide 2 mg/mL in 5 per cent glucose and in 10 per cent glucose. Hosp Pharm Pract 1991; 1:191–195.
85. Yahya AM, McElnay JC, D'Arcy PF. Photodegradation of frusemide during storage in burette administration sets. Int J Pharm 1986; 31:65–68.
86. Bhatia J, Stegink LD, Ziegler EE. Riboflavin enhances photo-oxidation of amino acids under simulated clinical conditions. J Parent Enter Nutr 1983; 7:277–279.

87. Dahl GB, Jeppson RI, Tengborn HJ. Vitamin stability in a TPN mixture stored in an EVA plastic bag. J Clin Hosp Pharm 1986; 11:271–279.
88. Dahl GB, Svensson L, Kinnander NJG, Zander M, Bergström UK. Stability of vitamins in soybean oil fat emulsion under conditions simulating intravenous feeding of neonates and children. J Parent Enter Nutr 1994; 18:234–239.
89. Shine B, Farwell JA. Stability and compatibility in parenteral nutrition solutions. Br J Parent Ther 1984; 5:44–46.
90. Allwood MC. The influence of light on vitamin A degradation during administration. Clin Nutr 1982; 1:63–70.
91. Kirk B. The evaluation of a new light-protective giving-set. Br J Parent Ther 1985; 6:146–151.
92. Smith JL, Canham JE, Wells PA. Effect of phototherapy light, sodium bisulfite, and pH on vitamin stability in total parenteral nutrition admixtures. J Parent Enter Nutr 1988; 12:394–402.
93. Schmutz CW, Martinelli E, Mühlebach S. Stability of vitamin K_1 assessed by HPLC in total parenteral nutrition (TPN) admixtures. Clin Nutr 1992; 11(suppl):110–111.
94. Billion-Rey F, Guillaumont M, Frederich A. Stability of fat-soluble vitamins A (retinol palmitate), E (tocopherol acetate), and K_1 (phylloquinone) in total parenteral nutrition at home. J Parent Enter Nutr 1993; 17:56–60.
95. Martens HJM. Stabilität wasserlöslicher Vitamine in verschiedenen Infusionsbeuteln. Krankenhauspharmazie 1989; 9:359–361.
96. Chen MF, Boyce W, Triplett L. Stability of the B vitamins in mixed parenteral nutrition solution. J Parent Enter Nutr 1983; 7:462–464.
97. Allwood MC. Compatibility and stability of TPN mixtures in big bags. J Clin Hosp Pharm 1984; 9:181–198.
98. Buxten PC, Conduit SM, Hathaway J. Stability of Parenterovite in infusion fluids. Br J Intravenous Ther 1983; 4:5,12.
99. Krämer I, Frank P. Lichtschutzbeduerftigkeit photoinstabiler Arzneimittel zur Infusion. Krankenhauspharmazie 1987; 8:137–144.

Index

Milton Keynes UK
Ingram Content Group UK Ltd.
UKHW020009071024
449327UK00031B/2717

9 780367 390358